Navigating an Evolving Sustainable Built Environment

Navigating an Evolving Sustainable Built Environment

Delivering Net Zero Carbon and Climate Resilience

Dr Mei Ren and Rachel Waggett

Published by Emerald Publishing Limited, Floor 5, Northspring, 21-23 Wellington Street, Leeds LS1 4DL.

ICE Publishing is an imprint of Emerald Publishing Limited

Other ICE Publishing title:
Sustainable and Resilient Cities: Insights into Urban Planning in US Cities and Communities
Cesar A. Poveda. ISBN 978-1-80686-688-5

A catalogue record for this book is available from the British Library.

ISBN 978-1-83662-853-8

Cover photo: Shutterstock.com/panuwat+phimpha

Commissioning Editor: Michael Fenton
Content Development Editor: Ryan Molyneux
Books Production Lead: Emma Sudderick

Typeset by KnowledgeWorks Global Limited
Index created by Lyn Nesbitt-Smith

Printed and bound by CPI Group (UK) Ltd, Croydon, CR0 4YY

Contents

Foreword

Few words in the English language are more ambiguous than 'sustainable'. Few phrases are more open to misinterpretation than 'net zero'. We humans understand that the world's climate is changing and that we are a major part of the problem. But there consensus ends. Even if it is agreed that something should be done, there is little agreement on what exactly can be done, who should do it, and how. This book gives some answers to the first and third of those questions. It is rooted in science, yes, good old-fashioned science. Not just the science that is 'settled', but the science that is constantly questioning, probing, analysing, putting up hypotheses so that they can be examined and either shot down or taken forward. The sort of science that has usually in the end solved the world's problems.

At heart, it is a practical guide for anyone who has to consider how to reduce carbon emissions or how to improve resilience to a changing climate. Architects, designers, planners, politicians, builders and more will all appreciate it. But it is also for you and me – the general public, who care about these issues and want to understand what's being done about them. It includes many practical examples of attempts to tackles sustainability challenges – what worked and what did not. It explains how advancing technology, not least artificial intelligence, can help us. It highlights how the natural world and its extraordinary biodiversity is key to adaptation – we damage it at our peril. The overall 'transformation capacities' of having a vision, rooted in technology, turning that vision into action; and good governance of the process underpin the book. As governments struggle with the cost of carbon reduction and climate resilience, this guide to how such work can be done most cost-effectively and with greatest chance of success could not be more timely.

Sir Tim Laurence GCVO, CB (Chair of the Science Museum Group Board)

Preface

Dr. Mei Ren

Dr. Mei Ren has a wide range of design, research and sustainability management experience from over 25 years of working with all stakeholders of built environment and construction industry. She is particularly interested in developing practical solutions of sustainability and climate resilience strategies. She started her career at the UK's Building Research Establishment before earning a PhD in building physics. She is a Chartered Engineer, a Fellow of The Chartered Institute of Building Services Engineers (CIBSE). She was awarded the inaugural NCE/ICE's Sustainability Champion of the Year in 2014 and was appointed a Royal Academy of Engineering Visiting Professor in Environmental Sustainability at the University of Manchester in 2015–2018. She is currently a Partner at Buro Happold. She is also a member of Science Museum Group's Masterplan and Estate Committee.

Rachel Waggett

Rachel Waggett is a Chartered Environmentalist and sustainability leader with a career spanning research, consultancy and the public sector. Beginning in environmental building services research, she moved into consultancy, working with both small and major firms to deliver innovative, practical sustainability solutions across diverse project scales, before moving into the public sector. Her expertise bridges strategic policy making and hands-on implementation, with a portfolio including advisory roles on both design and client sides. As Lead Officer for Environment at the Liverpool City Region Combined Authority, Rachel works with stakeholders, partners and peer authorities to accelerate sustainability action across the region. She is also closely involved in community energy initiatives in her local area, reflecting her commitment to grassroots climate action.

What inspired us to write this book?

Throughout our career, we've worked with talented professionals in the built environment who strive to embed sustainability in their work. Despite their commitment, many feel progress is slow and the scale of the challenge overwhelming. We want to show that meaningful progress is possible and our collective endeavour is not futile. Recognising what's already been achieved can inspire renewed action. This book aims to be a guiding compass, encouraging readers to reflect, refocus, and continue their sustainability journey with optimism and purpose. We want to demonstrate that with collective determination and responsible effort, innovative solutions are in our hands to ensure not only a habitable planet, but one where future generations can thrive.

Acknowledgements

We gratefully acknowledge our colleagues and collaborators, past and present, whose dedication, creativity, and expertise continue to enrich the built environment sector. We are especially grateful to those who generously allowed us to draw on their publications and project work, enabling us to share examples of best practice and inspiring case studies throughout this book. Thanks also to Rik Wilson for his help and creativity in reinterpreting and creating graphic images, and to the wonderful publication and editorial team who supported and encouraged us at every step.

Finally, we offer our heartfelt thanks to our families for their encouragement, patience, and unwavering support through this whole journey. This book would not have been possible without them.

Mei Ren and Rachel Waggett
ISBN 978-1-83662-853-8
https://doi.org/10.1108/978-1-83662-850-720261002

Introduction

The Brundtland Report, published in 1987 (Brundtland, 1987) – also called *Our Common Future* – by the World Commission on Environment and Development (WCED) introduced the concept of sustainable development: 'Sustainable development is development that meets the needs of the present without compromising the ability of future generations to meet their own needs'. Since then, there has been enormous progress made in our understanding of the concept of sustainability and how success can and should be measured, thanks to scientific discovery, technological advancement and, to a large degree, more and more climate-related natural disasters worldwide that have propelled increasing urgency.

Key principles that seek to promote understanding of the 'triple bottom line' – economy, environment and social needs – have laid a foundation for sustainability accountability. The SIGMA Guidelines (SIGMA Project, 2003) introduced five capitals (natural, social, human, manufactured and financial capital) as a means of enhancing accountability in a way that builds on the triple bottom line approach. However, in practice, there have been too many organisations making decisions using economic performance while excessively eroding environment or social wedges. By utilising the five capitals model and other methods of better valuing non-economic goods, it is possible to overcome some of the weaknesses of the triple bottom line concept. This shift to measure value created during the production of goods and services has established a strong direction to enable more tools and best practice to be developed (Forum for the Future, n.d.; Porritt, 2012).

However, recently, through the conversations with experienced engineers and architects who are leading on major projects delivering sustainability advice to their clients, the authors have noticed a real frustration and bafflement about why progress on sustainability has been seemingly so little, and why it is only lately that this subject of net zero and climate change has truly gained momentum. Their frustration about the inadequacy or limitation of existing assessment tools to drive change sufficiently quickly is clearly noticeable. The multitude of different options and issues can seem daunting and with a perceived lack of progress, the obstacles to action can seem insurmountable. Recognising the progress made to date can help mitigate the perception of being unable to advance. Framework thinking is required, and this book seeks to inspire readers to begin or continue to develop their own path to purposeful integration of sustainability in their practice.

Sustainability policy foundation

Chapter 1 explains the early sustainability initiatives with the evolving definition of sustainability principles, interconnected concepts and themes to set a context for further exploration of these topics.

The built environment system consists of real estate and infrastructure and touches all aspects of human life, from homes and offices to factories and highways. The built environment accounts for 14.4 metric gigatons of carbon dioxide equivalent ($GtCO_2e$) of emissions around the world annually. Approximately 26% of all the world's anthropogenic greenhouse gas emissions and 37% of combustion-related emissions come from the construction and operation of the built environment (UNEP, 2022).

As this impact became recognised, it is no surprise therefore that key policies and legislation began to be introduced in the late twentieth and early twenty-first century to promote sustainability in the built environment. A complex landscape of regulation now exists; these range from strategic documents to maintain consistency between trading partnerships, such as the European Energy Performance of Building Directive (EU, 2010), which was introduced in 2010, to national-scale building design codes, such as building regulations or energy performance certificates. Legislations such as the UK Energy Act 2011 (2011) enable increased implementation of energy efficiency measures for commercial and residential property and extend to entire energy supply chains. Chapter 2 provides an overview of the role that these national government policies and legislations play in laying the foundation of sustainable development and industrial initiatives in the journey addressing climate change challenges and promoting sustainability.

The Climate Change Act 2008 (2019) was the world's first legally binding carbon reduction legislation. It set out the UK Government's pathway to reduce greenhouse gas emissions by at least 80% (relative to 1990 levels) by 2050, since strengthened to 100% reduction (net zero carbon) by 2050. More recently, governments are attempting to streamline the complex legislative picture that has evolved in response to the rapidly evolving understanding of sustainability. The National Planning Policy Framework (MHCLG, 2024), originally published in March 2012, aimed to make the land use and planning system less complex and more accessible, to protect the environment and to promote sustainable growth, acknowledging that the planning system should support transition to a low carbon future due to the changing climate. The complexities of balancing so many aspects of the built environment proved to be exceptionally challenging, and the original aim of a simplified and consistent document has proved difficult to achieve in practice.

Nevertheless, since the introduction of these early policies and legislative requirements there has been real momentum and experience gained in designing and implementing sustainability measures. The built environment industry realised at a very early stage that navigating this complex field would be difficult, and industry-led methods to help guide good design emerged quickly in the form of environmental rating systems such as the Building Research Establishment Environmental Assessment Method (BREEAM) and Leadership in Energy and Environmental Design (LEED). A new profession of 'sustainability consultant' began to emerge from the 1990s, providing specialist technical advice as part of a professional team and helping to embed sustainability into standard practice. Despite this, in the early days, apart from a small number of pilot schemes genuinely adopting the first principles to push for higher standards, there have been abundant examples of 'box ticking' practices, using sustainability achievements merely to gain planning approval or meet the minimum regulations. Overinflated claims of sustainability achievement based on such practices led to accusations of 'greenwashing' and negative impressions or public disbelief in sustainability benchmark standards. For a long time, these standards were mistakenly believed to be synonymous with sustainable development. More recently, it is evident that the achievement of even the highest rating awarded under these schemes at design stage do not necessarily guarantee delivery of real low energy and sustainable buildings in practice.

Climate change

It is only reasonably recently that climate science and research have reached an overwhelming consensus that human activities are the main driver of climate change, primarily due to the burning of fossil fuels, which produces greenhouse gases. This was one of the most important outcomes from the IPCC *Fifth Assessment Report* in 2013 (IPCC, 2013). This unequivocal scientific evidence led directly to political agreement on a global scale.

With climate science now established and a strong consensus in place, efforts can shift from ambition to action. Advances in the field also allow climate models to better assess consequences and enable us to be better prepared to deal with the risks from human interference with the climate system. Our greater understanding means that the action is needed to cover multiple aspects, but falling broadly into two main areas: climate mitigation to reduce carbon emitted to the atmosphere, and climate adaptation to improve our resilience to extreme events and adapt to the increasingly warmer climate, as they are explored in detail as two parallel priorities in Chapter 3.

Framing purposeful sustainability

In recent decades, we are living through three noticeable epoch shifts: climate change, technological revolution and social change. The Covid pandemic has inadvertently become a catalyst for many of our deep global trends prompting new culture changes and innovation. Following the public launch of ChatGPT in November 2022, Artificial Intelligence (AI) is seemingly everywhere. How could the latest technology breakthrough be harnessed to accelerate our sustainability journey? Climate-related protests have been held in many world cities, and all global leaders' have participated in climate collaborations such as the now-annual Conference of the Parties (COP) (UNFCC, n.d.). The UK was the first nation in the world to declare a climate emergency with an objective to achieve net greenhouse gas emissions by 2050. Now 2366 jurisdictions in 40 countries have declared a climate emergency (CED, 2025).

Given such an immense amount of information on media and in the professional workplace, let alone a vast array of new terminology and concepts constantly emerging and perceived 'sudden' wakeup calls to sustainability and climate, it is natural to ask: 'What can I do? Should I feel doomed, or optimistic?'

Certainly, there is a need to apply practicality and realism to mixed emotions generated by the climate emergency. Practically, there is a gap between the essential optimism of the momentum and wider appreciation of the climate emergency, and the negativity and indeed the grimness of climate-related catastrophes that have happened already and how unprepared our education of young generations of professionals is to deal with these crises.

It is incredibly valuable to deepen our understanding of sustainability and climate as key significant factors that impact society and the environment – climate change is presently a major, global issue that impacts the environment and society in countless ways, and sustainable development aims to reduce the impacts of climate change that affect the environment and society. Sustainability is more than achieving net zero, as demonstrated by value-led framework and examples of framing purposeful sustainability integration introduced in Chapter 4. The value-led framework approach deals with not only the environmental indicators, but also social and economics that could be tailored to suit place-based development.

Responding to challenges

The built environment encompasses places and spaces created by people including buildings, parks and transportation systems. When we talk about sustainability, we generally mean making better operational choices to reduce the negative environmental impact on the world around us. There are, however, many options available to become more sustainable. Businesses and individuals must decide which methods are most accessible and impactful for them.

Starting from the bigger scale – our town and cities; urbanism is about the interaction between humans in towns and cities that create the built environment. Globally, an increasing number of people are leaving the countryside to live in cities and towns due to better employment opportunities and economic development. These trends have had a devasting effect on local rural communities. Despite the counter arguments of improved land efficiency, better resilience provision and economic opportunities for inhabitants in urbanisation, there are also large negative pressures on the environment of the towns and cities.

The drive to find a better way to manage these impacts led to eco-city and garden city examples in many parts of the world with various degree of success; an early example is Hammarby Sjöstad, Stockholm (C40 Cities, 2016). The 20-year masterplan development has changed Hammarby Sjöstad from a run-down, polluted and unsafe industrial and residential area in the 1990s to one of the world's most successful urban renewal districts. Sustainability was one of the primary areas of focus, with the high environmental sustainability ambitions being integrated into the planning process from the outset. Water sensitive design was adopted, with water, energy and waste being considered as a circular economy in an urban scale, a large portion of heating being produced from wastewater and municipal waste and all the electricity used coming from renewable sources. Hammarby Sjöstad is a successful demonstration of sustainable alternatives for urban infrastructure. However, despite such good practice examples, there are still few places aspiring to similar achievements.

In many parts of western countries since 2020, the pandemic has reversed the urbanisation trend, with more people moving to regional towns or rural areas where they have more space for a healthier lifestyle. This has only been made possible by technology advancement enabling many jobs to be done remotely from city offices and meetings to be held without travelling. Despite the reversed trend after the pandemic, in recent years, with more people moving back to cities, hybrid working is here to stay with employees' wellbeing being rated as one of the top priorities. The extent to which technology advancement drives social change, as emerging technology applications are used in the built environment, is explored together with social science findings on how physical and mental health of inhabitants can be measured, which could make a major impact on the future built environment. Chapter 5 reveals a host of sustainability processes, design matrices and tools.

For businesses, one of the biggest considerations in developing a more sustainable pathway is how to balance sustainable solutions with the cost of achieving them. This is not just unique to the private sector; public bodies making infrastructure plans or planning policies will need to consider the impact on economic and social aspects. Financial mechanisms such as carbon taxes can start to alter the cost/benefit assessment to make environmental and social considerations more financially valuable and redress the current imbalances. The use of a framework to capture essential considerations for a value-led development is explored in Chapter 4.

So, how we can account for 'E', 'S', 'G' in the sustainability journey of businesses?

E represents Environment. In practice, for an urban development, it means resource efficiency, environmental quality, biodiversity improvement and climate resilience. Resource efficiency includes energy efficiency, renewable energy, use of lower embodied carbon materials, waste reduction and water efficiency.

More and more detailed tools to empower professionals and real estate owners to adopt sustainability principles are available and this includes end use energy focused 'Design for Performance' standards, whole life cycle carbon assessment methodologies and others. Resource efficiency is best viewed through a lens of serving the needs of people in the most sustainable way. Nature-based design is viewed through a lens of protection of nature and promotion of regenerative solutions. These ambitions encourage actions to protect, manage and restore natural or modified ecosystems, which address societal challenges such as ecosystem degradation and biodiversity loss, effectively and adaptively, providing human well-being and biodiversity benefits (IUCN, 2020). Frameworks to achieve such principles provide strong evidence of co-benefits between protection of nature and biodiversity with people's health wellbeing.

S (social) and G (governance) criteria are less well-developed in terms of how to benchmark and assess their full value. Some common key performance indicators (KPIs) on corporate social responsibility reports include carbon and biodiversity, and associated governance and leadership for monitoring, verification and reporting. Social impacts include human capital, and social capital – that is, employment, job creation, education opportunity, wellbeing and health impact.

Environmental, social and governance (ESG) reporting especially for global companies is fast evolving starting with establishing a transparent data reporting and benchmarking framework. Chapter 4 outlines the different frameworks and guidelines for ESG reporting. The connectivity between sustainability-related factors and immediate financial viability is clearer than ever before. Transparent measurement and disclosure of sustainability performance is now considered to be a fundamental part of effective business management, and essential for preserving trust in business as a force for good. Many real estate companies and financial institutions are using ESG indices (or green indices) as the basis for institutional asset mandates, as well as exchanging traded funds, based on the underlying ESG data of listed real estate and infrastructure companies. Given the complexity surrounding sustainability disclosure, standard setters are joining forces to strive for consistency in corporate reporting.

Inspiring examples and endeavours worldwide

Each region of the globe faces different climate, environmental, economic and social challenges. Sustainability now means action. With this in mind, a range of international case studies across five continents of different scales have been selected and are included in Chapter 7 to showcase climate resilience and sustainable development endeavours. Urban development takes years to plan, develop and complete but the book has shown examples in as much detail as possible identifying their transformative impact. Three transformation capacities are used: *Vision and Technology* advancement, resulting in *Action* and *Governance* to showcase each case study's impact on sustainability transformation.

The examples aim to address climate change and environmental challenges, while not neglecting the other considerable epoch shift and advancements in technology, which have and will continue to affect how we tackle climate and social changes. The floods in Pakistan in 2023 claimed 1719 lives and engulfed a third of Pakistan, which has left the country with €40 billion in damages (BBC News, 2022a), while Europe's summer of drought in 2023 is reported to be the worst in 500 years (BBC News 2022b). In response, climate resilient floating structures and floating city examples have been considered as they are being implemented to tackle rising seawater or more extreme flooding events as one of the pressing priorities in South Asia and many parts of the world.

Indoor air quality plays a critical role on our health and wellbeing. Millions of global deaths are attributed to household air pollution every year (WHO, 2025), especially in African countries where women and children are the most affected. A Kenya female-led start-up enterprise, Mukuru Clean Stoves, won the Earthshot Prize in 2022 for development of an innovative and sustainable cook stove product to reduce unhealthy indoor pollution.

After decades of urbanisation and pollution, China is now faced with both water shortages and flooding – only made worse by the effects of climate change. Water resource has become limited and unevenly distributed. Climate change is causing heavier rainfall and storms, affecting large areas of southern China including the Yangtze basin and its tributaries. To combat this threat, 'sponge cities' have been introduced in the planning requirements in many cities in China using a nature-based solution to retain water at its source, slow down water flow and clean it throughout the process.

The Middle East region is at the cusp of a green renaissance, from green finance for large solar parks, to carbon capture, and Al Wadi public park to improve the blue infrastructure. Dubai's Museum of the Future was opened in 2022 marked World Environment Day by promoting eco-friendly behaviour. Museum of Future and Terra are among the inspirational buildings showcased at Expo City Dubai, which aims to be a net-zero-carbon city by 2050, integrating sustainability into its operations and events.

In the United States, the built environment industry is embracing the Living Building Challenge concept of building design to maximise positive environment gains rather than just reducing negative impact (which is the key underlying fundamentals for LEED and BREEAM). The Living Building Institute has introduced a Living Building Challenges rating allowing the flexibility for the real estate owners to choose Zero Energy, Zero Water or Healthy building certifications. A city hall building in Santa Monica was designed to meet the most stringent Living Building Challenges, having achieved Zero Energy and Zero Water standard.

In Australia, a pioneering scheme, National Australian Built Environment Rating System (NABERS) was launched to cutting down performance gaps in buildings. NABERS has shown that customers have reduced their energy use by an average of 30–40% over a 10-year period. NABERS case studies have been selected, as the rating system to realise Design for Performance goals, to drive financial growth and returns in portfolios across building sectors.

Europe is leading the way in transitioning from a linear to a circular economy where there is very little waste and wastage; products and raw materials can be reused or repurposed. The 'Gent Waste Brick for DING' was applied in the new wing of Design Museum Gent, whose brick was made from recycled municipal waste in the city. Oslo is currently the first city worldwide to adopt a climate budget; the budget includes several measures quantifying changes and emission cuts needed by 2030. Kristian Augusts gate 13, Oslo project is a case study to illustrate the circular economy practice as a key to achieving the emission cuts.

UK Green Building Council (UKGBC) trailblazed a series of guidelines and cross-sectoral partnerships including net zero definition, net zero building verification, health and wellbeing to support the UK built environment to plan their way toward meeting the net zero by 2050 target. The UK has been leading the world, especially in policy frameworks (e.g. planning frameworks to encourage adoption of sustainability practices), and methodologies to measure and benchmark sustainability performance of developments. As a developed economy, both ecological infrastructure and socio-economic infrastructure is well established in policies, research and development, instrumental in implementation of sustainability practices with a degree of success.

Future gazing

What is next for the built environment? Technology and our thinking are moving forward at a greater pace than ever and are prone to huge disruption. This book aims to empower and inspire readers to begin or strengthen their integration of purposeful sustainability into all aspects of their work. Chapter 6 provides examples of emerging concepts and near-term development opportunities where widespread adoption, success and impacts are now emerging, while Chapter 8 attempts to provide a synopsis of gazing to the future, exploring the twin major forces of change of climate and AI on the impact on future living, future energy and, indeed, our humanity at the end of this century.

Climate action is at a tipping point. To address growing climate risks, everyone has a role to play. World leaders and policy makers regularly debate the opportunities and actions that need to be taken. Just as in any project or programme, agreement is often hard to find with so many different perspectives and interests at stake. At COP28 in Dubai, countries reached an agreement to create a fund of losses and damages to help vulnerable communities to deal with the impact from the climate change (UNCC, 2023), but practical implementation can prove more difficult than policy agreement.

Nature is critical to meeting the Sustainable Development Goals and limiting global warming to 1.5 degrees. Our land and nature provides important human benefits supplying ecosystem services that help to safeguard the planet for the future of humanity. Since 2022, although there has been some headway on pledges to conserve 30% of Earth's land and seas by 2030 (Convention on Biological Diversity, 2022) by many international communities, progress must accelerate if biodiversity decline is to be stopped and reversed.

The agreed actions may not be enough, they could have been more ambitious, they could be delivered more quickly. But agreement to move forward with a common purpose has nevertheless been made. This agreement shows that action is possible, however challenging, and this sets important precedents in finding further steps to cut down carbon emissions, which are critical to limit the global temperature rise and minimise the climate disruption that will follow. New powerful technologies such as AI could accelerate the pace for changing built environment for better life quality and protecting our environment. There is every reason to be optimistic that we can achieve these ambitious goals.

As professionals, we have the opportunity to lead with purpose – championing sustainable choices and inspiring our clients and collaborators to do the same, so together we can shape a built environment that becomes a powerful legacy for future generations. This book sets out to show precisely how we arrived at this point in our journey – and how we must now move forward with clarity, optimism, and purposeful action.

REFERENCES

BBC News (2022a) *Pakistan Floods: One Third of Country Is Under Water – Minister.* https://www.bbc.co.uk/news/world-europe-62712301 (accessed 25/08/2025).

BBC News (2022b) *Europe's Drought the Worst in 500 Years – Report.* https://www.bbc.co.uk/news/world-europe-62648912 (accessed 25/08/2025).

Brundtland GH (1987) *Our Common Future: Report of the World Commission on Environment and Development.* UN-Dokument A/42/427. https://www.are.admin.ch/en/1987-brundtland-report (accessed 19/01/2026).

C40 Cities (2016) *C40 Good Practice Guides: Stockholm – Hammarby Sjöstad – C40 Cities.* https://www.c40.org/case-studies/c40-good-practice-guides-stockholm-hammarby-sjostad/ (accessed 05/05/2025).

CED (Climate Emergency Declaration) (2025) *Climate Emergency Declarations in 2,366 Jurisdictions and Local Governments Cover 1 Billion Citizens.* https://climateemergencydeclaration.org/climate-emergency-declarations-cover-15-million-citizens/ (accessed 05/05/2025).

Climate Change Act 2008 (2050 Target Amendment) Order 2019 (2019) https://www.legislation.gov.uk/ukdsi/2019/9780111187654/contents (accessed 10/05/2025).

Convention on Biological Diversity (2022) *COP15: Nations Adopt Four Goals, 23 Targets for 2030 in Landmark UN Biodiversity Agreement.* https://www.cbd.int/article/cop15-cbd-press-release-final-19dec2022 (accessed 10/05/2025).

EU (European Union) (2010) *Directive 2010/31/EU of the European Parliament and of the Council of 19 May 2010 on the energy performance of buildings (since revised)* https://eur-lex.europa.eu/eli/dir/2010/31/oj/eng (accessed 19/01/2026).

Energy Act 2011 (2011) https://www.legislation.gov.uk/ukpga/2011/16/contents (accessed 17/01/2026).

Forum for the Future (n.d.) *The Five Capitals – A Framework for Sustainability.* https://www.forumforthefuture.org/the-five-capitals (accessed 17/01/2026).

IPCC (Intergovernmental Panel on Climate Change) (2013) *Climate Change 2013: The Physical Science Basis. Contribution of Working Group I to the Fifth Assessment Report of the Intergovernmental Panel on Climate Change.* Cambridge University Press, Cambridge, UK; New York, USA, 10.1017/CBO9781107415324.

IUCN (International Union for Conservation of Nature) (2020) *Ensuring Effective Nature-Based Solutions.* https://iucn.org/resources/issues-brief/ensuring-effective-nature-based-solutions (accessed 09/05/2025).

MHCLG (Ministry of Housing Communities and Local Government) (2024) *National Planning Policy Framework.* https://assets.publishing.service.gov.uk/media/67aafe8f3b41f783cca46251/NPPF_December_2024.pdf (accessed 17/01/2026).

Porritt J (2012) The five capitals framework. In *Capitalism as if the World Matters.* Earthscan, London, UK, pp. 157–167, 10.4324/9781849770675.

SIGMA Project (2003) *The SIGMA Guidelines – Toolkit.* https://www.sustainabilityexchange.ac.uk/files/sigmaguidetosd.pdf (accessed 25/08/2025).

UNEP (United Nations Environment Programme) (2022) *2022 Global Status Report for Buildings and Construction: Towards a Zero-emission, Efficient and Resilient Buildings and Construction Sector.* https://www.unep.org/resources/publication/2022-global-status-report-buildings-and-construction (accessed 05/05/2025).

UNFCC (United Nations Framework Convention on Climate Change) (n.d.) *Conference of the Parties (COP).* https://unfccc.int/process/bodies/supreme-bodies/conference-of-the-parties-cop (accessed 25/08/2025).

United Nations Climate Change (2023) *Fund for Responding to Loss and Damage.* https://unfccc.int/fund-for-responding-to-loss-and-damage (accessed 17/01/2026).

WHO (World Health Organisation) (2025) *Household Air Pollution.* https://www.who.int/news-room/fact-sheets/detail/household-air-pollution-and-health (accessed 19/01/2026).

Mei Ren and Rachel Waggett
ISBN 978-1-83662-853-8
https://doi.org/10.1108/978-1-83662-850-720261003

Chapter 1
Background and definition of sustainability

1.1. Scope and aim of this book

With over half of the global population of almost 8 billion individuals living in urban areas, a figure anticipated to rise to almost 70% by 2050, the built environment is the frontline of the sustainability challenge (United Nations, 2022; United Nations Department of Economic and Social Affairs, 2019).

Responding to any challenge with purpose and ambition requires an element of consolidation, to consider the current position and how it has been achieved. Without this perspective it is difficult to gather the insights that are necessary to aspire to future progress.

Nowhere is this more vital than in the built environment where every aspect is impacted by global climate change and the urgent need to create developments that will be sustainable for current and future generations.

Yet, remarkably, this realisation has been recently achieved and most practitioners will have experienced instances where the real impacts of climate change are still not understood, or where the implications for built environment projects, both immediate and future, are not acknowledged.

All change is challenging, and the fact that achieving sustainable development requires such wholesale change across multiple areas and all disciplines is especially so. But considering how far built environment professionals have come in a short time can and should be inspiring. A solid foundation for progress now exists and practitioners have the platform to apply their skills and creativity to this undertaking.

This chapter sets out a summary of the early initiatives for sustainable development and sustainability in the built environment, as practitioners began to understand the need to act and to set in place mechanisms to do so. By looking at what has been accomplished and how the existing position has come about, a sense of perspective and renewed purpose for future steps can be achieved.

1.2. Definition of sustainability principles

The term 'sustainability', in the sense of a process that can be continued over a long time, was first used in the mid-1900s to describe economic mechanisms to do with commodities and resources. It was not until the 1980s that the term began to be used to describe processes that ensure the continuation of the natural world and natural systems, coinciding with raised awareness of the depletion of natural resources across the globe.

Degradation of natural environmental and ecological systems had been known and understood long before this time. Demand for resources and materials was gathering pace and was increasingly being satisfied by a global marketplace that obtained goods without due consideration of the consequences for the local systems that provided such resources. By the mid-twentieth century, the impacts of pollution from human technologies were clear (the first British Clean Air Act was passed in 1956); however, the concern was primarily for human health, rather than the wider environment. The emerging environmental movement of the time was galvanised by a number of publications raising awareness of the consequences of polluting activity on natural systems. One such book, *Silent Spring* (Carson, 1962) was fundamental to the understanding and eventual acceptance that there were unseen consequences of human technology. Increased access to global media and film of the destruction of natural environments such as forests enabled a raised awareness about degradation and destruction of natural systems. It was this more widespread documentation and acknowledgement that enabled the creation of formal environmental campaigning groups in the 1960s; their advocacy slowly led to increased global concern and eventually, a recognition that action was necessary.

The first United Nations Conference on the Human Environment was held in 1972, recognising globally for the first time the need for concerted gathering of information and action on environmental matters. The paper resulting from this conference uses terminology that now appears unfamiliar but reflects the emerging viewpoint that it is necessary to protect natural environmental systems while enabling social and economic development. For the first time, this need was globally accepted and articulated.

> A point has been reached in history when we must shape our actions throughout the world with a more prudent care for their environmental consequences. Through ignorance or indifference we can do massive and irreversible harm to the earthly environment on which our life and wellbeing depend. (Declaration of the United Nations Conference on the Human Environment, 1972 (United Nations, 1973: p. 4))

At the United Nations General Assembly 1983, it was determined that a special commission would be set up to provide an 'environmental perspective' to the year 2000. The report of this special commission (later known as the UN World Commission on Environment and Development), *Our Common Future*, was published in 1987 (Brundtland, 1987). It brought together a common definition of a sustainable future and set out principles for a way forward that still resonate today. The concept of intergenerational and international justice, although not articulated in quite this way, was a fundamental foundation of the report and the recommendations for action that it made.

> Humanity has the ability to make development sustainable to ensure that it meets the needs of the present without compromising the ability of future generations to meet their own needs. (Sustainable development definition from *Our Common Future* (Brundtland, 1987: para. 27)

The report did not underestimate the challenge of action across multiple aspects, multiple jurisdictions and sometimes conflicting priorities. The commission's work began to change the view of 'development' as purely economic, something to which all populations and nations strive to achieve, to a realisation that any development path that left future generations with a greater burden was by definition, undesirable.

The challenge laid down by this document was that of turning the wide-ranging and comprehensive call to action into achievable next steps in order 'to keep options open for future generations' (Brundtland, 1987; para. 105).

The concept of sustainable development is by no means perfect. Inherently broad and wide-ranging in its definition, it is capable of interpretation into a myriad forms. It has been interpreted as balancing economic development as equally important to environmental and social aspects (three 'pillars' of sustainable development), which many commentators feel has minimised the extent to which prosperity of human civilisations depend on environmental and social resources for survival. It also lacks a framework for demonstrating achievement in any quantifiable manner.

Nevertheless, the imperfection of the concept and the multiple interpretations to which it is subject does not make it redundant or unhelpful. As an aspirational concept and as a framework for discussion, debate and action determination, it has proved remarkably durable and effective. *Our Common Future* itself reminds readers that the recommendations are based 'on the realities of present institutions, on what can and must be accomplished today' (Brundtland, 1987). A framework for common action was the next challenge that the UN attempted to achieve.

At the 1992 UN Conference on the Human Environment, the 'Earth Summit', more than 178 countries adopted Agenda 21 (the *Rio Declaration on Environment and Development*), which comprised global strategies and actions around the principles set out in *Our Common Future* (United Nations, 1992b). Agenda 21 outlined global strategies for restoring the environment and encouraging environmentally sound development (United Nations, 1992c). The Commission on Sustainable Development was created to monitor and report on implementation.

For the first time, this new framework specifically targeted urban centres (objective 3.8 (k)) and while not specifically referring to design and construction, there was a clear requirement to address material, product and resource use (objective 4.20), including the need to develop a life cycle assessment of products and processes. The document also focuses on the need to provide information to assist individuals to make 'environmentally informed choices' (objective 4.22).

There is also an extensive section that focuses on a series of programmes around sustainable human settlement development. Programmes include the following.

- Provision of shelter.
- Improved management of settlements including infrastructure, land use decision making.
- Sustainable land use planning and management.
- Integrated provision of environmental infrastructure including water, sanitation, drainage and solid waste management.
- Sustainable energy and transport systems in human settlements.
- Human settlement planning and management in disaster-prone areas.
- Sustainable construction industry activities.
- Human resource development and capacity-building for human settlement development.

The advisory document also set out a range of objectives associated with the integration of these principles into decision-making, fiscal and economic methodologies, in the recognition that these aspects were necessary to incentivise and provide adequate support for the environmentally beneficial changes required.

The Agenda 21 report is a remarkable synthesis of the actions that we understand today to be necessary. Its content was non-binding and was not measured or quantified at that time, with an emphasis on locally applied solutions within the Agenda 21 framework. In common with many UN initiatives, the necessity of drawing conclusions and recommendations that apply to highly 'developed', high consuming nations as well as low consuming countries led to non-specific, broadbrush recommendations using general principles rather than specific action.

This led to conflict in implementing the recommendations in the policy frameworks of high consuming countries, with existing systems that required considerable change in order to achieve the aspirations set out in the report. Often, this was due to the rebalancing of economic factors that would be required – for example, the report recommends a move to renewable energy sources including solar, hydro and wind power; however, at that time, coal, oil and gas provided the incumbent energy supply in high consuming countries. Compared against the fossil-fuelled energy systems already in place, renewable energy was economically uncompetitive and perceived as higher risk to investors.

This was despite the recognition by all signatories that policies developed 'without addressing the sustainability of the resources on which production is based will sooner or later run into declining productivity, which could also have an adverse impact on poverty' (United Nations, 1992c: para. 3.2).

In 1997, 5 years after Agenda 21 and the Rio Declaration had been enacted, the first comprehensive review against the action plan was undertaken, concluding that little progress had been made. The resolution states that in many areas, environmental degradation was worsening, in areas such as transboundary air pollution where acid rain caused by industrial pollution had now spread into non-industrialised geographies (United Nations, 1997). In response, a strategy for a new 5-year programme was initiated for further implementation of Agenda 21.

The arrival of the millennium in 2000 encouraged global leaders to focus on the relationship between development, good governance and human rights and the plight of some of the world's most vulnerable people. The Millennium Declaration (United Nations, 2000) was signed at the 2000 UN Millennium Summit and reaffirmed the principles of Agenda 21 alongside other humanitarian principles. Out of this document came, for the first time, a global governance framework for the achievement of sustainable development. A series of goals were set out for achievement within a time period of 15 years that was felt to be ambitious but manageable. The Millennium Development Goals included targets mostly related to health and poverty, such as hunger, universal primary education, maternal health and combatting the spread of viruses such as HIV/AIDS (United Nations, 2015a). However, in an important step, goals related to environmental sustainability were included alongside human sustainability goals, albeit focused more on those aspects beneficial to humans, such as minimising loss of environmental resources, minimising biodiversity loss and increasing safe drinking water supplies. Nevertheless, these goals were a significant milestone, not least because they set in place time-limited specific targets with a regular monitoring schedule against which performance could be measured and monitored.

In 2012, 20 years on from the Rio Declaration, the UN again reiterated the common and familiar commitments (United Nations, 2012). The progress that had been successfully made, particularly in areas of policy development and national scale government commitments was celebrated; however, the lack of progress in many critical areas was acknowledged. The Millennium Development Goals had been in place for more than 10 years by this point, and, while progress was not significant across all areas, the concept of a suite of time-limited goals had been successful in driving and – importantly – in monitoring progress. With the Millennium Development Goals nearing the end of their 15-year period, a series of wider-ranging sustainable development goals was proposed, focused on priority areas for achievement of sustainable development and intended to serve as a driver for implementation and mainstreaming sustainable development. These UN Sustainable Development Goals (SDGs) that developed from this proposal took 3 years to develop and were adopted by the United Nations in 2015 in conjunction with an updated reiteration of principles and commitments (United Nations, 2015a). The result of this proposal was the development of the groundbreaking UN SDGs consisting of 169 indicators or targets across 17 different categories covering the full breadth of sustainability aspects (further discussion of the SDGs can be found later in this chapter and the inclusion of the SDGs in sustainability frameworks is covered in Chapter 4).

At the same time as the sustainable development frameworks were being developed, scientific understanding and global consensus was also emerging around the damaging effects of carbon emissions and their impact on the global climate.

Also initiated at the Rio 'Earth Summit' of 1992, the United Nations Framework Convention on Climate Change (United Nations, 1992b) clearly recognises the risks to the global climate system because of the release of greenhouse gases (GHGs) into the planetary atmosphere. Perhaps the most remarkable aspect of this convention was its global acceptance even though, at that time, the scientific consensus was by no means well developed. The convention sets ambitious carbon reduction targets (particularly for high consuming, historically developed nations) although risks associated with action or inaction were not fully understood at this time.

Commitments have been strengthened in the years since, with binding emission reduction targets proposed and accepted by many nations including industrialised high consuming countries in the Kyoto Protocol (United Nations, 1998). With the emerging scientific consensus enabling nations to better understand the risks of inaction and delay, more rapid decarbonisation targets were agreed in the Paris Agreement of 2015 (United Nations, 2015b), this time setting a clear goal to 'hold the increase in the global average temperature rise to well below 2 °C above pre-industrial levels' and to seek to 'limit the temperature increase to 1.5 °C above pre-industrial levels'.

These 'nationally determined contributions' (NDCs) rest on a concept adopted right at the start of the journey in *Our Common Future*, that of 'common but differentiated responsibilities and respective capabilities' (United Nations, 1992a). In other words, those who have more opportunity to take action, should do so, to achieve a common goal. Each of the signatory countries submits a national climate action plan containing their nationally determined contributions on a 5-year cycle, with each successive plan intended to reflect increasing ambition and achievement. This landmark agreement with its binding targets and individually developed goals is also accompanied by clear and transparent monitoring protocols to hold countries to account on the progress that they are making toward the common goal on GHG emissions.

The Paris Agreement marked a significant turning point in recognition of the urgent need for action on GHG emissions, but it is often forgotten that the Paris Agreement does not neglect the other environmental and social aspects of sustainability. The Agreement commits signatories to respecting and promoting social sustainability and climate justice. It also recognises and notes the importance of environmental matters such as ecosystem integrity.

The Paris Agreement was made possible because of the work of the Intergovernmental Panel on Climate Change (IPCC). Formed in 1988, the IPCC produces regular reports providing the scientific case and evidence to inform policy makers. Each report is assembled by teams of scientists utilising increasingly sophisticated climate modelling tools and insights to translate complex concepts into information for decision-making. This scientific consensus of the implications of global GHG emissions and the risk of failure to limit global heating was fundamental in enabling global political consensus to be reached.

1.2.1 Climate mitigation and climate adaptation

The IPCC was established by the United Nations to assess the science related to climate change, following publication of *Our Common Future*. The panel was tasked with assessing the scientific consensus around climate change, assessing and gathering knowledge on the potential impacts and considering potential response strategies.

At the outset of the panel's work, determining the global GHG emissions was by no means a straightforward task. From our current perspective, with readily available, peer reviewed global data translated into simple, easily accessible graphics, the volume of GHG emissions and their steady upward climb is evident. The scientific effort involved in observing, quantifying and modelling GHGs, their sources and the impact of change on the global climate cannot be underestimated. The reconstruction of past climatic factors and atmospheric GHGs from many different data sources has been fundamental, demonstrating beyond doubt that the planet is in an unprecedented crisis. The work of the scientific community to help policy makers to understand these complex matters has been crucial – presentation and communication has proven to be as vital as scientific rigour.

It was the IPCC reports that first divided their investigation work into two tasks: to consider climate mitigation and climate adaptation (IPCC, 1992).

Climate mitigation is a term used to group efforts to reduce the emission of GHGs and or increase their absorption into natural cycles (such as increasing forests to remove carbon from the atmosphere). The vast majority of efforts to address global climate change have focused on climate mitigation policies and strategies, from the move to reduce fossil fuel-based energy use, to efforts to reduce deforestation. In the context of the built environment, building regulation codes and frameworks that seek to limit energy consumption, include renewable energy technologies or reduce the need for transport energy use are all examples of climate mitigation.

Climate adaptation acknowledges that global climate changes resulting from historic and current GHG emissions will result in a range of impacts that will need to be accommodated and planned for. The impacts vary from country to country and region to region; however, the need for adaptation was recognised from the earliest stages of IPCC investigation. Even the first report in 1990 referred to the need to consider increased desertification, polar melting leading to sea level rise, a rise in prolonged drought and extreme weather events with potential impacts on human health. In the context of the built environment, adaptation has had a reduced focus to date in regulation and frameworks. Where these exist, the emphasis is on minimising exposure to risk (such as flood risk assessments regulated through urban planning and building codes) and providing services sufficient to cope with changes in weather conditions (avoidance of overheating, for example).

1.2.2 Planetary boundaries

While the nationally determined contributions of the Paris Agreement relate to GHG emissions, the agreement notes the importance of other environmental aspects. However, the mechanisms for measuring and monitoring global environmental impacts were less well developed. Crucially, the concept of 'thresholds' and the impact of exceeding those thresholds – the equivalent of the 1.5 °C boundary – had not been clearly defined at that time.

The development of planetary environmental boundaries means that this compelling consensus is now starting to be applied to wider environmental aspects. Planetary boundaries are the subject of a conceptual framework developed by the Stockholm Resilience Centre. In this framework, nine 'boundaries' are proposed within which humanity can continue to develop – a 'safe operating space' for humans. Breaching the boundaries increases the risk of major disruption to earth systems and irreversible change in the global environment. Since the conceptual framework was first published in 2009 (Rockström *et al.*, 2009), data improvements have meant that all nine planetary boundaries can now be quantified, as well as the global status against each boundary. Unfortunately, the latest assessment indicates that six of the nine boundaries have already been breached (Richardson *et al.*, 2023) and unless this balance can be reinstated, major disruptive change is likely.

The planetary boundaries are a methodology for quantifying current environmental performance both globally and locally. It is possible to assess the extent to which proposed or existing developments fall within the planetary boundaries and where necessary, design interventions that will move the development closer to a balanced situation within the human 'safe space' (see Chapter 4 for more information on planetary boundaries and discussion of interaction with social aspects).

1.2.3 Triple bottom line (the '3 Ps')

The 'triple bottom line' is a phrase that uses language of business and economy to describe the interrelated nature of environmental, social and financial aspects to successful operation. In operating a business or corporation, directors are accustomed to considering the 'bottom line' – or the profit (or loss) made by an organisation over a period. Triple bottom line applies the same methodology to social and environmental (and ecological) concerns.

The term was created by John Elkington in 1994 (Elkington, 1994) as a response to *Our Common Future* and the need to reconcile the concepts within that document with business language and frameworks to inspire action from organisations and economy-driven boards of directors. The ability of this concept to inspire business professionals that a 'win:win:win' aspiration was possible cannot be understated.

This framework is also sometimes referred to as the '3 Ps': people, planet and profit (also coined by Elkington), a term now used in a synonymous manner to the triple bottom line, with the same concept of integrating benefit or harm to social, environmental and ecological systems alongside financial concerns. The intention is to widen consideration of the 'hidden' cost of decisions and processes (externalities) that are often not borne by the organisation responsible.

Methodologies for assessing the impact on social, environmental and ecological impact can be challenging to measure as not all impacts can be monetised. It is difficult, therefore, to demonstrate parity in assessing the relative impacts of each of the three 'bottom lines' and concluding to what extent all three have been satisfied.

1.2.4 The Five Capitals

The Five Capitals approach (Figure 1.1) provides a way of understanding sustainability based on the value or 'wealth' defined by five types of capital

- natural capital
- human capital
- social capital
- manufactured capital
- financial capital.

Figure 1.1 The Five Capitals (image reproduced with kind permission of Association for Consultancy and Engineering (2020) based on concept and development by Forum for the Future (Porritt, 2012)).

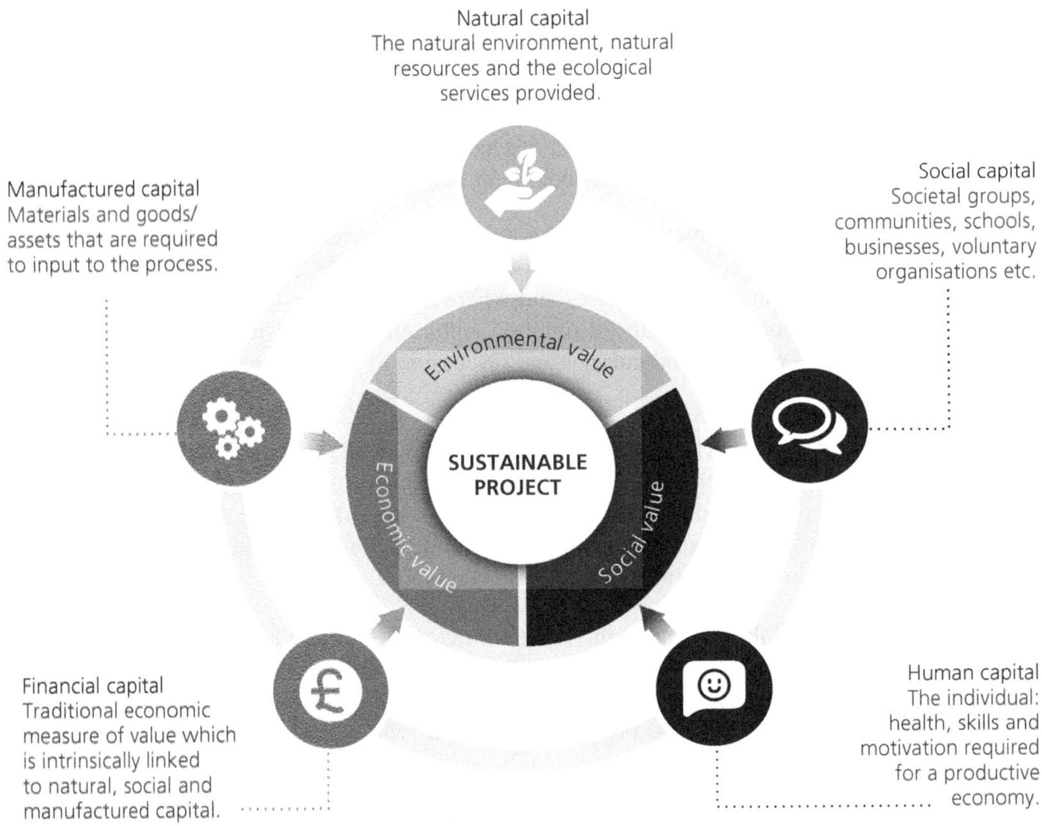

The approach was formalised and promoted by Forum for the Future (Porritt, 2012) as a means of helping businesses to understand their role in enhancing the stocks of wealth in all of these capital areas, rather than solely financial or manufactured capital. Use of economic terminology to consider natural, human or social 'stocks' that can be depleted or exhausted, is a deliberate effort to communicate in a language understandable to business and financial institutions.

The approach has been picked up and built upon by many other organisations, demonstrating its appeal and wide applicability to different sectors and geographies.

Natural capital

Natural capital can be defined as that part of nature which directly or indirectly underpins value to people, including ecosystems, species, freshwater, soils, minerals, the air and oceans, as well as natural processes and functions (Natural Capital Committee, 2019). In combination with other types of capital (including manufactured capital and financial capital), natural capital forms part of our wealth; that is, our ability to produce actual or potential goods and services into the future to support our wellbeing.

Natural capital assets such as soils, species, minerals, freshwater and so forth, provide 'ecosystem services' such as pollination, protection from flooding, alleviation of heat and so on, which can be measured and 'valued', albeit imperfectly, in terms of their importance to humans. The term implies a purely economic concept and is used to enable these important services to be captured in business case and economic assessments.

Human capital – knowledge and skills

Human capital can be defined as the stock of knowledge, skills, trained accomplishments and aptitudes of people. In economic and political terms, it is often linked to productivity, since a workforce with high levels of human capital has the ability to be more productive. The term is most often applied, therefore, at the scale of an organisation or workforce, rather than on a population scale.

Social capital – levels of trust and connections among people

Social capital can be defined as the stock of products resulting from human interactions; this can include personal relationships and networks, the ability to obtain resources, assistance and to share endeavours. High social capital will tend to foster trust and respect between groups or individuals, thereby increasing the likelihood of success for any shared project.

1.2.5 Corporate level environmental, social and governance (ESG) reporting

In the corporate world, ESG processes are used to enable organisations to measure and assess the overall sustainability of their operation. Originally the concept of measuring and managing an organisation's (often regulated) environmental, health and safety impacts, this concept has grown over time to encompass social responsibility and the broader impacts of an organisation's operations. Ultimately, the ESG concept is now intrinsically linked to the perceived risk status of an organisation, and hence its investibility. Financial markets respond favourably to organisations that are actively measuring and managing their ESG obligations.

In many countries, legislation now requires that corporate and certain other organisations disclose data against specific sustainability indicators or key performance indicators (KPIs) publicly as part of annual reporting. Again, the underlying reason for this is to enable investors to assess the relevant risks that companies face and how they are managing those risks. Public disclosure also has the effect of enabling customers to understand the principles by which an organisation operates and, increasingly, this is a factor in customer perception, choice and brand loyalty.

This is an area of great development in recent years, with many countries seeking to establish standards and protocols whereby organisations are required to formally report on ESG matters. More information about these new and emerging standards and protocols can be found in Chapter 4.

1.2.6 Circular economy

Agenda 21 may have set a 'clear requirement' to address material, product and resource use with life cycle assessment in 1992, but the complexity of carrying out this objective stymied implementation for decades. While life cycle assessment slowly gained ground, it has only been in recent years that the mechanisms and models for comprehensive adoption have evolved. The concept of the circular economy has significant benefits in this respect with its intuitive methodology and wide applicability to many different sectors.

There are many definitions of the circular economy, similar in underlying principle there are differences depending on the industry or process to which the concept is being applied. The dictionary definition focuses on the economic aspects of the concept: 'a system based on using renewable resources, eliminating waste, and reusing and recycling material goods'. The Ellen MacArthur Foundation defines the circular economy as 'a system where materials never become waste and nature is regenerated', taking a more broad and holistic definition of the concept (Circular Economy Introduction, n.d.).

Our current economic systems overwhelmingly rely on a linear economy. Materials are extracted, processed into components and then at the end of their useful life, they are disposed of; their value at an end. Where products are recycled or reused, this is often as low value components that have little remaining worth in future systems.

In the context of the built environment, the circular economy is generally applied with relevance to material use, particularly construction materials. Application in this context becomes a process by which the value of components in a system is maintained, so that components at the end of their usefulness in one function can be reused, forming components of multiple future systems. This process maintains the value of the components over time and through multiple potential uses.

The circular economy system also implies the value in prolonging the use and longevity of a component, prolonging the usefulness of components for as long as possible.

Following this process to its conclusion, the preservation of longevity and flexibility in the built environment (Gallego-Schmid *et al.*, 2020) retaining materials in use in their original form, is a key underlying principle. The importance of refurbishment (and construction for ease of future refurbishment) is understood as a fundamental aspect in applying circular economy thinking to the built environment that has been an ongoing topic of discussion since the 1990s (Brand, 1995). Although the terminology and language has changed, the concepts remain consistent with the longevity and lasting relevance of built environment constructions being key (Cheshire, 2019).

The broad definition of the circular economy supported by the Ellen MacArthur Foundation emphasises planetary benefits of slowing resource loops and retaining materials in use for longer, highlighting the benefits to nature and the natural environment of avoiding material extraction. Many commentators avoid this complexity when assessing value chains in the built environment, partly because such benefits are difficult to assess and quantify. Nevertheless, this is an important component that will gain importance in future application of circular economy thinking, which is still rapidly evolving.

1.3. An overview of the current status of sustainability and climate practice – 40 years since *Our Common Future*

1.3.1 UN SDGs

While Agenda 21 had been successful, the progress review in 2012 (United Nations, 2012) made it evident that more rapid and time-defined actions would be needed. The scientific case for rapid action was solidifying and there was an increasing global consensus that actions needed to speed up.

As a result, in 2015, the UN adopted the 2030 Agenda for Sustainable Development (United Nations, 2015a), which included 17 SDGs covering an incredibly broad range of aspects related to environmental and social global equity (Figure 1.2). At its core, this initiative was similar to previous declarations, recognising the need for global social and economic betterment, but emphasising the urgent requirement to do so while protecting the stability of the global climate.

Figure 1.2 UN Sustainable Development Goals (SDGs) (reproduced by kind permission of https://www.un.org/sustainabledevelopment/. The content of this publication has not been approved by the United Nations and does not reflect the views of the United Nations or its officials or Member States)

Unlike previous documents, a target end date of 2030 was set to aim for achievement of the 17 goals and their component targets. The setting of specific goals was a recognition that voluntary, less prescriptive calls to action had failed to achieve the necessary change since the publication of *Our Common Future*.

The highly ambitious nature of these goals and the rapid implementation that would be necessary was acknowledged but the more quantitative approach that this framework represented appeared to many to present a more credible method of driving change by holding participants to account on their actions and progress. Nevertheless, there was a recognition that sustainable development remained a collective endeavour and that countries needed to recognise that progress needed to be prioritised for those countries with the greatest challenges.

Annual progress reports have been published in each year since the adoption of the 17 SDGs; during this time, the indicators have shown a lack of progress across the globe, with the 'world's poorest and most vulnerable people bearing the brunt of the consequences'. The language used in the summary report has grown increasingly more strident over the years and with only 6 years (at time of writing) to the target end date, it is acknowledged that the achievement of the SDGs is slipping out of reach.

Many of the SDGs and their targets are of relevance to the built environment, the most explicit being SDG 11: Sustainable Cities and Communities, which deals with urbanisation. All the 17 SDGs are fully integrated to account for multiple overlapping and complementary impacts, and much of the detail that sits behind the headline goals are of relevance to urban areas and the built environment. Importantly, tracking these indicators over time allows demonstration of progress toward overarching sustainable goals, allowing areas of success and areas that need more focus to be reported and acted upon.

Since publication of the SDGs, their acceptance by the international community has led to their acknowledgement in the policies and strategies of many global industries and private organisations, as well as the public sector. All signatory nations report annually on progress against the SDGs, and, increasingly, municipalities and organisations are adopting the goals as a mechanism to frame their actions across this wide-ranging area. The SDGs are therefore influencing national and sub-national strategy and policy, as well as decision making on a range of matters concerning the built environment by clients and occupiers.

Embedding action – albeit at varying degrees of commitment – on sustainable development into built environment matters is now commonplace at every level of implementation. Chapter 4 has further discussion on embedding SDGs into built environment frameworks.

1.3.2 Development level action (environmental rating and award systems)

The publication of *Our Common Future* in 1987 marked a turning point for the built environment. The authors spoke of the need to create a common framework for action, which a few years later resulted in the publication of Agenda 21 (the Rio Declaration). In the interim, however, the design and construction industry in the UK immediately rose to the challenge, setting out to create a framework for the sustainable design of buildings under the auspices of the then Government-owned Building Research Establishment.

'Eco-labels' for products to reach a particular standard of performance in environmental matters were not unknown at this time. There was growing consumer interest in purchasing products that reduced environmental impact, especially pollution in their manufacture (Elkington and Hailes, 1988). Eco-labels were a method of simply labelling products to signify that they met a set of defined, third-party verified standards, to help inform consumers. The German Blue Angel standard was launched in 1978 for a small number of products, focusing on avoidance of environmental damage compared with competitor products (Figure 1.3). Over the next decade, many countries and trading blocs developed similar ecolabels for a range of different products. Issues considered in these early ecolabel schemes tended to focus on global scale damages such as deforestation (particularly tropical hardwoods) and ozone-depleting chemicals such as chlorofluorocarbons (CFCs), which were widely used in consumer products and building systems at the time.

Figure 1.3 Examples of successful eco-labels providing a recognisable brand identity ('FSC Logo' by Precisetti79 is licensed under CC BY-SA 4.0. 'BLUE ANGEL Logo' by Publicgarden GmbH is licensed under CC BY-SA 4.0)

The innovative concept behind such eco-labels centred on an independently judged award against a set of stringent criteria, to form a trusted brand recognition and enable consumers concerned about minimising environmental impact to confidently select a product rather than its competitors. The concept is built on the idea of consumer pressure being a significant driver of change, and the need for a trusted award that would combat over-exaggerated claims of environmental performance.

In providing a single 'award' identity, all eco-labels, even these early incarnations, hide a significant level of complexity and detail that necessarily underpins the award. A successful eco-label must be sufficiently stringent to ensure that not all products can achieve the standard, yet enable at least some products to do so. An eco-label that no producers can achieve, will not stimulate competition among the producers and will fail in its purpose. The eco-label criteria must therefore be continually updated to drive progress while rewarding only a portion of the producers.

Eco-labels in the built environment

The Building Research Establishment Environmental Assessment Method (BREEAM) was first published in 1990, focusing on the design of new office buildings. The creators aimed to produce an aspirational 'eco-label for buildings' to drive competition and widen the scope of designer and client ambition away from solely energy-related concerns prevalent at the time.

The BREEAM scheme rated individual office building designs against various environmental achievements including a wide range of issues from component materials (timber and recycled materials, for example) and occupant health and wellbeing (noise levels, ventilation). Each achieved issue was awarded one 'credit' up to a total (of 27 in the first edition, 42 in the second edition); the total number achieved resulted in an overall 'BREEAM Rating', the top rating at that time being 'Excellent' (now 'Outstanding'). Like the eco-labels then available for consumer products, the award was made by a separate, independent organisation using third party assessors.

It is difficult to convey, more than 30 years later, quite how innovative the BREEAM scheme was at that time. The 'conservation of fuel and power', familiar now as Part L of the UK Building Regulations, was only introduced in 1985 (just 5 years before the publication of the first BREEAM document) and focused on minimum insulation standards in common elements of buildings (Davies, 2013). From its inception, BREEAM required calculation of carbon dioxide emissions, with benchmark targets leading to higher 'credit' scores. BREEAM also went beyond the environmental performance of the building itself, to consider wider environmental impacts of operating an office development, awarding credits for provision of cyclist facilities, avoidance of wind effects and overshadowing and provision of ecologically beneficial landscaped areas.

The prescriptive nature of the 'credits', which was required to ensure the criteria were sufficiently precise and challenging, meant that the BREEAM methodology for office buildings could not be easily applied to other types of building. In the following few years, the BREEAM methodology was published for food retail stores (supermarkets), new homes, industrial buildings and non-food retail buildings. The need to ensure that the methodology remained sufficiently challenging also meant that regular updates to the methodology were required.

Framework models versus award schemes

While BREEAM set out to achieve an 'eco-label' approach, in the UK at that time there were a number of environmental frameworks being created by the construction industry and research bodies. A framework, unlike an award scheme such as BREEAM, sets out a range of issues and considerations that an exemplar development should consider and aspire to achieve. It does not, however, *require* a particular standard or compliance, and use of the framework does not result in a branded 'award' or validation by a third party.

This approach has both opportunities and limitations. Without the need to strike a delicate balance between achievability and achievement, a framework can explore emerging issues where consensus has not yet been reached, it can encourage consideration without determining what level of achievement indicates 'success'. Of course, the limitation to this approach is that use of the framework does not result in an independent 'brand' award that a client can use to demonstrate their commitment to environmental performance, with the associated competitive advantage.

One example of such a framework model is the Environmental Code of Practice for Buildings and their Services (BSRIA, 1999) first published in 1994, emerging at a similar time to the early BREEAM schemes. As a framework model, it explores issues beyond the scope of the contemporary BREEAM schemes, including life cycle costing, supply chain issues and design for ease of maintenance, for example, all of which were concepts in their infancy at that time. An excellent example of the ability of frameworks to challenge current accepted standards, this document also widened the scope of consideration to construction, commissioning, occupation/maintenance and deconstruction rather than simply building design, foreshadowing the circular economy and life cycle assessment concepts. Subsequent iterations of BREEAM and other building 'eco-labels' have embedded these issues into the award scheme format.

Framework models, due to their less prescriptive nature, can be adopted more readily by other countries, as they can be more easily applied to different regulatory environments and regional variations. Nevertheless, the unique benefits of an award scheme proved to have wider appeal, by providing a recognised award that provided a competitive aspect, and these award schemes formed the basis of much international effort in the years that followed.

Following the launch and success of BREEAM in the UK, other countries began to consider their own need for similar systems, including Green Star (Australia and subsequently New Zealand), Comprehensive Assessment System for Built Environment Efficiency (CASBEE) (Japan), Canada's Building Environment Performance Analysis Club (BEPAC) and UAE's Estidama (Reed *et al.*, 2009).

The USA Green Building Council was formed in 1993 and very shortly after determined that a US version of the award scheme would be beneficial. The Leadership in Energy and Environmental Design (LEED) scheme was launched in 1998 (Richards and Institute for Environmental Entrepreneurship, 2012) and shared many characteristics with the BREEAM scheme, including third party verification leading to an awarded rating. Being aligned to US building codes and standards, it was more applicable to countries utilising those standards.

One of the perennial criticisms of these award schemes has been the gap between design intent and actual performance, often (but not exclusively) related to energy consumption. In response, various 'in use' accreditations were introduced by scheme operators. National Australian Built Environment Rating System (NABERS) (originally launched in Australia but now available elsewhere, including the UK) rates energy efficiency (and water efficiency) using an independent award mechanism, but there are other examples that achieve similar or overlapping goals, including BREEAM In-Use. Further information on the current built environment award schemes available across the globe can be found in Chapter 2.

The award scheme 'eco-labels' and sustainable development guiding frameworks tend to contain their objectives to individual buildings and single developments and only peripherally consider the wider impacts of the development in its context – for example, provision of bicycle storage, changing facilities and safe routes for cyclists within the boundary of the development might be rewarded, but there is no consideration given to the provision or otherwise of infrastructure to support cycling in the wider neighbourhood.

This is a pragmatic approach that recognises the limited opportunities of influence available to those developing buildings, where the wider environment into which the development fits, lies beyond their control. There is little or no opportunity to influence 'upward' to the city or urban-scale environment. However, this means that the surrounding built environment and the interactions with the building project are not well articulated.

While understandable and practical, this limited scale application of sustainable development in the built environment inevitably risks missing opportunities for more impactful action. In recognition of this consequence, scheme operators launched versions that sought to take a wider perspective, such as BREEAM Communities, which sets standards and targets for masterplanning and strategic development.

The BREEAM scheme 'eco-label' for buildings of different types separates buildings into different sectors to enable specific criteria to be used that are common across all buildings of that type. Recognising this, in the early 2000s the Building Research Establishment set out to create a methodology that combined their successful BREEAM scheme with a framework methodology for design of new communities. Intended to support and guide decision-making at masterplanning, outline design and detailed design stages, it aimed to link the design of a neighbourhood to the design of individual buildings. This is an important benefit of a framework approach that allows the wider considerations of a neighbourhood to be taken into account.

The methodology also widens the scope of the BREEAM methodology to include more social and economic matters relevant to community development and a high value is placed on community consultation to design around the needs of the existing community. While BREEAM Communities has been used very successfully for larger scale projects, it is more difficult to apply to smaller sites or those that are developed over long periods of time. It also applies a top-down, linear delivery assumption that is not necessarily the case even in a country with a highly structured planning system. A framework that encompasses a scoring and rating system is inherently inflexible both in terms of application to different places as well as in terms of continuing to push design boundaries at a time of rapid change. Nevertheless, BREEAM Communities was a recognition that individual buildings, however sustainably designed, are only a part of the solution in achieving sustainable development.

In recent years, the project-scale schemes have been subject to several changes that have begun to move toward a more holistic and ambitious scope. BREEAM Infrastructure and recent BREEAM updates have begun to include wider implications of a development on the surrounding neighbourhood and urban environment, encouraging the consideration of how an individual development can widen their positive influence and the maximise the scope of the sustainable benefit being delivered.

This process is challenging to reconcile with existing construction methodologies and therefore the use of more holistic, wider-ranging urban development ecosystem frameworks has begun to emerge and aim to overcome this challenge. Further discussion on these emerging frameworks can be found in Chapter 4.

1.4. Urban development ecosystems

Since the publication of *Our Common Future*, with its broad and encompassing viewpoint, action has tended to coalesce around smaller, more focused opportunities. Valuable though this action is, it often leads to a perceived neglect of other, more broad and arguably more fundamental areas that require equal focus. A wider viewpoint in sustainable urban development does not neglect these ecosystems and can weave specific actions into a wider context to support larger, more coherent improvement.

1.4.1 Connected to nature – nature networks

As discussed earlier, the environmental and sustainability movement as understood today was underpinned initially by a focus on the understanding that human action can harm the natural environment, either directly or indirectly. The fundamental role of nature in sustainability has been somewhat degraded in recent years in a focus on carbon and climate change, to the detriment of sustainable development principles.

Focus has turned more recently to the need to recover nature in all parts of the globe, not just in those areas of particular fragility. *Making Space for Nature*, an authoritative report on England's ecological status and recovery opportunities (Lawton *et al.*, 2010) recommends development of a 'resilient and coherent ecological network' to counter habitat loss, fragmentation and loss of quality. While it is acknowledged that the urban and built environment seldom functions primarily as a landscape for nature, there is no doubt that it can play a vital role in connecting core site habitats that surround the built environment. Pockets of rich biodiversity can and do co-exist alongside urban functions, where concerted efforts have been made to enable this function.

There is a need, therefore, to be aware of the wildlife supported by these habitats that may exist outside of the built environment and how they can be better connected by 'wildlife friendly' design to help movement of species between these habitats. Sites to accommodate nature within the built environment can be viewed as 'stepping stones' that help connect core habitats and make hard landscapes and built-up areas permeable and hospitable to a range of species. The frequency and type of such sites within a built environment will depend on the particular species; investigation of the likely species that would benefit from increased movement through the space to surrounding habitats is essential – for example, spaces to support pollinators will require more frequent stepping stone sites than those to support bird species.

Nature supportive development has been shown to deliver considerable benefits for communities, such as climate change adaptation and health benefits. The World Economic Forum (World Economic Forum, 2020) estimates that US $44 trillion – over half the world's total GDP – is moderately or highly dependent on nature.

Indeed, businesses, investors and governments can now integrate a quantified value for these nature services into business case development (Natural Capital Coalition, 2016). Building design can benefit from integration of nature, utilising trees to shade spaces and reduce the urban heat island effect or green roofs to mitigate rainfall runoff into urban drainage systems, for example, creating a nature network within the built environment context therefore delivers benefits for people and nature. In designing a sustainable built environment, the natural processes that existed on the site before development should be considered and, where possible, design should work to accommodate such natural systems, preserving resilience and the natural capital benefits that such systems deliver.

Practitioners are often urged to consider 'green infrastructure' in designs for the urban environment. While beneficial, the non-specific characteristics of green (and blue) infrastructure mean that it will often be designed without focus on the outcomes for specific species requiring support. Green and blue infrastructure networks, therefore, do not necessarily deliver nature networks. In a sustainable built environment, they can be designed to do so alongside other benefits. Forward-thinking developers and professionals are now beginning to embrace this wider inclusion of nature networks (see Chapter 5 for a discussion of nature and green infrastructure design tools).

Mobility

The mechanism for people and goods to move through and between urban environments is often considered separately to the design of the buildings and immediate surroundings. In urban developments of the past, pedestrians were the primary users of streets, with horses and wheeled vehicles accommodated on slightly wider routes for goods deliveries and passenger transport. As car ownership increased, many urban areas were redesigned to place the needs of motorised vehicles first and pedestrians were confined to a narrow strip at the side of the street (where accommodated at all). In many parts of the world, it is almost impossible to travel through the urban environment safely without being in a motor vehicle.

A sustainable mobility framework considers the most beneficial movement of people and goods. A high proportion of people living in cities and the urban environment do not own or have access to car transport, yet their ability to move around their community can be significantly impeded by infrastructure designed for rapid car travel through the neighbourhood. In current design exercises, even now, it is common to focus on the mechanisms by which motorised vehicles may enter and move through a site, rather than an initial focus on people. Yet, in order for people to live sustainably in the urban environment, accessing employment, goods and everyday services is essential. Therefore, mobility infrastructure underpins the sustainable built environment and its long-term relevance.

Mobility choices are changing exceptionally rapidly with many urban areas seeking to redress the balance toward pedestrians, cyclists and slower-moving vehicles. Many municipalities are also seeking to increase public transport linkages and use, diverting away from privately owned transportation modes. For those who continue to require a private vehicle, change is still inevitable, as ownership modes and fuel source adapt to the need to move away from fossil fuels.

Mobility frameworks that have until recently focused on 'traffic' modelling and enabling continuous movement of vehicles are now being re-designed to utilise multiple mobility modes and a more nuanced modelling system. Dealing with rapidly evolving mobility systems will be a challenge for built environment practitioners, but the result of more broad, flexible and permeable networks will provide resilience to future change as well as current benefits.

1.4.2 Enabling systems – energy planning, smart cities, digital connectivity

Underpinning every urban environment is the hidden network of systems that function to support the operation of the urban space. In many urban areas, developed areas conceal decades or centuries of enabling systems, including water supply and sewerage systems, gas and electrical infrastructure and, more recently, data and information cabling and mobile connectivity networks. Developed with long design lives due to their often sub-surface nature and the disruption associated with replacement, modernising such systems to deal with emerging challenges or new technology is exceptionally complex with long lead times.

Enabling system design now has to deal with increasing complexity, uncertain inputs and outputs as well as accelerating technological change. In modern times, designers have relied upon benchmark and 'reference' building codes to design suitable risk-balanced system capacities. Such codes are renewed periodically to ensure that they remain relevant to new technologies and demands across the majority of built environment needs. However, it is increasingly evident that benchmark and reference codes, based on average demands such as weather patterns, are struggling to cope with increasing uncertainty. Designing for uncertain demands puts additional emphasis on the need for flexibility and enabled change to be built in to all such systems at the point of installation.

Smart cities that use information and communication systems to manage infrastructure in real time, responding to changing demands, are one response to this need for flexibility – for example, smart controls can respond to heavy rainfall forecasts and empty storage vessels or channels ahead of time, in preparation for excessive water flows that would otherwise cause flooding. Such systems are increasingly being piloted and implemented in urban environments, from public transport management to air quality interventions. The evolution of artificial intelligence systems will further accelerate opportunities in this area; this subject is explored further in Chapters 6 and 8.

Smart cities can also work in a participative manner with members of the community, by allowing data to be shared in real-time and in collecting and utilising data provided by the community to improve the predictive results. This enables a much more nuanced approach to design and operation, which is genuinely locally sourced and therefore more responsive to community need – for example, Google Maps predict journey times accurately by combining historical journey times across a route with the real-time location data of users also utilising that route. Users willingly share their anonymised location data in exchange for a smoother and more predictable journey.

1.5. Healthy cities

Today around 56% of the world's population live in cities – approximately 4.4 billion people. By 2050, this trend will continue with the urban population more than doubling its current size with nearly seven out of ten people living in cities. Ensuring that our urban areas support health and wellbeing is both vital and urgent. In recognition of increased urbanisation, the World Health Organisation (WHO) Healthy Cities project was launched in 1978. Its ambition is to promote health and wellbeing through action at the level of individual local authorities. WHO Healthy Cities is a global movement working to put health high on the social, economic and political agenda of city governments (WHO, n.d.).

WHO (WHO, 2020) defines a healthy city as being a process, not an outcome. This means that a healthy city is not one that has achieved a particular health status, rather it is conscious of health and striving to improve it. The goal is specified as healthy cities are places that deliver for people and the planet. They engage the whole of society, encouraging the participation of all communities in the pursuit of peace and prosperity, promoting good governance and leadership for health and wellbeing (Figure 1.4).

Figure 1.4 Healthy cities key themes

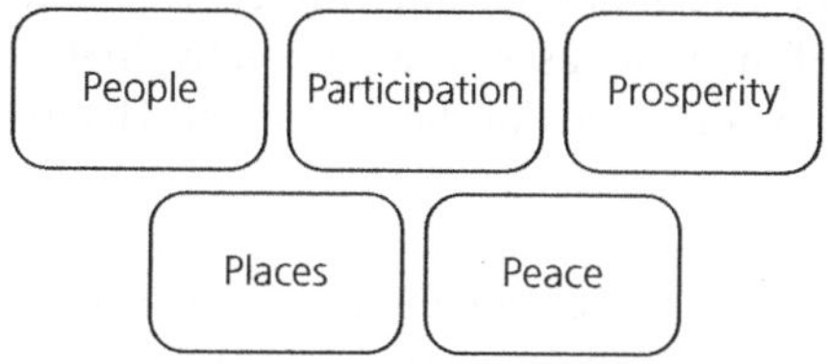

The publication of the UN SDGs built on the work already underway to recognise and support the health of populations in cities. The areas of health promotion, governance for health and social determinants of health (SDH) already developed as part of the healthy cities project underpin many of the social aspects of the SDGs.

One of the first steps that cities take in the WHO Healthy Cities project is to develop a comprehensive city health profile, a public health report that describes the health of the city's population, bringing together key pieces of information on health and its determinants in the city.

This profile usually uses health indicators to define the population's health and presents information on the lifestyles, environmental and social factors in the city that affect health. This is an important element in the process of improving health and thus moving closer to the reality of a healthy city.

The WHO European Healthy Cities Network is a coalition of some 100 flagship cities and approximately 30 national networks, working to engage local governments in political commitment, institutional change, capacity-building, partnership-based planning and innovation. Despite the many challenges that exist in developing cities that support rather than undermine the health of their occupants, it was recognised that the challenges of creating and sustaining healthy cities have many commonalities. Working together to create mutual support networks and investigating opportunities together is a key strength of networks of this type.

One important aspect is the recognition that moving toward healthier cities is a political choice requiring recognition of health as a core value (for more discussion on value-setting for social considerations, see Chapter 4). Alignment with political agendas at all levels of governance structures creates legitimacy and long-term recognition of the value of healthy communities, thereby emphasising the need for the urban fabric to enable that value to be realised. Starting more than 30 years ago, the Healthy Cities movement has attempted to drive this change to create healthier urban settings that support the health and wellbeing of the people that use them. Steady expansion of the network to regions around the world will help new city regions and urbanised areas, which must grow to accommodate their increasing populations, to do so in a way that supports rather than undermines health indicators, in line with the SDGs.

1.5.1 C40 Green and healthy streets

Underlining the importance of health as a political and strategic matter, some of the most ambitious initiatives have come from political leaders of cities around the world. C40 is a global coalition of city mayors who have declared their determination to meet the challenges of the Paris Agreement to cut their emissions and thereby play their part in limiting global temperature increases. C40 is supporting cities to transform their cities into greener, healthier, and more prosperous places to live. In 2017, C40 launched the Green and Healthy Streets Declaration, which includes a pledge to establish a significant area of their city as zero-emission by 2030. The Zero Emission Area Programme works with the 36 signatory cities of the Green and Healthy Streets declaration specifically on this commitment.

Three-quarters of C40 cities are now decreasing their per capita emissions at a faster rate than their own countries, while high-impact actions delivered by C40 cities – such as introducing low-emission zones, increasing green spaces and planting more trees to improve air quality and to reduce urban heat risk, and achieving universal waste collection – have tripled in the past decade, which is vital if global heating is to be kept under 1.5 °C (C40, 2023).

1.5.2 Planning for health and wellbeing

Worldwide health issues such as obesity, mental illness, growing health inequalities and climate vulnerability cannot be solved solely by healthcare provision and medicines; cities must take actions of tackling the social, economic and environmental determinants. In a time when unhealthy and unsustainable conditions are being built into the physical fabric of many cities, a new awareness and strategy is urgently needed to putting health and wellbeing at the heart of city and urban planning (Barton *et al.*, 2015). Spatial distributions of built and landscape form and transport impact on diet, physical activity, social interactions, air quality and access to nature, affecting physical and mental wellbeing. Barton (2016) called for putting people back at the heart of planning. Barton's 'City of Well-Being' research recommends urban planners and decision makers think strategically and be 'explicit about values' to truly make genuinely sustainable places that are not generating a future burden for health services.

The recent Covid pandemic and the increased awareness of population vulnerability to infectious disease in an urbanised scenario has added further weight to the need for health consideration in urban planning. It became evident that systems and local capacity to deal with infectious disease of this type was not keeping pace with the rate of urbanisation. New systems, new technologies and above all, a change in planning methodologies will be necessary if future populations are to be protected and resilience built (Capolongo *et al.*, 2020).

This is especially urgent given the exponential increase in urban populations expected over the next few decades across the world. Ageing municipal systems and lack of strategic planning regimes based on a business-as-usual methodology are not necessarily fit to achieve social and health equality, without significant change of focus. In this context, the coalition of political city leaders determined to achieve change through policy and strategic intervention is exceptionally timely.

REFERENCES

Association for Consultancy and Engineering (2020) *Measures for Successful Outcomes: The Five Capitals Approach: A Discussion Paper*. www.acenet.co.uk (accessed 16/01/2026).

Barton H (2016) *City of Well-Being: A Radical Guide to Planning*. Routledge, London, UK/New York, USA.

Barton H *et al.* (2015) *The Routledge Handbook of Planning for Health and Well-Being: Shaping a Sustainable and Healthy Future*. Routledge, London, UK/New York, USA.

Brand S (1995) *How Buildings Learn: What Happens After They're Built*. Penguin, New York, USA.

Brundtland GH (1987) *Our Common Future Report of the World Commission on Environment and Development*. UN-Dokument A/42/427, https://www.brundtland.co.za/wp-content/uploads/2022/08/Brundtland-Report-1987-Our-Common-Future.pdf (accessed 16/01/2026).

Building Services Research and Information Association (BSRIA) (1999) *Environmental Code of Practice for Buildings and Their Services*. Ed. 2 (COP6/99). https://www.bsria.com/uk/product/QnPXkn/environmental_code_of_practice_for_buildings_and_their_services_edition_2_cop_699_a15d25e1/ (accessed 29/09/2024).

C40 Cities (2023) *Green and Healthy Streets Accelerator*. https://www.c40.org/wp-content/uploads/2024/03/C40_Green__Healthy_Streets_Accelerator_Progress_Report_2023.pdf (accessed 16/01/2026).

Capolongo S *et al.* (2020) COVID-19 and cities: from Urban Health strategies to the pandemic challenge. A decalogue of public health opportunities. *Acta Bio-medica Dell'Ateneo Parmense/Acta Biomedica* **91(2)**: 13–22, 10.23750/abm.v91i2.9615

Carson RL (1962) *Silent Spring*. Houghton Mifflin Company, USA.

Cheshire D (2019) *Building Revolutions: Applying the Circular Economy to the Built Environment*. RIBA Publications, Newcastle upon Tyne, UK.

Circular Economy Introduction (n.d.). https://www.ellenmacarthurfoundation.org/topics/circular-economy-introduction/overview (accessed 28/09/2024).

Davies H (2013) Tracing the continuing development of Part L. *Modern Building Services* [preprint]. https://modbs.co.uk/news/fullstory.php/aid/12062/Tracing_the_continuing_development_of_Part_L.html#:~:text=As%20long%20ago%20as%201962,extended%20to%20non%2Ddomestic%20buildings (accessed 16/01/2026).

Elkington J (1994) Towards the sustainable corporation: win-win-win business strategies for sustainable development. *California Management Review* **36(2)**: 90–100, 10.2307/41165746.

Elkington J and Hailes J (1988) *The Green Consumer Guide: From Shampoo to Champagne: High-street Shopping for a Better Environment*. Gollancz, London, UK.

Gallego-Schmid A *et al.* (2020) Links between circular economy and climate change mitigation in the built environment. *Journal of Cleaner Production* **260**: 121115, 10.1016/j.jclepro.2020.121115.

IPCC (Intergovernmental Panel on Climate Change) (1992) *Climate Change: The 1990 and 1992 IPCC Assessments*. https://www.ipcc.ch/report/climate-change-the-ipcc-1990-and-1992-assessments/ (accessed 16/01/2026).

Lawton JH *et al.* (2010) *Making Space for Nature: A Review of England's Wildlife Sites and Ecological Network*. Report to DEFRA. https://webarchive.nationalarchives.gov.uk/ukgwa/20130402170324/http:/archive.defra.gov.uk/environment/biodiversity/documents/201009space-for-nature.pdf (accessed 16/01/2026).

Natural Capital Coalition (2016) *Natural Capital Protocol*. https://naturalcapitalcoalition.org/wp-content/uploads/2018/05/NCC_Protocol_WEB_2016-07-12-1.pdf (accessed 16/01/2026).

Natural Capital Committee (2019) *Natural Capital Terminology*. https://assets.publishing.service.gov.uk/government/uploads/system/uploads/attachment_data/file/909202/ncc-terminology.pdf (accessed 16/01/2026).

Porritt J (2012) The five capitals framework. In *Capitalism as If the World Matters*. Routledge, London, UK, pp. 157–167, 10.4324/9781849770675.

Reed R *et al.* (2009) International comparison of sustainable rating tools. *Journal of Sustainable Real Estate* **1(1)**: 1–22, 10.1080/10835547.2009.12091787.

Richards J and Institute for Environmental Entrepreneurship (2012) *Green Building: A Retrospective on the History of LEED Certification: Background Briefs for Better Sustainability Decisions in Collaboration with Sustainable Industries Magazine*. Institute for Environmental Entrepreneurship. https://enviroinstitute.org/wp-content/uploads/2012/09/GREEN-BUILDING-A-Retrospective-History-of-LEED-Certification-November-2012.pdf (accessed 29/09/2024).

Richardson K *et al.* (2023) Earth beyond six of nine planetary boundaries. *Science Advances* **9(37)**: 10.1126/sciadv.adh2458.

Rockström J *et al.* (2009) A safe operating space for humanity. *Nature* **461(7263)**: 472–475, 10.1038/461472a.

United Nations (1973) *Report of the United Nations Conference on the Human Environment.* https://documents.un.org/doc/undoc/gen/nl7/300/05/pdf/nl730005.pdf (accessed 16/01/2026).

United Nations (1992a) *Report of the United Nations Conference on Environment and Development*, A/CONF.151/26 (Vol. I). https://www.un.org/en/development/desa/population/migration/generalassembly/docs/globalcompact/A_CONF.151_26_Vol.I_Declaration.pdf (accessed 16/01/2026).

United Nations (1992b) *United Nations Framework Convention on Climate Change.* https://unfccc.int/resource/docs/convkp/conveng.pdf (accessed 16/01/2026).

United Nations (1992c) United Nations Conference on Environment & Development. Rio de Janerio, Brazil, 3 to 14 June 1992; Agenda 21. https://sdgs.un.org/sites/default/files/publications/Agenda21.pdf (accessed 16/01/2026).

United Nations (1997) *Resolution Adopted by the General Assembly, Nineteenth Special Session.* https://documents.un.org/doc/undoc/gen/n97/774/73/pdf/n9777473.pdf (accessed 16/01/2026).

United Nations (1998) *Kyoto Protocol to the United Nations Framework Convention on Climate Change.* https://unfccc.int/resource/docs/convkp/kpeng.pdf (accessed 16/01/2026).

United Nations (2000) *United Nations Millennium Declaration.* General Assembly resolution 55/2. https://www.ohchr.org/en/instruments-mechanisms/instruments/united-nations-millennium-declaration (accessed 16/01/2026).

United Nations (2012) *The Future We Want, Rio De Janeiro, Brazil.* Report, pp. 1–3. https://wedocs.unep.org/bitstream/handle/20.500.11822/13662/N1238164.pdf?sequence=1&%3BisAllowed= (accessed 16/01/2026).

United Nations (2015a) *The Millennium Development Goals Report 2015.* https://www.un.org/millenniumgoals/2015_MDG_Report/pdf/MDG%202015%20rev%20(July%201).pdf (accessed 16/01/2026).

United Nations (2015b) *Paris Agreement.* https://unfccc.int/files/essential_background/convention/application/pdf/english_paris_agreement.pdf (accessed 16/01/2026).

United Nations (2022) *World Population Prospects 2022. Statistical Papers – United Nations. Series A, Population and Vital Statistics Report*, 10.18356/9789210014380.

United Nations Department of Economic and Social Affairs (2019) *World Urbanization Prospects: The 2018 Revision*, 10.18356/b9e995fe-en.

World Economic Forum (2020) *Nature Risk Rising: Why the Crisis Engulfing Nature Matters for Business and the Economy.* World Economic Forum.

WHO (World Health Organisation) (2020) *Healthy Cities Effective Approach to a Rapidly Changing World.* World Health Organisation.

WHO (n.d.) *What is a Healthy City?* https://www.who.int/europe/groups/who-european-healthy-cities-network/what-is-a-health-city (accessed 29/09/2024).

Navigating an Evolving Sustainable Built Environment: Delivering Net Zero Carbon and Climate Resilience

Mei Ren and Rachel Waggett
ISBN 978-1-83662-853-8
https://doi.org/10.1108/978-1-83662-850-720261005

Chapter 2
Sustainability policy foundation

2.1. The evolution of sustainability policy in the built environment

The previous chapter set out the context of the sustainability challenges that face policy makers and practitioners in the built environment, the full extent of which have only been realised in recent years. From the early understanding of the planetary damage that was being caused by human civilisations, industrialisation and urbanisation, those working in the built environment have sought to find mechanisms to respond to these challenges. These responses have come from both practitioners (design teams and constructors) and from policy- and strategy-makers, demonstrating the need for all parties to work together to achieve a common goal of a sustainable built environment.

This chapter considers how government and other policy-makers and the built environment industry have responded in the face of a clearly accelerating need for action. The chapter is written from the English (and UK where appropriate) point of view but the main interventions and their objectives are similar across the globe, with a short summary of the practices in different countries in Section 2.5.

2.2. National government policy and legislative action

The previous chapter sets out the evolution of global understanding regarding the need for action on sustainable development. This global consensus has strengthened over the past decades, underpinned by an increasingly robust scientific consensus and this has led to an acknowledgement of the need for each country to take their share of the necessary action through Nationally Determined Contributions. Global agreements do not set actions for each individual country; instead, each country determines their own proposed course of action and is then responsible for setting out the legal and policy framework that will support delivery of these actions.

In all countries, the passing of legislation that binds government and populations to a particular action or suite of actions is the most ambitious form of commitment. Once enshrined into law, even governments can be held to account for any breach of that commitment and any changes require a formal alteration to be made (in the UK, this requires the Act of Parliament to be repealed). A legislative instrument that includes binding quantified commitments is the most stringent and enforceable form of legislation and is rarely used even in highly developed parliamentary democracies. This kind of outcome-based legislation can be helpful in dealing with big, strategic challenges such as climate change. Such complex challenges require long-term commitment and action by present and future governments. Legislating for future action helps to indicate a shared determination to act that can hope to be sustained even through changes in political direction or parliamentary control. However, there are risks to governments in enshrining specific targets in legislation and for that reason, they are seldom used.

2.2.1 National legislation – the Climate Change Act

In response to the international consensus on a need for action, the UK Government wished to show clear leadership in climate action. The UK Climate Change Act was passed in 2008 (Climate Change Act, 2008) and was the first legally binding climate change mitigation target set by a country anywhere on the globe. As well as setting a legally binding target for the total greenhouse gas (GHG) emission reductions to be achieved by 2050, it set out a series of 5-year interim carbon 'budgets' to ensure progress toward the overarching target was adequately monitored. To achieve this monitoring, it also set up a new independent Committee on Climate Change to carry out forecasting and monitoring. As well as these landmark commitments, the Act also makes provision for adaptation to climate change and adds legislation for supportive mechanisms such as carbon trading schemes and waste recycling among other provisions.

The original Act sets a target to achieve 80% reduction in GHG emissions against a baseline year of 1990. According to the provision of the Act, it was possible to alter this target in light of new scientific advice or in response to international commitments that demonstrate a need for alteration. In 2019, a significant amendment was made, legislating for achievement of 'net zero' (at least 100% of 1990 emission levels) GHG emissions, including methane, by 2050 (Climate Change Act 2008 (2050 Target Amendment) Order, 2019).

This strengthening to 'net zero' was initiated as the result of the agreements arising from the UN Climate Change Conference (COP21) in Paris in 2015. The agreement, known as the 'Paris Agreement' made at this conference (United Nations, 2015a) was a true landmark, with an overarching goal to hold 'the increase in the global average temperature to well below 2 °C above pre-industrial levels' and to pursue efforts to 'limit the temperature increase to 1.5 °C above pre-industrial levels'. This global agreement was possible because the scientific evidence presented by the Intergovernmental Panel on Climate Change clearly indicated the implications for the global climate that were likely to occur if the 1.5 °C global average temperature threshold was crossed. The sophistication of the climate models and the overwhelming scientific consensus shown by the multiple modelling techniques meant that the evidence of the impact of different decarbonisation trajectories was clear and the need for global action incontrovertible. The headline message was that to limit global temperature rise to 1.5 °C, GHG emissions must peak before 2025 at the latest and decline 43% by 2030.

As part of the agreement, all signatory countries agreed to produce a national climate action plan of nationally determined contributions to enable these global aspirational carbon reduction targets to be shared across all nations. The UK's Climate Change Act was strengthened based on the scientific evidence for accelerated action, the UK's Nationally Determined Contribution as well as the global agreement for action.

Each signatory to the Paris Agreement reports their plan and performance against the plan on a regular basis, allowing global progress to be calculated and future trajectories estimated. In the UK, annual reports are produced by the independent, expert Climate Change Committee (formerly called the Committee on Climate Change), setting out progress against the targets, to which government must respond.

The Climate Change Act is a UK-wide legally binding instrument; however, the devolved administrations in the UK of Scotland, Wales and Northern Ireland have developed their own strategies and, in some cases, their own legally binding targets.

The Climate Change Act (and subsequent amendments) does consider adaptation as well as mitigation of GHGs. A Climate Change Risk Assessment is required to be produced every 5 years assessing current and future risks and opportunities in the UK. In addition, the UK Government is required to produce a National Adaptation Programme (for England; the devolved administrations have their own programmes) that responds to the risks identified. Unlike mitigation activity, this risk assessment is somewhat subjective and does not have a measurable target and an end point against which performance can be appraised and a pathway drawn. As with performance on carbon emissions, the Climate Change Committee produces a regular (bi-annual) monitoring report on performance against adaptation indicators. However, assessment of performance is somewhat hampered by a lack of quantitative data and a wide margin of uncertainty in predicting the impact of future changes to climate and the associated risks. Inevitably, perhaps, this has resulted in less focused attention being paid to action on adaptation.

The global consensus on the need for action made it simpler to introduce legislation as ambitious as the Climate Change Act 2008. Without the global consensus, this multinational agreement would have been incredibly difficult to progress. There are always risks in changing political attitudes from different countries and, over time, commitment has wavered from some individual countries. But the fact of the agreement and its broad consensus makes it less vulnerable to political uncertainties. Decades after the UK Climate Change Act being entered into statute, the Act is still driving progress on climate and the mechanisms for budget-setting have proven to be remarkably resilient to the fast pace of change needed (OECD, 2021).

2.2.2 Multinational action and common legislative frameworks

The growing consensus on the need for climate action made it possible for existing trading blocs or groups of countries to make common multinational agreements across a wider geography than an individual country. The extent to which legislative decisions are binding on all members of the group varies depending on the type of legislation enacted. In the case of the European Union (EU), for example, the European Climate Law 2021 (Regulation (EU) 2021/1119) binds all EU members to achieve climate neutrality by 2050. As a Regulation, this is binding on all members (unless they have agreed a specific exemption) and is therefore the most ambitious form of multinational bloc agreement possible.

More commonly, the less stringent Directive is agreed. Such legislation sets out targets that member countries must achieve by a particular date; however, each individual country produces their own legislation to implement the Directive in their own preferred manner. Despite the differences in implementation between member states, the general consensus on the need for action generally leads to considerable strides forward. The European Directive on the Energy Performance of Buildings (Directive 2010/31/EU) is a good example of legislation that was enabled by the broad political consensus. The Directive was implemented in stages from 2007, with each European country (including the UK, at the time) transposing the Directive into local regulation, with the UK legislation (Energy Performance of Buildings Regulations (England and Wales), 2012) coming into effect in early 2013. The Directive was also incorporated into the Building Regulations for England and Wales and the devolved administrations of Scotland and Northern Ireland.

The Directive has been updated and revised several times since first adopted to progressively tighten performance requirement. Even in its initial form, it sought to make the energy efficiency of buildings transparent by provision of a certificate that displayed the energy performance of a building against an easily understood scale. The objective was to hold developers and landlords accountable to prospective purchasers and to enable energy performance to be used as a differentiator when selecting property for purchase or occupation. In addition, similar certificates were mandated to be displayed prominently in all public buildings, again with the intent of holding the building managers and owners publicly to account on the performance of their property. The underlying intent was to incentivise building owners and managers to invest in the improvement of energy performance of their buildings over time, thereby providing a competitive advantage for the most efficient properties and stimulating improvement.

The methodology for displaying the energy efficiency performance was to create a recognisable 'ecolabel' similar to that used for white goods, in order to be readily understood by customers without technical knowledge. Known as an Energy Performance Certificate (EPC), the document also includes information on the approximate running cost of the building given its current energy efficiency and recommendations on how efficiency could be improved (with consequent improvements to running cost). All of this is intended to further incentivise action on the part of occupiers and managers.

Setting standards and replicable targets across a wide geography including multiple nations is a complex business but the advantages, particularly within trading blocs where organisations can use identical standards in trading their products and services, can be considerable.

2.2.3 National legislation enabled by multinational action

The implementation of the European Directive resulted in the UK embedding a mandatory system of energy rating to each property, not reliant on voluntary action by tenants or owners. In itself, this did not require any specific energy performance to be met by the property owners. Once in place as a requirement, the existence of a methodology for building energy efficiency rating enabled the introduction of further regulation to be implemented that set a particular standard as a minimum requirement. In the UK, the Energy Act 2011 (among other provisions) implemented a minimum energy efficiency requirement for England and Wales – for example, all tenanted properties need an EPC rating of E or above (Energy Act, 2011). In future, it has been proposed that the requirement will be strengthened to require a rating of C or above. One could envisage progressive raising of these targets over time, further improving the energy efficiency of the rented building stock. The requirement could also be widened to apply to owner occupied properties, becoming required on the sale of the property, for example. This implementation moves beyond incentivisation into compulsion but is often accompanied by incentive schemes for early movers in the market (such as subsidised costs for insulation projects, changes to high-carbon heating systems or similar).

Naturally, such assessment and labelling of buildings only applies to a single building unit, rather than a wider urban development. The strengths and weaknesses of this type of approach are discussed in Chapter 1 and apply equally in the case of policy and legislation.

2.2.4 National legislation and dealing with complexity

Primary legislation, such as UK Acts of Parliament, are the principal mechanism for implementing mandatory standards. But their inflexibility and inability to react to rapid changes mean that they are often implemented through secondary legislation (such as regulations, rules and orders) to provide more detailed compliance requirements. There are many examples of such secondary legislation that is influential on buildings and the built environment, because the detail required to be effective is often best delivered at the secondary legislation stage, where it can be more easily updated and where any more complex methodologies can be explained.

In the built environment sector, both primary and secondary legislation has an enormous impact on design and construction. Land use planning and in particular town planning is one of the key regulatory instruments available to governments who seek to embed desirable action into the built environment. Surprisingly, even in the UK with its long history of urbanised settlement, use of regulation to control and enforce desirable designs of settlements is relatively recent, with primary legislation only enacted at the start of the twentieth century. Of course, cities and human settlements have always been 'planned' in terms of their favourable location – the earliest large settlements were situated close to sources of food, fuel and freshwater. Those most favourably situated could expect to attract more inhabitants for this reason, growing the settlement organically. Where new towns and settlements were founded by rulers or private landowners, settlements began to be more formally laid out, at least with respect to the main civic centres, principal buildings and entrance points to the settlement. In the UK, with many existing towns and cities dating to hundreds and even thousands of years, it is clear that this intuitive approach has the potential to be highly successful, at least in low population densities.

With the industrial revolution in the eighteenth century, people were required to settle in tightly confined areas determined not by the natural means of supporting themselves, but by proximity to the buildings and manufactories in which they needed to work. Poorly built, cramped and unhealthy settlements were the result and in the nineteenth century it was realised that proper planning codes were required to avoid significant health impacts on residents. The first UK national code came in the first few decades of the twentieth century (Housing and Town Planning &c Act, 1909, 1919) with several subsequent Housing Acts. These early legislative acts sprang more from concerns about human health rather than wider, more strategic considerations (Booth and Huxley, 2012). The UK planning system as it exists today, with a requirement to obtain 'planning permission' from a locally based planning authority before a development could commence and the need to prepare local development plans, did not begin formally until the mid-twentieth century (Town and Country Planning Act, 1947).

The Act was superseded in 1990 and at this point, the social and enabling infrastructure costs of development were recognised and a mechanism set up to enable those to be borne by the developer. Sustainable Development as a duty, rather than an aspiration was not included in Planning legislation until 2004 when a duty was placed on planners to contribute to the achievement of sustainable development (Planning and Compulsory Purchase Act, 2004). Subsequent to this, many amendments have sought to improve the ability of planning practitioners to integrate sustainable development aspects into their planning activities – for example, setting out ways in which renewable energy projects are dealt with, or encouraging the setting out of developments to encourage walking and cycling.

The nature of legislation and of a plan-driven regulatory approach does have considerable advantages, principally the assurance of compliance and the embedding of community consultation and participation in planning decisions (however imperfectly executed). The disadvantage of this approach is the pace with which alterations can be achieved in this type of system to respond to changing pressures and demands.

The rapidity with which new technologies are emerging (and can be viably included in development) and the speed with which a changing climate is demanding a response, are now causing significant difficulties with the planning process. The plan-led nature of local development, whereby a proposed plan is put forward, consulted upon, examined in public and then eventually adopted, is inevitably a slow process taking years rather than months. Making a plan that can respond to the rapidly changing need for new thinking is not aligned with this system. During the 1990s and early 2000s, the UK Government responded to this need by publishing a number of planning policy statements and planning policy guidance documents, effectively making these semi-mandatory 'directional' documents. In an effort to draw all of these multiple documents together and simplify and consolidate the planning picture, the National Planning Policy Framework (NPPF) was published in 2012 (Ministry of Housing Communities and Local Government, 2024).

Despite this laudable aim of simplification, there have been revisions to the NPPF several times since publication, which again underscores the immense difficulty of making long-term policy in a rapidly changing environment with conflicting viewpoints, demands and needs. Indeed, the frequency of published updates is accelerating, which is adding to the complexity of managing this wide-ranging legislation. At the time of writing, a further update to the NPPF is anticipated in 2026.

> This revised Framework replaces the previous National Planning Policy Framework published in March 2012, revised in July 2018, updated in February 2019, revised in July 2021 and updated in September 2023. The National Planning Policy Framework was revised...on 12 December 2024...and amended on 7 February 2025. (National Planning Policy Framework, Ministry of Housing Communities and Local Government, 2024)

The multiple changes to legislation, regulation and guidance in recent years illustrates that even where considerable secondary legislation and guidance exists, policy-makers have a very difficult job to keep pace with the rapidity of change.

Even throughout all of these updates, the fundamental objective of the planning system to contribute to the achievement of sustainable development in terms of economic, social and environmental objectives is clearly reiterated, with specific reference to the UN sustainable development goals (SDGs) (United Nations, 2015b). This policy foundation provides a level of certainty onto which further policy and more local decision-making can rely.

Town and land use planning is a devolved matter in the UK, and in this respect, administrations in Scotland, Wales and Northern Ireland have the opportunity to move at a different pace and with different priorities. This has led to considerable differences in the detail of sustainability planning in these different administrations. While planning authorities at smaller geographies with England are required to set Local Plans, these are constrained by national frameworks, and authorities have limited scope to act independently.

Other countries embrace a less centralised planning system, with planning powers more devolved to municipalities who can identify specific land areas and local planning codes. While this is also possible in the UK, through the creation of a Supplementary Planning Document (SPD), as the name implies, these are supplementary to the agreed Local Plan and compliance is not necessarily mandatory. The UK planning system relies on negotiation of an equitable level of compliance allowing for 'special circumstances', which can lead to extended timescales for decision making and, in some cases, permits lower sustainable design performances to be justified.

2.2.5 Secondary legislation mandating action

In some areas of influence, secondary legislation can be remarkably successful. Financial mechanisms are one of the key ways in which governments can incentivise changes desirable for sustainable development. In effect, financial incentives do not preclude damaging actions, but they put a financial burden on organisations who carry out such undesirable actions. In time, as decision-makers realise the financial benefits of making a change, the new, desirable action or behaviour is normalised in the marketplace. Such financial incentives are often easier to introduce than mandatory legislation, as they are seen as less constraining on economic and business operations. Implementation is more complex as it requires management of the incentive and a robust form of oversight, as well as regular review to ensure it continues to deliver the right incentive.

One example is the use of renewable energy generation certificates that incentivise energy generators to utilise renewable power generation rather than fossil fuel generation. In the UK (and separately in Scotland and Northern Ireland), the Renewable Obligation Orders enabled Renewable Obligation Certificates (ROCs) to be issued to renewable energy generators for every MWh (mega-watt-hour or large unit) of energy produced by their renewable technology. The number of ROCs issued varied depending on the type of technology and the year in which the technology began operation (to take account of decreasing generation technology costs over time). Certificates have a marketplace value as they can be purchased by energy supply companies who wish to demonstrate the level of energy generated from renewable energy or by organisations who may wish to offset their carbon emissions. There are therefore financial incentives to a generation company to invest in renewable energy technology, over and above the financial benefit of the energy retail income.

The benefit of a national scheme is that it can be agile and respond to changing technology cost, consumer demands and political need. As a response to the maturing renewable energy industry, the ROC scheme began to be replaced from 2014 with Contracts for Difference (CfD), a capacity market scheme that could respond in a more agile way to the rapidly decreasing capital cost of delivery renewable energy generation and electricity cost volatility.

CfDs differ in that they provide an agreed unit price for the energy generated over the period of the contract, secured by Government on a best-value basis using a series of 'auctions' to obtain the best offers on price from the prospective generators. The ROC scheme was fully closed and replaced with the CfD scheme from 2017, demonstrating maturity in the market and showing the flexibility of the secondary legislative process in responding to rapid change.

2.2.6 Secondary legislation – trading mechanisms

Carbon trading mechanisms are an example of secondary legislation used to financially incentivise beneficial action – in this case, more broadly across large organisations rather than focusing on the energy generation sector.

Carbon trading is a method of financially incentivising reduced carbon emissions and efficiency projects. Carbon trading can be participated in voluntarily by individual organisations or can be mandated by governments. Mandated carbon trading is particularly used at nation or state level to incentivise action in the industrial and commercial sectors, by means of a carbon 'cap and trade' system.

At the national level, a carbon 'cap' is set for specific industries or sectors by government, who then issue permits up to the 'cap' level. Any organisation or industry who can reduce their carbon emissions below the agreed cap, can then 'trade' their surplus permits to other organisations who have failed to meet or have exceeded their cap. This 'cap and trade' system incentivises efficiency and can significantly penalise those trying to purchase permits, whose value can be volatile and considerable.

It is essential that the cap is set at an optimal level, however – too high and there would be no need for trading; too low, and there would be few permits available for trade. Setting capped limits is a delicate business if the organisations and industries subject to the requirement are not to be financially impacted in a damaging way. Partly for this reason, it has not been possible for a global system of cap and trade carbon permits to emerge, and most schemes work at the national or sub-national (state) level. The European Union has a large-scale cap and trade system that has met with success despite the complexity of application across many national and sectoral types.

In the UK, carbon trading was mandated through secondary legislation in 2020 (Greenhouse Gas Emissions Trading Scheme Order, 2020). This Order created a UKETS (UK Emissions Trading Scheme) applying to energy intensive industries (primarily industrial), power generation and certain aviation routes. From 2026, the UKETS also applies to domestic maritime transport and, from 2028, to waste incineration and energy sector waste. Again, this shows the flexibility of secondary legislation to adequately manage rapid change in a variety of sectors – in this case, the industrial sector, once the relevant regulations and orders have been put into place.

2.2.7 Key secondary legislation – building regulation

In terms of secondary legislation with a key influence on the built environment and development sectors, building regulation is surely the most prominent and effective. Building regulations were originally conceived to deal with issues of safety, recognising that constructing buildings that were unsafe or built from inappropriate materials posed a risk not only to the occupants of the dwelling but also the surrounding buildings, particularly in urban areas where building density was high. In fact, the earliest building regulations in the UK were developed on a municipal rather than national scale and related to issues such as prevention of fire (an immense threat in urban areas with open cooking fires and predominantly timber buildings) and, later, adequate drainage and ventilation.

While these regulations could be applied by a municipality, they were not applied consistently and often only applied to urban areas. In fact, the first national approach to dedicated building regulation was not implemented until the late 1950s, in Scotland, with England and Wales's regulation following shortly afterward. Building regulations as we know them today in the UK with the detailed guidance provided in the Approved Documents system, was enabled by the Building Act 1984, which gave national government the ability to issue building regulations that would be enforced by development controllers (Building Act, 1984). Shortly afterward, the now-familiar Part L: Conservation of Fuel and Power in Buildings was included in the regulations, responding to the realisation of the costs and impacts of fossil fuel use since 2002. While these early regulations appear very simple and rudimentary in comparison with the current regulation, this was the start of formal recognition of the need to reduce energy consumption in buildings, setting out the minimum standard across the country.

Initially, regulatory standards were set for a limited number of thermal elements of a building, such as roof, walls, windows and so on. As understanding of energy use and its complexity grew, and the desire to factor in the different carbon emissions of various fuel types, it was necessary to develop a model capable of more detailed representation.

Over time, this simple building representation (of a 'box' comprising replicated area/volume and fabric thermal characteristics) has matured into a complex national calculation methodology and requires building modelling software to complete.

As buildings are mandated to be more thermally efficient and better constructed for air tightness, regulators find inevitably that this can create other concerning issues, such as insufficient ventilation, leading to condensation or overheating. Further regulation is then required to ensure that these unwanted impacts are not inadvertently caused by the drive to reduce energy consumption. As a result, the building regulations inevitably become more complex and varied in nature – for example, all parts of the UK now (at time of writing) have building regulations that include minimisation of energy consumption, ventilation standards and, most recently, avoidance of overheating, following updates to the detailed Approved Documents in 2022.

It is anticipated that additions will be made in future, such as inclusion of embodied and life cycle carbon (or energy), water efficiency and further provision for sustainable drainage (which is also dealt with through separate legislation). Adoption of standards that go considerably beyond current practice in new construction is challenging in any legislation (even secondary legislation) due to its mandatory nature and its inherent inflexibility, trading off the potential to inspire innovation to ensure minimum compliance across a wide geography.

Any innovation and beyond-compliant construction are therefore entirely voluntary in such a system.

2.2.8 Building sustainability frameworks purposed as codes and standards

Building framework codes and standards occupy somewhat of a middle ground between legislation and voluntary, choice-based action. They can be used in a regulatory capacity, whereby governments (either national or local) can insist upon compliance with a particular building code as a condition of consent. They can also be used on a voluntary, choice-based basis, whereby a developer or client could determine that they wish to work to a particular code, without there being a clear requirement to do so. Both have an important place in driving sustainable development and their potential for inclusion where there is a lack of clear legislation can be transformative.

The development of building sustainability frameworks and 'eco-labels' is covered in other chapters describing how such systems were created and evolved over time. Here, we are focusing on the purposing of such frameworks as supplementary to, or in place of, legislated sustainability action. In the UK, this was first taken forward by certain local governments and forward-thinking clients, who began requiring a certain level of Building Research Establishment Environmental Assessment Method (BREEAM) standard (for example) in development proposals, as a method of demonstrating that the wide range of sustainability aspects had been considered and to drive continual improvement across their geography or portfolio. The level of achievement required to be met was invariably higher than that required by mandatory planning standards and, therefore, adoption of these frameworks acted to drive improved sustainability performance.

National government departments soon realised that they could use a similar practice to drive up sustainability performance in their own estate and it became common for public buildings such as schools, prisons and hospitals to have a certain level of performance mandated as a client requirement. This grew familiarity with the sustainability frameworks and the practice was eventually adopted by national government.

The Code for Sustainable Homes was launched in 2006 and subsequent iterations followed (Department of Communities and Local Government, 2010) replacing a predecessor 'BREEAM style' assessment methodology called 'EcoHomes' (Building Research Establishment (BRE), 2006). The Code for Sustainable Homes was similar to other BREEAM methodologies in that it set out a range of sustainability measures, achievement of each being rewarded by a score that translated into an award level. Not all of the sustainability measures were required, even to achieve the highest award level, thereby allowing an element of flexibility that was considered valuable.

The UK Government made assessment against the Code for Sustainable Homes mandatory in 2008 for all new homes, as part of a Home Information Pack that was required to be prepared for all homes before sale. In 2010, Government determined that this requirement was no longer necessary, unless it was required as part of locally determined planning consent (note that this was no longer permitted after 2015, when Government removed the ability for local government to set such requirements locally) or as a requirement of funding (Pickles, 2015). Instead, the carbon aspects of the Code for Sustainable Homes would form part of the Building Regulations.

However, energy (carbon emissions) from operational energy use was only one part of the Code scheme with other energy-related actions (from related issues such as provision of renewable energy, to less obvious practical measures including provision of space to dry laundry). The Code, like BREEAM, incentivised other environmental actions which were not within the remit of Building Regulations, including water use, materials, flood risk and surface water attenuation, ecology and actions to improve health among other aspects.

This focus on carbon/energy use coincided with the adoption of the Climate Change Act and its specific focus on carbon dioxide equivalents. While this ground breaking legislation was instrumental in driving forward action on decarbonisation, this is an example of how emphasis on carbon can manifest as a very narrowly focused interpretation of sustainability, to the detriment of other equally valuable aspects. The Code for Sustainable Homes was by no means exhaustive in consideration of environmental impact, but it did widen the suite of consideration beyond energy/carbon and had a considerable impact on construction design choices.

With their removal from national legislation, such building design codes reverted to their original, entirely voluntary status. Nevertheless, the flexibility of approach and their ability to be regularly updated (and continue to demonstrate best practice in the built environment) meant that many developers and clients continued to see benefit in their use.

2.2.8.1 The emerging Future Homes and Buildings Standard

At time of writing, the UK Government has launched plans to develop a Future Homes Standard and a separate Future Buildings Standard that are intended to complement (and potentially, replace) Building Regulations and lead to large reductions in carbon emissions from new construction when it comes into effect.

Unlike the Code for Sustainable Homes, the Future Homes and Buildings Standards is likely to focus on energy (carbon) with a proposed suite of measures that can be selected to decarbonise new properties and render them fit for future ‘net zero’ operation without retrofit or modification.

The focus is squarely on operational energy use; embodied carbon in construction materials is, at present, omitted from the proposed standard, as is the carbon implication of other operational measures (such as proximity to good quality public transport, goods and services, electric vehicle charging etc.). Similarly, the opportunity to address other important sustainability measures that can be best incorporated at design stage (such as water efficiency and future climate resilience) are omitted in current draft versions.

Other selected sustainability measures are being included as a mandatory requirement in other legislative documents, such as the inclusion of Biodiversity Net Gain and (a planned) requirement for sustainable drainage systems for surface water into planning regulation. This indicates that Government is still committed to inclusion of wider sustainability aspects into legislation, where a clear benefit and a suitable mechanism can be found.

2.3. Government strategy and policy directives

Most countries operate a two-tier system of governance whereby legislation sets a formal, mandatory framework within which all parties must operate. The secondary level of governance is the production of strategy and policy directives, which while not legally binding (since they do not form legislation) set out Government intent. Government can use such strategies and policies to determine investment priorities and to raise awareness in private markets of the desirability of action in line with strategic objectives set out. Strategy publications often signal a direction that Government is considering for future legislation, indicating a direction of travel that prudent organisations evaluate when making decisions. Strategy can also be used to set fiscal incentives to action, helping to build a business case for private action without the need for legislation and regulation. Because of the rapidly evolving nature of the climate and sustainable development agenda, strategy and policy has had a raised profile and impact, particularly where it applies to the intersection between economic and environmental policy.

2.3.1 Clean Growth Strategy

In recent decades, many countries have assembled and published variations of clean growth strategy, bringing together their targets and aspirations for decarbonisation together with their ambition and desire to maintain and improve economic growth. In the UK, a new Clean Growth Strategy was published in 2017 (HM Government, 2017) that responded to and fulfilled relevant aspects of the Climate Change Act. Despite global acceptance of the need for decarbonisation, one of the major barriers to taking action at this time was the belief that changing from the current fossil-fuel dominated system to transition to clean energy sources would significantly damage economic growth and thus, prosperity in a country.

The UK's Clean Growth Strategy provided evidence that since the ambitious decarbonisation target had been set, it had been possible to make considerable reductions in carbon emissions whilst maintaining economic growth. The reduction in carbon emissions referred to in the strategy was almost wholly due to a decarbonising electrical energy grid, driven by reducing prices in renewable energy technology (and supported by financial incentives to take-up).

This evidence, and similar evidence from other countries including the Intergovernmental Panel on Climate Change (IPCC, 2014), supports a view that it may be possible to maintain economic growth while simultaneously reducing GHG (carbon) emissions. This viewpoint is supported by various quantifications and forecasts of the cost of allowing continuing GHG emissions globally, resulting in climate changing impacts. In the UK, the first authoritative estimates of the potential economic consequences of inaction were published in 2007 in the Stern Review (Stern, 2007) (commissioned by the UK Government of the time), and the author has made clear in subsequent articles that, in fact, a business as usual (fossil fuel consuming) trajectory will result in considerable damage to economic growth, to a greater extent than a decarbonised trajectory, due to the costs of inaction (Stern, 2016).

Alternative viewpoints contending that continuing economic growth with disregard to the environmental consequences is not possible or supportable if a sustainable global future is to be achieved, have growing numbers of supporters. This view is further explored in Chapter 4 on the inclusion of social value into frameworks.

These arguments make clear that whatever viewpoint is held, decarbonisation is necessary to the industries and businesses that support economic prosperity. The Clean Growth Strategy and others of its kind begin to set out mechanisms by which this decarbonisation can occur. Such strategies indicate clearly to businesses and investors that decarbonisation is necessary if their organisation is to prosper in forthcoming regulatory frameworks. The most successful companies will be those that can be most resilient to future change, and therefore such strategies that clearly signal political and legislative trajectories can be very helpful in moving board decisions in desirable directions.

In common with many other strategies published in the years following the Climate Change Act, the Clean Growth Strategy portrays a targeted and narrow focus on 'decarbonisation' (reduction and elimination of GHG emissions) and while natural landscapes, the wider environment and social aspects of sustainability are mentioned, it is within the context of associated benefit to decarbonisation, or the promotion of carbon-storing environments. More recently, a more holistic and broad interpretation of climate action is being built into strategy and policy. The UK's Modern Industrial Strategy (Department for Business and Trade, 2025) moves beyond a discussion of the importance of net zero, to specifically supporting sectors in their efforts, such as electrification of industrial processes. This document also embeds a wider interpretation of climate action, including life-cycle impacts, recycled materials and resource extraction. It goes beyond carbon to emphasise the importance of climate adaptation and nature recovery, demonstrating how policy-led thinking has evolved beyond carbon considerations.

2.3.2 25-year Environment Plan

Strategies and policies are also helpful because they can draw together areas of focus that may be spread among multiple different areas of government and potentially, multiple areas of legislation. This is particularly true of environmental aspects of sustainability that, unlike carbon emissions, do not have a single metric that can be used to set targets and track progress.

The UK Government's 25-year Environment Plan sets out a series of actions and targets across multiple areas of intervention ranging from air and water quality, biodiversity and species recovery, agriculture and land use (HM Government, 2018). The plan also signals the need to address the impacts of climate change that will result from global temperature rise, including flooding. The document draws clear links between economic and community benefits and achievement of environmental goals. The premise set out in the UN SDGs (United Nations, 2015b) that economic, social and environmental challenges are interlinked and indivisible is reiterated by the strategy. Inevitably, the breadth of aspects considered means that such a strategy contains high-level targets and objectives, leaving the detail of delivery and intermediate targets to supporting documents. The benefit of utilising a strategy document rather than primary or secondary legislation is the flexibility of this approach and the ability to adopt high ambitions for achievement without risking legislative enforcement if high-level, long-term targets are missed.

Nevertheless, strategies of this type set clear direction and ambitions that can be built upon and expanded in more detailed documents and do have a very helpful function in making the long-term vision clear. In the case of the Environment Plan, a revision of the document five years after publication (HM Government, 2023a) set out achievements to date and in some areas, set out shorter term actions that legislation would be brought forward to deliver. The benefit of such documents for organisations and investors is the ability to understand the overarching direction that legislation will take in the future. Organisations can therefore seek to set their own strategies to align with this vision. Since such strategies often form a precursor to legislation – the 25-year Environment Plan led to an Act of Parliament (Environment Act, 2021) setting out various mandatory requirements – for example, this is a prudent step for businesses whose activities may be impacted by such legislation.

2.3.3 Gear Change – cycling and walking strategy

For specific environmental aspects, strategies can introduce bold and transformative change that would be difficult to agree if introduced as legislation. In 2020, the UK Government introduced a 'vision' document that set out their strategy for increasing cycling and walking, with a view to achieving a shift away from private motor vehicles (Department for Transport, 2020). This document envisaged a step-change in 'active travel', partly in response to an identified need to improve people's health and levels of activity. The document proposed a range of actions and measures to achieve the objectives, including investment in infrastructure and measures to encourage cycling and walking. Importantly, the document indicated a clear direction in terms of highway infrastructure funding, which would henceforth only be able to access funding if adequate active travel infrastructure, such as dedicated cycling lanes, were incorporated. The indication of future funding formulae and specific actions to achieve compliance ensured that practitioners working in infrastructure design and place-making took immediate notice. Whether the transformative effects envisaged in the strategy are realised in actuality, this example illustrates how valuable a clear and deliverable strategy can be in influencing action by organisations and individuals working in affected sectors.

2.4. Market-led corporate action

The global nature of climate change and the international consensus on the need for action has prompted many corporate sectors to act to prepare their own responses, even without legislation or a clear national strategy picture. Globalisation of corporates and the interconnected nature of financial and trading markets now mean that disruptive change can have widespread global impacts.

2.4.1 Climate related financial disclosures

In recognition of this, the financial sector took an early lead to assemble a group focused on examining vulnerability and stability in peer organisations and processes. The Financial Stability Board (formerly the Financial Stability Foundation) counts central banks and treasuries among its members and therefore has a high level of credibility and influence; essential in this type of non-legislative, international effort. The board set up a task force on climate-related financial disclosures (TCFD) in 2015, recognising that

> globally consistent and comparable disclosures by companies of their climate-related financial risks are increasingly important to market participants and financial authorities, in order to . . . manage risks and seize opportunities stemming from climate change. (FSB, 2020: p. 1)

The role of the task force was to define a set of disclosure recommendations for companies to ensure that relevant information was made available in a consistent manner to investors, lenders and insurance underwriters. The task force recommended that companies disclose the nature of the climate risks to which their business was exposed, the impact of these risks and the resilience of the business to such risks, including different climate future scenarios. The long timescales associated with businesses investing in property and real-estate, added to their exposure to extreme climate events, has made this global drive toward risk and exposure reporting particularly important.

In the UK, mandatory reporting is required for certain organisations, particularly larger organisations and those listed on the UK Stock Exchange. This includes major investors and owners in real-estate markets in the UK, raising awareness among the boards of these companies of the importance of integrating sustainability and resilience into their asset holding.

With the publication in 2023 of the Green Finance Strategy, central government departments and other associated investment organisations of importance (such as public sector pension funds) are now also required to implement the TCFD disclosure recommendations, which will further drive the transparency of sustainability implementation in asset holdings (HM Government, 2023b).

The disclosure guidelines produced by the task force have formed a foundation for international financial sustainability standards including International Financial Reporting Standards (IFRS) and the Foundation's International Sustainability Standards Board (ISSB), which uses the TCFD recommendations as a basis but with additional sustainability (not just climate-related) issues considered (IFRS, n.d.).

Although these recommendations and guidelines began as international collaborations on the part of larger corporates and sectoral coalitions (for further details and comparison of these standards, see Chapter 4), they are now being utilised as a basis for mandatory standards. While legislation is a time-consuming business requiring wide consultation and iteration, the authoritative nature of an industry-compiled standard and its widespread acceptance as 'best practice' in relevant sectors makes it simpler for governments to utilise such a standard as a basis for legislation. This can speed up the process of regulation and, in addition, streamlines the process of updating regulatory standards at a time of accelerating change in market places.

This market-led approach is now being championed in other sectors particularly relevant to the urban environment, such as the emerging (at time of writing) UK Net Zero Carbon Buildings Standard. Similar to the financial disclosures frameworks, this standard is being developed by an ambitious coalition of actors in the built environment sector and aims to set a rigorous and science-based approach to claims of 'net zero' for buildings in the UK.

2.4.2 Voluntary market-based financial mechanisms

Corporate action can be incentivised by consumer choice and reputation, particularly where mechanisms exist that makes choice easy to navigate. Market-based mechanisms create a system whereby customers or organisations who wish to demonstrate their commitment to sustainable goals can do so in a way that financially benefits suppliers.

Similar to the renewable energy certificates, Renewable Energy Guarantees of Origin (REGOs) are a market-based mechanism that allows customer choice to drive a beneficial market for renewable energy. In this mechanism, renewable energy generators are issued with a REGO for each unit of renewable energy generated, proving that the electricity generated is from renewable energy sources. When renewable energy is purchased from such a source, the REGO is purchased alongside as proof of origin. Any surplus REGOs can also be traded as an additional income source for the energy generator. Consumers seeking to reduce carbon from their energy supply can choose a supplier offering renewable energy, backed by a REGO guaranteeing that the supply is genuinely renewable (and ensuring no 'double counting' of renewable energy supply). Importantly, the UK energy regulator Ofgem issues and regulates the REGO market, providing confidence in the market that is fundamental to participation.

Many countries and trading blocs have set up Guarantee of Origin schemes, responding to consumer demand for guaranteed renewable energy supplies and also to assist companies that are seeking to improve their performance in annual carbon reports (discussed in more detail in Chapter 4). Government regulation that requires energy supply companies to declare the fuel mix of their energy supply incentivises participation in this market also, although participation is not compulsory.

2.4.3 Voluntary carbon trading

It is also useful to note that while not mandated by Government, voluntary carbon trading is increasing in prominence in organisations and industries keen to demonstrate that they are playing their part in decarbonisation efforts. By purchasing carbon 'credits' to offset some or all of their carbon emissions, organisations can demonstrate that they are participating in global decarbonisation. Types and cost of carbon 'credit' can vary widely, depending on the type of decarbonisation project that they are enabling. Projects can range from tree-planting to efficiency projects that avoid carbon emissions through to direct air capture of carbon, for example. Because voluntary carbon trading does not require purchase of a particular carbon permit type, they are not constrained by national boundaries and this has enabled carbon credits to become a global commodity trade that is now growing in size and importance.

The voluntary carbon trading market has been legitimised and strengthened by approval of Article 6 of the Paris Agreement (at COP26 in Glasgow 2021), which sets out a framework for voluntary cooperation between countries to achieve emission reduction targets (United Nations, 2015a). This enables countries to transfer carbon credits earned by decarbonisation projects to another country, thus helping to meet the receiving country's decarbonisation targets. Of course, if such trading is to be credible and robust, it must be properly governed to ensure that there is no double counting, and that the decarbonisation projects do indeed deliver the carbon reduction claimed. This is an emerging opportunity that offers considerable scope for benefit for participant organisations and nations but (at time of writing) is still evolving.

The participation in voluntary carbon trading is assisted by Government mandates to publicly report specific environmental metrics, primarily carbon emissions – for example, in 2019, the UK required all large companies to report publicly on energy use and carbon emissions (HM Government, 2019). The main driver behind such reporting requirement is to ensure that companies are quantifying their carbon emissions but also reporting them publicly to allow the organisation to be held to account by customers or investors where appropriate. Purchase of traded carbon credits can allow an organisation to report reduced carbon emissions and can be reputationally beneficial, thus mandated environmental reporting can have a genuine benefit for global decarbonisation efforts.

2.4.4 Voluntary impact reporting

Just as the mandated financial disclosures followed voluntary efforts by companies, carbon reporting (and reporting of other environmental and social indicators) was being done voluntarily by many companies, before being made compulsory. While some companies are mandated by Government to report on their carbon emissions, it is generally only the larger organisations or those listed on various stock exchanges that are required to do so. Smaller organisations or those based in countries where such mandates do not exist are currently under no obligation to publicly report on their impacts. Nevertheless, many companies choose to do so; their reasons range from a perception that their customers and investors demand this type of transparency, to a deliberate decision to brand their organisation as carbon and climate conscious. Similarly, many larger organisations choose to report not only their carbon emissions but also widen this reporting to other environmental and social impacts. Support for such reporting has been growing and is now widespread across many sectors.

Environmental, social and governance (ESG) reports are now a regular part of annual reporting for many companies of all sizes, with popularity of such approaches growing year on year. In order to avoid perceptions of 'greenwash', several organisations such as the Global Reporting Initiative (refer to Chapter 4, Section 4.6.), the International Organisation for Standardisation (ISO) and others offer standardised methodologies. These help to align reporting processes so that companies can more easily benchmark themselves and their progress in a robust and believable manner.

As the name implies, these reporting standards cover a broad range of reporting criteria including environmental aspects (and, increasingly, natural environment such as biodiversity impacts), social aspects to do with community and people and organisational governance matters.

2.5. Sustainability benchmark assessment methods in different countries

Clients seeking to apply sustainability frameworks to their building and asset portfolios find it particularly helpful to utilise sustainability benchmark assessment methods that allow a summarised set of indicators and goals that respond to the sector and geography. Originating as a method for UK portfolio holders to benchmark their various assets, the concept of an 'off the shelf' indicator set, allowing comparative assessment (thus providing a competitive element) for buildings was rapidly taken up in other geographies.

Since BREEAM was launched in the 1990s, many countries developed their own version of 'BREEAM' or equivalent that is suited to country-specific climate and culture. These frameworks are designed to evaluate a building's environmental impact and performance and operate as green building 'rating systems' that compare the performance of assets. In some cases, they include new and existing buildings as well as extending to public realm or infrastructure.

As discussed in Chapter 1, in response to climate change and wider environmental and energy conservation challenges, a holistic assessment method became essential to help guide built environment professionals in their desire to improve the overall environmental performance of buildings to promote sustainable design, and best practices in construction and operations. It also provided government bodies with an industry-led policy lever in benchmarking and steering towards more sustainable practices. Benefits of using these assessment methods to developers and asset owners include the following.

- Meaningful differentiation of their buildings in the real estate market.
- Positive effect on corporate brand, leasing and resale value of buildings.
- Reduction in energy, water and material resource usage.
- Reduction of potential environmental impact.
- Improvement in indoor environmental quality for better health and wellbeing.
- Provision of clearer direction for continual improvement.

The intended audience for these assessment methods are developers, asset owners, government agencies as well as architects and designers. Indeed, a study (Cushman and Wakefield, 2021) showed that an office building certified under the Leadership in Energy and Environmental Design

Figure 2.1 LEED-certified against non-LEED full service rent (per square foot) (created using data from Cushman Wakefield CoStar (Cushman and Wakefield, 2021))

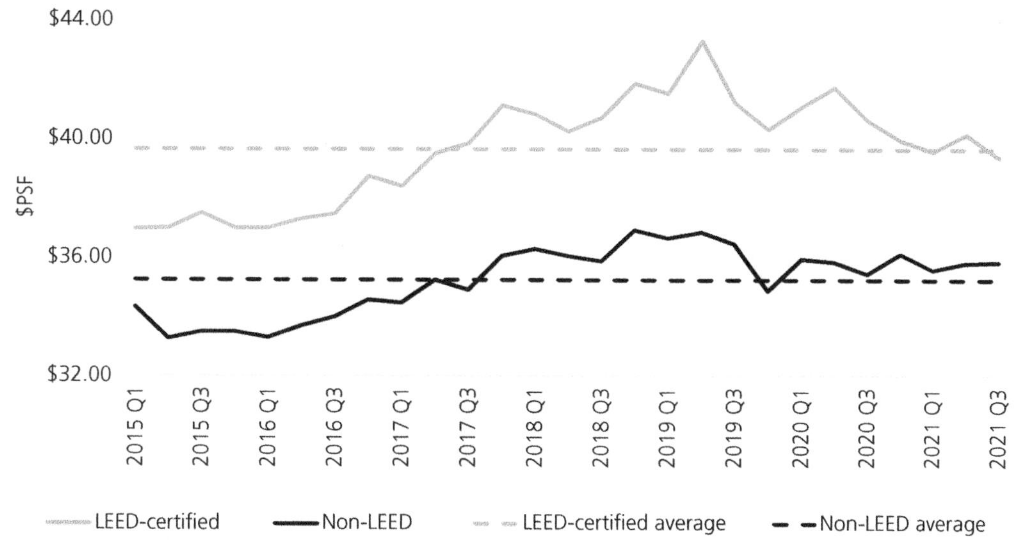

(LEED) scheme (a rating system widely used in the USA, see below) will likely secure higher rental yield than a non-sustainability certified building (Figure 2.1). The commercial real estate company CBRE's assessment report in 2022 re-confirmed that tenants benefit from energy savings and increased productivity, while owners benefit from higher asset value (CBRE, 2022).

A range of leading assessment methods and rating systems across Europe, the United States of America, Asia and Australia is shown in Figure 2.2 below. These green building rating systems are designed to evaluate a building's environmental impact and performance. Common criteria among these systems are thematically based and include natural resource efficiency practices such as energy and water, indoor environment performance as well as wider issues

- energy, carbon
- water use
- materials
- land use
- impacts on site ecology
- transport
- indoor air quality, lighting, acoustics
- waste and adaptability, flexibility
- management – construction, operations and maintenance.

As the first generation of sustainability assessment methods, schemes such as BREEAM and LEED have not tended to cover social, economic and urban planning issues. In the UK, these issues are usually addressed by Environmental Impact Assessment (EIA) and Sustainability Appraisal, commonly required for major developments by the UK town planning regulation systems. EIA complements rating systems such as BREEAM in a number of areas. EIA focuses on the impact of

Figure 2.2 Global sustainability assessment methods or rating systems

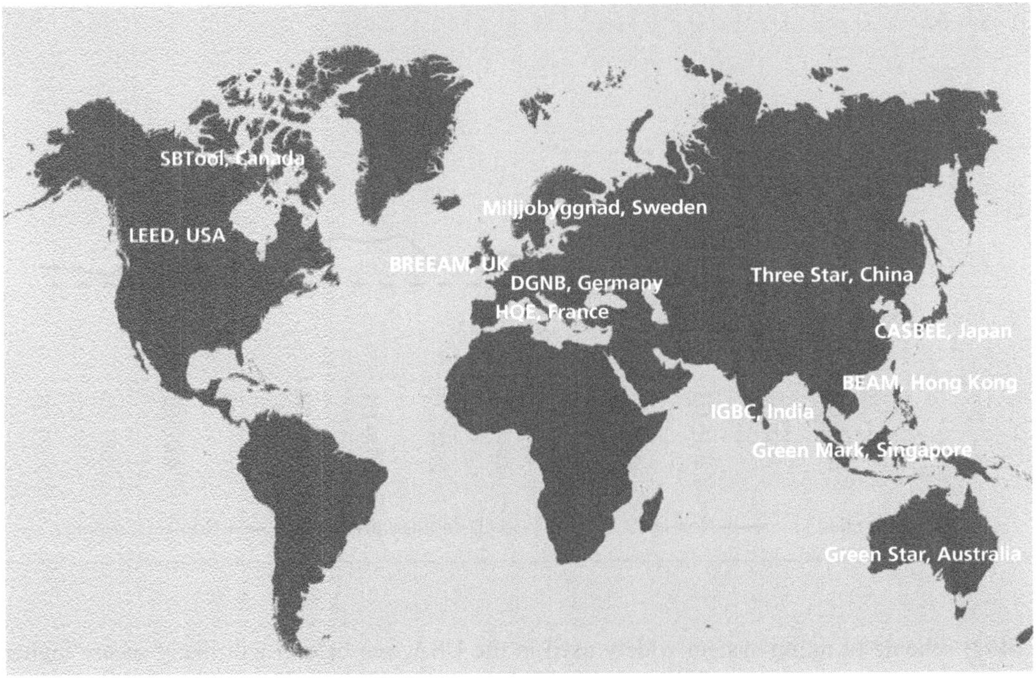

BREEAM: Developed by BRE, UK

LEED: Leadership in Energy and Environmental Design, developed by US Green Building Council

DGNB: developed by German Sustainable Building Council

HQE: French rating system

SBTool: set up by International Initiative for a Sustainable Built Environment (iiSBE) in Canada

Miljobyggnad: Scheme by Swedish Green Building Council

Green Star: Rating scheme developed by Australian Green Building Council

CASBEE: Comprehensive Assessment System for Built Environment Efficiency developed by Japan Green Build Council and Japan Sustainable Building Consortium

Green Mark: Rating scheme developed by Singapore Building and Construction Authority (BCA)

HK-BEAM: Hong Kong Building Environmental Assessment Method (HK-BEAM)

Three Star System: introduced Chinese government agency, Ministry of Housing and Urban-Rural Development (MOHURD)

IGBC: Indian Green Building Council rating system

a proposed development on the potential receptors such as local economy, heritage, nearby neighbourhood environment quality and transport issues, whereas BREEAM mainly addresses sustainable design practice of the proposed building itself. Taking an example of daylighting, BREEAM requires the indoor daylight factor and view to outside to achieve a certain standard for a workplace, while EIA requests the proposed development does not cause a negative reduction in access to daylight and sunlight of a neighbouring home due to the new building proposed. EIA can be time-consuming to apply and is relatively inflexible; a proposal to replace the EIA process with a new Environmental Outcomes Framework to help overcome these limitations has been proposed but not yet developed (at time of writing).

Many of these methods have numerous similarities but nevertheless differ in various respects. They differ in the way assessment is carried out and the way information is prepared, the legal obligation for code compliance and certification, the criteria they contain and their methods of scoring performance, the cost of certification and the possibilities of assessing different building types and different planning stages. Markelj *et al.* (2013) provides a comprehensive comparison across the most popular methods including BREEAM, LEED and Deutsche Gesellschaft fur Nachhaltiges Bauen (DGNB) (German Sustainable Building Council). It is important to note that these methods are constantly evolving to reflect the latest advancement and practice in sustainable development, with new updates to the content and indicator standards published every few years.

2.5.1 Tools for large scale development

Due to the limitations associated with single building sustainability assessment, BREEAM and LEED and other established schemes like Comprehensive Assessment System for Built Environment Efficiency (CASBEE) launched neighbourhood sustainability assessment (NSA) tools (see Table 2.1), which examine the buildings together with their environments and evaluate the topics like society, land usage, transportation, water, air, energy and bio diversity as a whole; economically, environmentally and socially (Yıldız *et al.*, 2016). A comparison of these NSAs (Yildiz *et al.*, 2016) found that each of these schemes reflect local conditions and priorities – for example, the dependence of the transportation on cars in the USA and its position as a significant natural resource consumer led to giving the highest weighting to transportation in LEED among all the systems. BREEAM gives more weight to the protection of the environment including land usage, resources and energy usage. Whereas social development in CASBEE and economic development in DGNB clearly stand out when compared with the other systems.

Further similar neighbourhood assessment methods such as LEED SITES and BREEAM Infrastructure have been introduced to assess the public realm, parks and infrastructure related developments.

Indicators and composite indicators are increasingly recognised as a useful tool for policy making and public communication in conveying information on countries' performance in fields such

Table 2.1 Neighbourhood assessment tools

BREEAM Communities	LEED ND	DGNB NSQ/NUD	CASBEE UD	Green Mark Districts
Building Research Establishment Environmental Assessment Method for Communities (BRE, 2012)	Leadership in Energy and Environment Design – Neighborhood Development (USGBC, 2014)	Deutsche Gesellschaft fur Nachhaltiges Bauen Neubau Stadtquartiere New Development (DGNB, 2020)	Comprehensive Assessment System for Built Environment Efficiency Urban Development (JSBC and IBEC, 2014)	Green Mark for Districts (BCA, 2013)

as environment, economy, society or technological development. City assessment tools help the users to understand the real status of their own cities. Today, many city rankings are published by various institutions such as the 'Liveability Ranking' of Economist Intelligence Unit, 'Quality of Living Survey' of Mercer Human Resource Consulting and 'Global Power City Index' of the Mori Memorial Foundation. However, these tools are usually focused on a high level for benchmarking the chosen indicators.

By incentivising the built environment marketplace to continually update their concept of 'best practice' in sustainability, building assessment and rating systems have been very successful. However, more recently it has become increasingly recognised that building assessment systems on their own are neither sufficient nor adequate to address the sustainability challenges faced by urban societies. The science is now showing with alarming clarity how quickly time is running out to avoid catastrophic and irreversible changes to the globe. There is now scientific consensus of the need to take urgent action to almost halve global emissions by 2030 and eliminate them completely by the middle of the century. Since the 2015 Paris Agreement, the development of dedicated indicators/indexes for specific areas has accelerated. This change complements the existing sustainability rating systems and to a large degree, provides an imperative for the existing rating systems to enhance and evolve to take account of the complexity and urgency of the challenge.

REFERENCES

Building Act 1984 (1984). https://www.legislation.gov.uk/ukpga/1984/55/contents (accessed 17/01/2026).

Building and Construction Authority (BCA) (2013) *BCA Green Mark for Districts*. https://www1.bca.gov.sg/docs/default-source/docs-corp-buildsg/sustainability/gm_district_v2_2.pdf?sfvrsn=81b80c4a_0 (accessed 17/01/2026).

Booth P and Huxley M (2012) 1909 and all that: reflections on the Housing, Town Planning, etc. Act 1909. *Planning Perspectives* **27(2)**: 267–283, 10.1080/02665433.2012.646776.

BRE (Building Research Establishment) (2006) *EcoHomes 2006 – The Environmental Rating for Homes* (*Issue 1.2.*). https://tools.breeam.com/filelibrary/Technical%20Manuals/EcoHomes_2006_Guidance_v1.2_-_April_2006.pdf (accessed 17/01/2026).

BRE (2012) *BREEAM Communities Standard*. https://breeam.com/standards/communities (accessed 17/01/2026).

CBRE (2022) *Green Is Good: The Enduring Rent Premium of LEED-Certified U.S. Office Buildings*. https://www.cbre.com/insights/viewpoints/green-is-good-the-endurance-of-the-rent-premium-in-leed-certified-us-office-buildings (accessed 23/10/2024).

Climate Change Act 2008 (2008). https://www.legislation.gov.uk/ukpga/2008/27/contents (accessed 17/01/2026).

Climate Change Act 2008 (2050 Target Amendment) Order 2019 (2019). https://www.legislation.gov.uk/ukdsi/2019/9780111187654/contents (accessed 17/01/2026).

Cushman and Wakefield (2021) *Green Is Good: Sustainable Office Outperforms in Class A Urban Markets*. https://assets.cushmanwakefield.com/-/media/cw/americas/united-states/insights-pdfs/green-is-good-spotlight_final1.pdf?rev=e26315797d7d49faa6c58ca7762f91a6 (accessed 17/01/2026).

Department for Business and Trade (2025) *The UK's Modern Industrial Strategy*. CP 1337. UK Government. https://assets.publishing.service.gov.uk/media/68595e56db8e139f95652dc6/industrial_strategy_policy_paper.pdf (accessed 05/09/2025).

Department of Communities and Local Government (2010) *Code for Sustainable Homes: Technical Guide*. https://assets.publishing.service.gov.uk/media/5a74b01b40f0b619c86599ff/code_for_sustainable_homes_techguide.pdf (accessed 17/01/2026).

Department for Transport (2020) *Gear Change: A Bold Vision for Cycling and Walking*, pp. 2–50. https://assets.publishing.service.gov.uk/media/5f1f59458fa8f53d39c0def9/gear-change-a-bold-vision-for-cycling-and-walking.pdf (accessed 17/01/2026).

Directive 2010/31/EU of the European Parliament and of the Council of 19 May 2010 on the energy performance of buildings (since revised) (2010). https://eur-lex.europa.eu/eli/dir/2010/31/oj/eng (accessed 17/01/2026).

Energy Act 2011 (2011) https://www.legislation.gov.uk/ukpga/2011/16/contents (accessed 17/01/2026).

Energy Performance of Buildings (England and Wales) Regulations 2012 (2012) https://www.legislation.gov.uk/uksi/2012/3118/contents (accessed 17/01/2026).

Environment Act 2021 C.30 (2021) https://www.legislation.gov.uk/ukpga/2021/30/contents (accessed 17/01/2026).

DGNB (Deutsche Gesellschaft für Nachhaltiges Bauen) (2023) *DGNB System for Districts Version 2020*. https://www.dgnb.de/en/certification/districts#:~:text=In%20all%20certification%20phases%2C%20the,not%20serve%20the%20general%20public (accessed 17/01/2026).

FSB (Financial Stability Board) (2020) FSB response to the IFRS Foundation's Consultation Paper on Sustainability Reporting, 22 December 2020. https://www.fsb.org/uploads/P221220.pdf (accessed 17/01/2026).

Greenhouse Gas Emissions Trading Scheme Order 2020 (2020) https://www.legislation.gov.uk/uksi/2020/1265/contents/made (accessed 17/01/2026).

HM Government (2017) *The Clean Growth Strategy: Leading the Way to a Low Carbon Future*. https://assets.publishing.service.gov.uk/media/5ad5f11ded915d32a3a70c03/clean-growth-strategy-correction-april-2018.pdf (accessed 17/01/2026).

HM Government (2018) *A Green Future: Our 25 Year Plan to Improve the Environment*. https://assets.publishing.service.gov.uk/media/5ab3a67840f0b65bb584297e/25-year-environment-plan.pdf (accessed 17/01/2026).

HM Government (2019) *Environmental Reporting Guidelines: Including Streamlined Energy and Carbon Reporting Guidance*. https://assets.publishing.service.gov.uk/media/5de6acc4e5274a65dc12a33a/Env-reporting-guidance_inc_SECR_31March.pdf (accessed 17/01/2026).

HM Government (2023a) *Environmental Improvement Plan 2023: First revision of the 25 year Environment Plan*. https://assets.publishing.service.gov.uk/media/64a6d9c1c531eb000c64fffa/environmental-improvement-plan-2023.pdf (accessed 17/01/2026).

HM Government (2023b) *Mobilising Green Investment: 2023 Green Finance Strategy*. https://assets.publishing.service.gov.uk/media/643583fb877741001368d815/mobilising-green-investment-2023-green-finance-strategy.pdf (accessed 17/01/2026).

Housing, Town Planning &c Act 1909 (1909) https://www.legislation.gov.uk/ukpga/Edw7/9/44/enacted (accessed 17/01/2026).

Housing, Town Planning &c Act 1919 (1919) https://www.legislation.gov.uk/ukpga/1919/35/enacted (accessed 17/01/2026).

IFRS – IFRS Sustainability Standards Navigator (n.d.) https://www.ifrs.org/issued-standards/ifrs-sustainability-standards-navigator/ (accessed 17/01/2026).

IPCC (Intergovernmental Panel on Climate Change) (2014) *AR5 Climate Change 2014: Mitigation of Climate Change IPCC Working Group III Contribution to AR5*. https://www.ipcc.ch/report/ar5/wg3/ (accessed 17/01/2026).

JSBC and IBEC (Japan Sustainable Building Consortium and Institute for Building Environment and Energy Conservation) (2014) *CASBEE for Building (New Construction)*. https://www.ibecs.or.jp/CASBEE/english/download/CASBEE-BD(NC)e_2014manual.pdf (accessed 17/01/2026).

Markelj J, Kuzman M and Zbasnik-Senegacnik M (2013) A review of building Sustainability Assessment Methods. *AR: Architecture, Research* **(XIV)**: 22–31.

Ministry of Housing Communities and Local Government (2024) *National Planning Policy Framework*. https://assets.publishing.service.gov.uk/media/67aafe8f3b41f783cca46251/NPPF_December_2024.pdf (accessed 17/01/2026).

OECD (2021) *The United Kingdom's Pioneering Climate Change Act*. https://www.oecd.org/en/publications/ipac-policies-in-practice_22632907-en/the-united-kingdom-s-pioneering-climate-change-act_c08c3d7a-en.html#:~:text=of%20climate%20action.-,Approach,This%20kind%20in%20the%20world (accessed 17/01/2026).

Pickles E (2015) *Planning Update Statement* 25 March 2015. *Statement UIN HCWS488*. DCLG, UK. https://questions-statements.parliament.uk/written-statements/detail/2015-03-25/HCWS488 (accessed 17/01/2026).

Planning and Compulsory Purchase Act (2004) https://www.legislation.gov.uk/ukpga/2004/5/contents (accessed 17/01/2026).

Regulation (EU) 2021/1119 of the European Parliament and of the Council of 30 June 2021. Establishing the framework for achieving climate neutrality and amending Regulations (EC) No 401/2009 and (EU) 2018/1999 ('European Climate Law') (2021).

Stern N (2007) *The Economics of Climate Change: The Stern Review*. Cambridge University Press, Cambridge, UK. https://assets.cambridge.org/97805217/00801/frontmatter/9780521700801_frontmatter.pdf (accessed 17/01/2026).

Stern N (2016) Economics: current climate models are grossly misleading. *Nature* **530(7591)**: 407–409, 10.1038/530407a.

Town and Country Planning Act 1947 (1947) https://www.legislation.gov.uk/ukpga/1947/51/enacted

United Nations (2015a) *Paris Agreement*. https://unfccc.int/files/essential_background/convention/application/pdf/english_paris_agreement.pdf (accessed 17/01/2026).

United Nations (2015b) *Transforming Our World: The 2030 Agenda for Sustainable Development | Department of Economic and Social Affairs*. https://sdgs.un.org/2030agenda (accessed 17/01/2026).

USGBC (US Green Building Council) (2014) *Guide to LEED Certification: Neighborhood Development*. https://www.usgbc.org/tools/leed-certification/neighborhood (accessed 17/01/2026).

Yıldız S *et al.* (2016) Neighborhood sustainability assessment tools and a comparative analysis of five different assessment tools. *Journal of Planning*, 10.14744/planlama.2016.05914.

Navigating an Evolving Sustainable Built Environment: Delivering Net Zero Carbon and Climate Resilience

Mei Ren and Rachel Waggett
ISBN 978-1-83662-853-8
https://doi.org/10.1108/978-1-83662-850-720261006

Chapter 3
Climate change initiatives

3.1. Reaching a global scientific consensus on climate change

As early as 1896 when Svante Arrhenius (1859–1927) first claimed that fossil fuel combustion may eventually result in enhanced global warming, he proposed a relationship between atmospheric carbon dioxide concentrations and temperature (Arrhenius, 1896). Heated debate regarding the science underpinning climate change and especially, the extent to which human combustion of fossil fuels are a major cause has existed ever since. The climate change debate has been one of flaring points among politicians and scientists, especially since the 1990s, even though the scientific consensus is clear that humans have caused the warning trend of the last century (Myers *et al.*, 2021).

The Intergovernmental Panel on Climate Change (IPCC) stated in its 5th Assessment Report (AR5) in 2013 that it is 'extremely likely that most of observed increase in global surface temperature since 1951 is caused by human influence' (IPCC, 2013). That statement was based on the use of climate models to investigate what the world's climate would have been like without human emissions of greenhouse gases (GHGs) and land use change. AR5 published further guidance on impacts, adaptation and vulnerability in 2014 to help evaluate how patterns of risks and potential benefits are shifting due to climate change (IPCC, 2014).

The Paris Agreement, signed in 2015, was a turning point in international climate negotiations (United Nations, 2015). Almost every country agreed to take collective global action to tackle climate change. Under the Paris Agreement

- countries have committed to limit global warming to 2°C above pre-industrial averages
- if possible, limit warming to 1.5°C
- increase measures to adapt to climate change
- provide enhanced support to developing countries.

The UK was one of the early signatories of the Paris Agreement, and the UK Government set a legally binding target of net zero emissions by 2050. Without the development of sophisticated climate models in the last 50 years, the attribution of global warming to human factors could not have been made. Climate scientists have since changed their focus and motivation from verifying Arrhenius' hypothesis to the investigation of the effects of all associated factors to improve climate projections and inform decision making (Slingo, 2016).

The predictive power of climate models enables better preparation to deal with the risks that will be caused by the changing climate system. It is important to note the different focuses on climate mitigation and adaptation; mitigation refers to measures to reduce carbon (and other GHGs) emitted to the atmosphere as much as possible, ideally to be net zero carbon, whereas adaptation refers to measures to adapt to climate change already built-in because of carbon already emitted. As carbon continues to accumulate in the atmosphere despite reductions in carbon emissions, the planet will continue to warm. Climate resilience and adaptation therefore addresses the natural and built environment and communities to enable them to be more resilient to extreme climatic conditions and adapt to the changes. This chapter examines the important concepts of climate mitigation and adaptation methodologies and provides a good practice guide with examples.

3.2. Understanding of climate science and climate projection models

Climate change policy and action initiatives have been largely informed by the advancement of climate science and its models of planetary systems. Since the 1950s, the principal tools of climate science have been computer simulation models of the global general circulation moving towards increasingly comprehensive coupled models of the entire climate system (Edwards, 2010). Climate model evaluation and intercomparison is changing modelling into a more standardised, modular process, presenting the potential for unifying research and operational aspects of climate science.

The UK's Met (Meteorological) Office Hadley Centre is a global leading centre for climate science and services that played a significant role in the compilation of all six IPCC Assessment Reports.

Climate models are based on the components of a global coupled climate model that include an atmospheric model, an ocean model, a land model and a sea ice model. In addition, some recent versions of models include a land ice model to simulate ice sheets and mountain glaciers. Led by the World Climate Research Programme, a range of different research centres run the same set of experiments on their models aiming to create a better understanding of climate change. Attribution of climate change now goes beyond just considering the global mean surface temperature to address other components, with more regional aspects of the climate system and even extreme events such as flooding, drought and heatwaves.

Although the scientific community's understanding and ability to simulate the climate is advancing all the time, these models are not able to represent all the features seen in the present-day real climate. Weather conditions have natural variability over months and years, whereas these models seek to understand and simulate the underlying pattern of change over time, to predict future outcomes. It is important to note the commonly quoted phrase 'Climate is what you expect, weather is what you get' that seeks to promote understanding of the differences. It is in fact the same science underpins them both. The climate science clearly indicates that in a warming world, the greatest impacts of climate change on society will be felt through changes in the weather, especially hazardous weather such as floods, storms, heatwaves and drought. The transformation of climate science to a science that is rooted in local weather prediction is the next big step; however, this is beyond the discussion of this book.

A few important versions of Hadley models are summarised as follows.

- HadCM3 model – Hadley Centre Coupled Model version 3, developed in 1999, was the first unified model of climate configuration not to require flux adjustments (artificial adjustments applied to climate model simulations to prevent them drifting into unrealistic climate states). HadCM3 was one of the major models used in the IPCC 3rd and 4th Assessments, and also contributed to the 5th Assessment.
- HadGEM1 stands for the Hadley Centre Global Environment Model version 1. It was developed in 2006 and used in the 4th Assessment Report.
- HadGEM2-AO – a coupled climate model and was selected as standard scenarios for the IPCC 5th Assessment Report (AR5).

- HadGEM3-GC3.1 is the third generation of Hadley climate models. With a host of advances over previous models, UKESM1 (an Earth model) and HadGEM3-GC3.1 are two state-of-the-art climate models, and have been used to form the basis of UK contributions to IPCC 6th Assessment Report AR6.

CMIP6 is the latest phase of collaboration under the Coupled Model Intercomparison Project (CMIP), which represents the most current global climate model data available. This climate data provides the foundation for the IPCC's 6th Assessment Report. Murphy *et al.* (2023) compared the difference in climate change projections from the UK Met Office's Climate Projections (UKCP) land scenarios and CMIP6 models.

The UKCP programme has published a series of climatic projections to inform policy making, including UKCIP02, UKCP09 and lately UKCP18 and Shared Socio-economic Pathways (SSP), based on the aforementioned Hadley climate models.

3.2.1 Representative Concentration Pathways (RCPs)

In order to assess climate responses, RCPs using the coupled climate model HadGEM2-AO were selected as standard scenarios for the IPCC 5th Assessment Report (IPCC, 2013) and these scenarios include time paths for emissions and concentrations of GHG and aerosols and land-use/land cover. The RCPs represent different assumptions about how global human society will respond to the threat of climate change, as they are a set of scenarios that help to predict future outcomes. RCP2.6, 4.5, 6.0 and 8.5, considered in the IPCC 5th Assessment Report, range from a scenario assuming aggressive mitigation measures (RCP2.6) to one in which carbon emissions are assumed to rise substantially (RCP8.5). RCP8.5 is the least optimistic scenario, assuming little in the way of reduction in carbon emissions throughout the twenty-first century. RCP2.6 is a much more optimistic scenario, assuming that very stringent reductions in carbon emissions would be achieved in the near future, with other scenarios representing intermediate positions between these two scenarios.

The commonly used RCP8.5, 6.0, 4.5 and 2.6 W/m^2 (Caesar *et al.*, 2018; Fung, 2020) radiative forcing scenarios correspond to the range from pessimist to the optimistic emission scenarios. Table 3.1 shows the key headline difference in these models.

The UKCP provides a range of projections: UKCP global (60 km), regional (12 km) and local (2.2 km) – for example, 2.2 km means modelling grids of 2.2 km with more accurate predictions for the local UK geographical coverage. Global climate model projections include 15 from the UK Met Office Hadley Centre (PPE-15) and 13 models from other international climate modelling

Table 3.1 Key differences in different climate change scenarios for the RCP pathways (created from data within IPCC (2013, AR5 WG1: Table 12.3))

	Global mean surface temperature averaged over 2081–2100 compared with the pre-industrial period (average between 1850–1900)
RCP2.6	+1.6°C (Likely range 0.9 to 2.3°C)
RCP4.5	+2.4°C (Likely range 1.7 to 3.2°C)
RCP6.0	+2.8°C (Likely range 2.0 to 3.7°C)
RCP8.5	+4.3°C (Likely range 3.2 to 5.4°C)

centres (CMIP5-13). It represents a mitigation scenario aiming to limit the increase of global mean temperature to around 2°C above pre-industrial levels for mid-range climate sensitivity.

UKCP Global (60 km) allows users to explore the impacts of different emissions scenarios using the outputs of a global climate model – that is, RCP8.5 and RCP2.6. As expected, the projections for RCP2.6 (Fung *et al.*, 2020) show less global warming than RCP 8.5 (see Figure 3.1). Similar increases in the global mean surface air temperatures are seen for both emissions scenarios up to around 2040. The projections for RCP8.5 then increase further, whereas RCP2.6 projections level off from the 2050s onwards.

UKCP18 provides an updated generation of national climate projections for the UK and provides users with recent scientific evidence on projected climate changes with which to plan. It is a cornerstone principle of resilience preparation that risk planning should be undertaken for a wide range of possible future changes, in parallel with taking actions to reduce the likelihood of the worst scenario becoming reality. The UK Government's intention is that while continuing to spearhead international efforts to keep the global temperature rise to well below 2°C and pursue efforts to limit it to 1.5°C (by reducing the UK's own emissions, supporting the less-developed world to do the same and contributing to global diplomatic and scientific efforts), adaptation plans should be developed to accommodate the potential occurrence of more severe climate change risk.

It is noted that general climate change trends projected over UK land for the twenty-first century in UKCP18 are broadly consistent with earlier projections (UKCP09) showing an increased chance of warmer, wetter winters and hotter, drier summers along with an increase in the frequency and intensity of extremes (Met Office, 2022). This is seen in all of the probabilistic (25 km), global (60 km), regional (12 km) and local (2.2 km) projections.

Figure 3.1 Modelled temperature projections for RCP 8.5 and RCP 2.6 (UKCP Global 60 km) (© Crown Copyright 2020, Met Office (Met Office *et al.*, 2019))

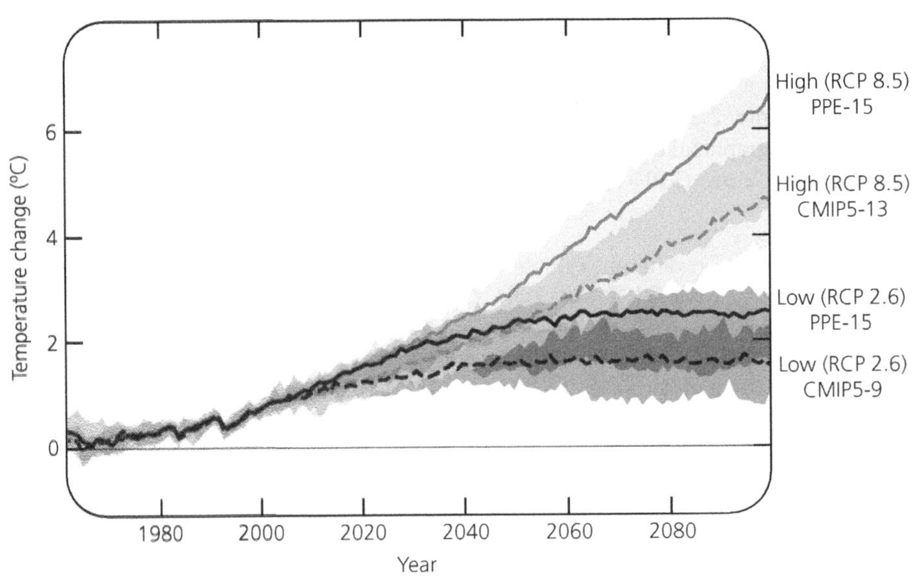

There is a great deal of overlap between UKCP09 and UKCP18 despite some differences. The previous set of projections, UKCP09, were used to assess climate risks for the 2017 UK Climate Change Risk Assessment and inform policy-making regarding response to these risks by including them within a National Adaptation Programme (HM Government, 2024). UKCP18 provides the most up-to-date assessment of how climate in the UK is expected to change over the coming century and was used to inform the updated CCRA in 2022 (HM Government, 2022c).

3.2.2 Shared Socio-economic Pathway scenarios (SSPs)

SSP scenarios, launched in 2021, differ from RCP emission scenarios in that they incorporate socioeconomic factors over the next century to provide scenario projections. SSP is now used in the IPCC's 6th Assessment Report (AR6).

SSPs predict that the global temperature will reach the 1.5°C warming threshold (identified as an aspirational limit by the Paris Agreement) at some point in the middle of the twenty-first century. Various SSP projections are shown in Table 3.2.

The AR6 report concluded in early 2023 that if the world does not begin to drastically cut emissions by the time of the next report of the IPCC, then it would no longer be possible to prevent 1.5°C of warming (IPCC, 2023).

In keeping with the global agreement to strive for a maximum 1.5°C warming limit, SSP1-1.9 provides a new pathway with a rather low radiative forcing of 1.9 W/m^2 in 2100 to model a scenario consistent with the 1.5°C threshold. However, even in this scenario, the global temperature peaks at 1.6°C in the years 2041–2060 and declines after.

The remaining carbon budget consistent with limiting warming to 1.5°C is estimated at ~500 $GtCO_2$ from the start of 2020 for a 50% likelihood. This magnitude of carbon budget implies the

Table 3.2 SSP and temperature projections (IPCC, 2023)

SSP	Scenario	Estimated warming (2041–2060)	Estimated warming (2081–2100)	Very likely range in °C (2081–2100)
SSP1-1.9	Very low GHG emissions: CO_2 emissions cut to net zero around 2050	1.6°C	1.4°C	1.0–1.8
SSP1-2.6	Low GHG emissions: CO_2 emissions cut to net zero around 2075	1.7°C	1.8°C	1.3–2.4
SSP2-4.5	Intermediate GHG emissions: CO_2 emissions around current levels until 2050, then falling but not reaching net zero by 2100	2.0°C	2.7°C	2.1–3.5
SSP3-7.0	High GHG emissions: CO_2 emissions double by 2100	2.1°C	3.6°C	2.8–4.6
SSP5-8.5	Very high GHG emissions: CO_2 emissions triple by 2075	2.4°C	4.4°C	3.3–5.7

necessity of reaching global net-zero CO_2 emissions before mid-century. Higher likelihood thresholds correspond to smaller budgets: ~400 $GtCO_2$ for a 67% (likely) chance and ~300 $GtCO_2$ for an 83% (very likely) chance of limiting warming to 1.5°C.

SSPs projections were introduced to provide greater clarity in policy analysis by using different socioeconomic factors than the RCPs. There is a shift towards using the SSPs to assess climate resilience, although it is likely that climate projections from RCPs and SSPs may not predict significant difference in climate projection for near-term future dates such as the 2030s. Differences are more marked from the 2050s projections onwards. For modelling some specific climatic variables such as sea level rise or regional scale projections, the scientific literature still largely uses modelling results based on RCPs. Both climate models are often seen in UK risk assessments with regards to climate adaptation and future impacts in the UK.

3.3. Climate mitigation practices

Climate mitigation aims to reduce GHG emissions to atmosphere as much and as quickly as possible. This has been the main emphasis so far of global climate actions.

As discussed in Chapter 2, in the UK, the majority of policy levers promote the transition to low carbon practices through legislation, planning policy and incentives. The National Planning Policy Framework (NPPF), released in 2012 and recently updated in 2024 (MHCLG, 2024), with a further review pending (at time of writing) promotes that the planning system should support transition to a low carbon future. It should help towards enabling radical reductions in GHG emissions, improving neighbourhood resilience and encouraging the reuse of existing resources, such as existing buildings and associated infrastructure. It requires that any proposed development must demonstrate provision of a positive strategy for maximising the potential of energy from renewable and low carbon sources, and opportunities for the development to be supplied from decentralised, renewable or low carbon energy systems, as well as for co-locating potential heat customers and suppliers.

It is recommended that when reviewing planning applications for renewable and low carbon development, the local planning authorities should recognise that even small-scale projects provide a valuable contribution to cutting GHG emissions.

The UK Climate Change Act 2008 committed the UK to a 57% reduction in emissions by 2032 and an 80% reduction by 2050, subsequently raised to target 100% reduction by 2050 (Climate Change Act 2008 (2050 Target Amendment) Order 2019). Further policy interventions and strategic documents have set out the implications for different built environment sectors since the zero carbon target was set. Three illustrative pathways are included in the UK's Clean Growth Strategy (CGS) to meet the long-term emissions reduction target (HM Government, 2017). These pathways do not represent the most likely or preferred routes to meeting that target but show that the modelled pathways would leave open a wide range of options for reaching the 2050 target.

The Government's National Infrastructure Assessment sets out the UK's long term infrastructure plan, with at least 50% renewable electricity generation by 2030; buildings that require less energy to heat and progress to zero carbon heat. This could be achieved through developing the evidence base on the different options, identifying areas for potential future cost reduction and progressing towards trialling low carbon hydrogen supply and manufacture at scale, including carbon capture and storage. The plan also sets out the UK's intention to pilot test hydrogen and heat pumps as low carbon heating options.

The Ministry of Housing, Communities and Local Government released in January 2022 an updated version of Building Regulation Part L – Conservation of fuel and power, which came into force in June 2022 (HM Government, 2022a). The update to Part L is designed to provide a stepping-stone to more stringent requirements to meet the Future Homes Standard in 2025 and focuses on a combined approach of improving fabric thermal performance and mitigating the carbon emissions from using fossil fuels.

Broadly speaking, a series of energy and carbon targets have been set for the progressive energy improvement policies towards achieving net zero carbon standard for built environment in the future.

- Ac.30% reduction in carbon emissions over the previous Part L 2013 standard, achieved through a more minor increase to fabric standards, alongside low carbon heating and/or renewables – for example, photovoltaic systems (PV) systems.
- The intention for an average home built to Future Homes Standard 2025 will have 75–80% fewer carbon emissions than one built to Part L 2013.

Due to the decarbonisation of the national electrical grid, the 'new' Part L will lead to heat electrification solutions in the UK. The majority of UK heat sources traditionally come from natural gas fired boilers; this represents more than 50% of the space heating and hot water in building carbon emissions. The UK Government has introduced a range of incentives to encourage both the public and private sectors to invest in low-carbon heat infrastructure.

Energy Act 2023

The Energy Act became law in October 2023. It represents the most significant energy legislation in decades and establishes the framework for a future-ready energy system. Electricity supply in the UK has been regulated by a national regulator (Ofgem) but the Energy Act 2023 (Energy Act, 2023) expands Ofgem's remit to cover heat networks. The aim is to allow for more innovative heating systems that require regulation and oversight to ensure protection of consumers. Ofgem will be responsible for setting rules on excessive pricing and improving quality of service for the growing number of heat network consumers across the country. The Act also contains powers to enable the Government to bring forward heat network zoning in England.

Taken together, these measures aim to reduce the cost of living by regulating prices (given the UK's dependence on imported natural gas supplies) and ensure this sector can contribute to delivering net zero in the most cost-effective way.

With these policy levers, legislation and incentives, climate mitigation measures in the UK comprise carbon reductions and renewable energy generation. This is broadly consistent with approaches taken in other countries seeking to reduce greenhouse gas emissions in accordance with global commitments.

3.3.1 Carbon reduction

The building and construction sector is responsible for approximately 40% of global energy-related carbon emissions and 50% of all extracted materials. Released at the latest round of climate talks in Egypt, COP27, it finds that the sector accounted for over 34% energy demand and around 37% of energy and process-related carbon dioxide emissions in 2021 (UNEP, 2022; United Nations Environment Programme and Yale Center for Ecosystems + Architecture, 2023). In the UK, emissions from buildings account for around half of total UK GHG emissions.

It is important to break carbon reduction efforts down to two types: operational carbon due to energy consumption and embodied carbon due to the use of materials used to construct the built environment. For a typical office building, operational carbon due to energy use contributes 30–40% of the total carbon emissions (see Figure 3.2). This energy use mainly comprises heating, ventilation and air conditioning (HVAC) consumption, domestic hot water and lighting as well as from other uses such as IT and other equipment.

In the context of offices, embodied carbon, contributing up to 60–70% of building carbon footprint is the carbon produced through the extraction, transportation, manufacturing and assembly of the materials used to build, fit out and furnish an asset. Design choices on structures and facade can make significant impact on the embodied carbon.

Not all the energy uses in a building are controlled by legislation; in England and Wales, only 'regulated' energy is considered within the Building Regulation Part L Conservation of fuel and power and mandatory Energy Performance Certificates (EPC). 'Regulated energy' is the component in a building where the designers and developers' design can make a real impact on potential energy use. This component includes HVAC, domestic hot water and lighting, whereas the cooking or small power energy use due to plug-in loads are usually called 'unregulated' energy, which falls within the remit of energy-use behavioural patterns.

Despite regulations intended to achieve carbon reduction, energy efficient designs may not lead to real carbon savings – that is, EPC and Part L only leads to energy efficiency in theory but not in practice. Office buildings often consume more than twice the energy in operation than estimated using design simulations based on the Part L regulated energy methodology. This has created a 'performance gap' between the designed energy consumption and the actual consumption.

To rectify this issue, Display Energy Use Certificates (DEC) were introduced to measure actual building energy use. These are required to be produced annually for public buildings and those

Figure 3.2 Examples of typical building types carbon emission distributions (created from data adapted from RICS (2017))

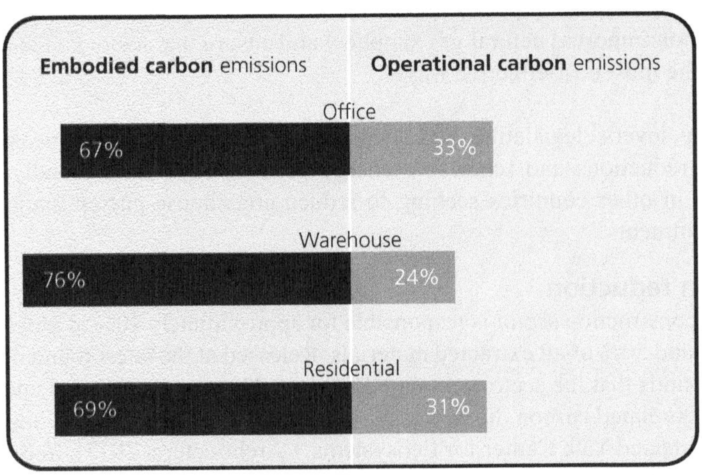

occupied by public authorities that have a total useful area greater than 250 m^2. While a step forward in closing the performance gap, DECs may not provide sufficient detail to help close the performance gap for a private sector building, or a building with multi-tenant occupation, for example. Recently, a building rating system known as NABERS (National Australian Built Environment Rating System) has been introduced to the UK from Australia where it was first established. This rating system aims to adopt Design for Performance methodology to provide a simple, reliable system for rating the energy efficiency of office buildings across England, Wales, Scotland and Northern Ireland.

Similar to the star-rating system applied to white goods (such as washing machines), NABERS awards office buildings a rating from one to six stars. This helps building owners to understand their building's performance compared with other similar buildings, providing a benchmark for progress.

The UK Carbon Budget produced by the Climate Change Committee (CCC) estimates that its balanced net-zero pathway – when based on fuel-switching, carbon capture and storage (CCS) and improvements to resource and energy efficiency – sees manufacturing and construction emissions reduced by 70% by 2035 and 90% by 2040 from 2018 levels. This means significant effort on energy efficiency improvement, together with significant deployment of renewable energy generation is needed to meet the target.

Embodied carbon is not currently mandated by legislation in the UK and its assessment is being driven by voluntary reporting through corporate policy or planning requirement. This may change; a recent consultation has proposed introduction of a 'Part Z' to the Building Regulations that proposes introduction of mandatory embodied carbon assessments for construction projects. At present, Part Z is an industry led proposal for an amendment to the Building Regulations advocating mandatory whole-life-carbon assessment and reporting to reduce embodied carbon in construction.

3.3.2 Operational carbon

Operational carbon is the emissions of carbon dioxide and other global warming gases during the in-use operation of a building.

In 2007, the UK Government announced its aspiration for new non-domestic buildings to be zero carbon in operation by 2019, and domestic buildings by 2016 (Code for Sustainable Homes Level 6). Between 2007 and 2010, the Government consulted on the definition of 'zero carbon' for non-domestic buildings. As a minimum, it was stated that the zero-carbon definition for non-domestic buildings would cover 100% of regulated emissions – that is, a Building Emissions Rate (BER) of zero. Regulated energy typically represents 60–70% of a building's total energy use, as shown in the example in Figure 3.3.

This concept is underpinned by an energy hierarchy used to achieve zero carbon standards for buildings, as shown below. The hierarchical approach is widely adopted by planning policy as a mechanism to demonstrate that a low carbon design has been proposed.

Reduce energy demand and energy efficiency measures – to ensure that buildings are constructed to very high standards of fabric energy efficiency and use efficient heating, cooling, ventilation and lighting systems. This usually entails improving the thermal performance of the building

Figure 3.3 Breakdown of carbon dioxide emissions from a typical mixed-use building (created using data from Target Zero (2012))

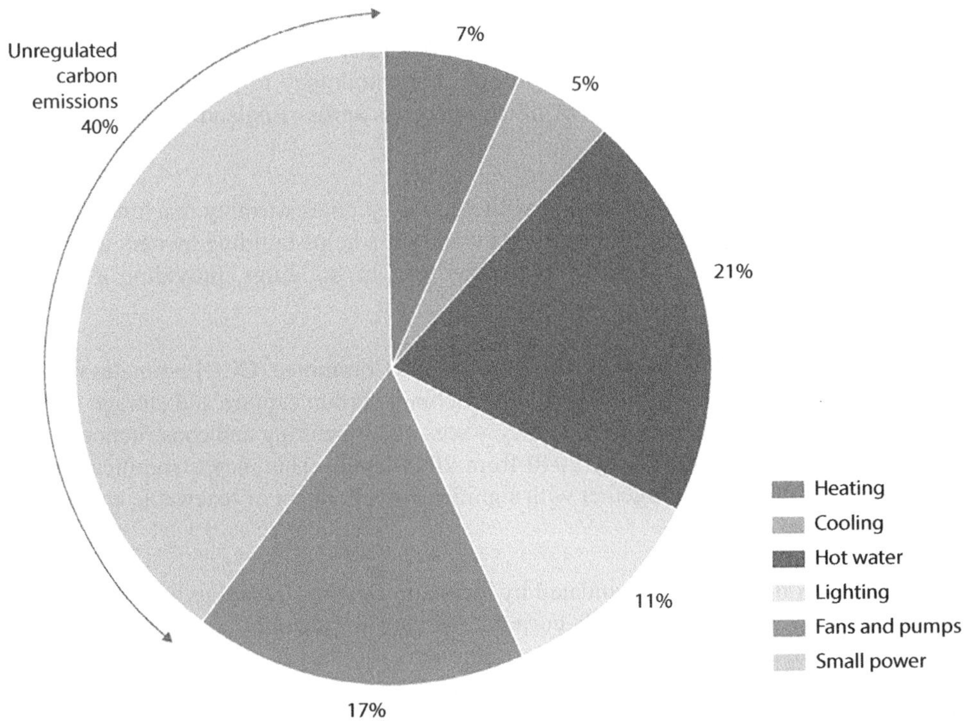

envelope, such as enhancing U values, air permeability target and g value for solar gains limit and so on; system efficiency standards such as lumens per watt for lighting and specific fan power (W/L/S) for ventilation and so on.

The current mechanism for non-domestic buildings sets minimum energy-efficiency standards based on delivered heating and cooling energy (kWh/m^2/yr), as defined in the Building Regulations for regulated energy. The upcoming (in draft at time of writing) UK Net Zero Buildings Standard (UK Net Zero Buildings, 2024) will likely introduce more matrices such as peak demand for heating or cooling.

Renewables on or near site – this represents the minimum carbon abatement required using energy efficiency measures and on-site low and zero-carbon (LZC) technologies, or through directly connected heat or coolth sources, such as heat networks. The required standard is largely driven by Building Regulations compliance, planning policy, Building Research Establishment Environmental Assessment Method (BREEAM) energy requirements, or local planning obligations for renewable-energy integration.

Allowable solutions – Allowable solutions are supplementary measures intended to offset residual emissions, for example, supplying low-carbon or renewable heat to nearby developments or supporting LZC community-heating schemes. This step has seen limited uptake in the UK and was ultimately withdrawn in 2015 as part of a policy change prioritising housebuilding productivity.

In the UK Government's *Construction 2025* strategy, published in 2013, a target was set to reduce built-environment GHG emissions by 50% by 2025 from a 1990 baseline, consistent with the 4th Carbon Budget under the Climate Change Act. Despite these high-level commitments, there has been limited policy or regulatory action to drive operational carbon reductions from buildings. In the absence of stronger policies and regulation by the Government, leading organisations have published their own strategies and recommendations outlining how the built environment could deliver the necessary operational carbon reductions needed to help meet the UK Climate Change Act commitments. These include the following.

UK Green Building Council (UKGBC)

UKGBC produced leading guidance such as the Net Zero Carbon Buildings Framework and the Net Zero Whole Life Carbon Roadmap. In the context of building operational energy, UKGBC defines net zero carbon as

> when the amount of carbon emissions associated with the building's operational energy on an annual basis is zero or negative. A net zero carbon building is highly energy efficient and powered from on-site and/or off-site renewable energy sources, with any remaining carbon balance offset. (UKGBC, 2019)

This effectively means any net zero carbon building must consider not only regulated energy component but also the unregulated component as an actual energy consumption during operation in use. UKGBC subsequently published a series of guidance on Renewable Energy Procurement and Carbon Offsetting Guidance for Net Zero Carbon Buildings. Chapter 4 provides more details.

London Energy Transformation Initiative (LETI)

LETI was established in 2017 to support the transition of the London built environment to net zero carbon, providing guidance that can be applied to the rest of the UK.

In order to meet national 2050 reduction targets, LETI believed that by 2025, 100% of new buildings must be designed to deliver net zero carbon, and the whole construction industry will need to be equipped with the knowledge and skills necessary (LETI, 2020a). In terms of operational energy, LETI recommends a series of indicative design measures for offices, residential properties as well as for naturally ventilated schools – for instance, measures to meet the operational energy consumption target for commercial offices of 55 kWh/m^2/yr including building fabric design standards, glazing ratio, natural ventilation or low carbon HVAC systems.

RIBA 2030 Climate Challenge (Royal Institute of British Architects)

In 2019, in response to the UK Climate Change Act, RIBA published their 2030 climate change targets for operational energy, embodied carbon and water use reduction. See Section 4.2 for more details. RIBA's operational energy targets for non-domestic buildings for 2020, 2025 and 2030 are shown below. The RIBA 2030 Climate Challenge has now been re-issued as Version 2 (RIBA, 2021), with refined and updated targets that reflect developments in embodied carbon practice and align with LETI and the latest jointly authored guidance. Version 2 of the Climate Challenge presents ambitious but achievable forward-looking performance outcomes, moving beyond business-as-usual compliance approaches.

CIBSE TM54 Energy Assessment (The Chartered Institution of Building Services Engineers)

In an effort to overcome the UK's design-for-compliance culture, which has contributed to the 'performance gap' between the original design intent and the actual in-use performance, CIBSE TM54, *Evaluating Operational Energy Use at the Design Stage*, has been recently revised (CIBSE, 2022). The document aims to encourage better prediction of actual building performance to inform design decision making. It is a guidance document providing practical steps for how to undertake predictive energy modelling of buildings. The methodology can be applied to any type of building, such as new or existing buildings. It is now referenced in BREEAM and several institutional sustainability standards such as University College London's (UCL's) Sustainable Building Standard as well as in the latest revision to Building Regulations Part L 2021 edition.

Part L 2021 requires that buildings above 1000 m^2 should include a forecast of actual energy use in kWh/year, which is broken down by fuel type as part of the handover information. CIBSE TM54 (CIBSE, 2022) is an accepted approach of meeting this requirement. Operational energy modelling should reduce the gap between predicted and actual performance, but some level of performance gap will always remain. It is important that the client and all those involved in the design and operation of buildings understands this aspect.

Focus area

There has been a significant amount of effort in addressing operational carbon and 'performance gap' issues in buildings. For a building over 60 years, operational carbon emissions make up more than 60% of the whole life carbon emissions; therefore, operation energy and carbon remains the priority for reduction, with the focus development in the following areas.

- Low-carbon heat strategy and plans to phase out fossil fuels from buildings.
- Policies to improve energy efficiency for all buildings.
- New build standards to ensure all new homes are ultra-efficient and use low-carbon heating.
- Closing of the performance and compliance gaps.

3.3.3 Embodied carbon

Embodied carbon emissions that are contained within the materials used in construction and operation of the built environment have only reached prominence more recently. Typical breakdowns of whole life carbon emissions for different building types were shown in RICS's Whole Life Carbon Assessment for the Built Environment, regularly updated with the latest information (RICS, 2024). The definition and calculation are in line with the modular structure of BS EN 15978 (BSI, 2011) (see Chapter 4 for more information and detail) and account for all components making up the finished building covering a cradle to grave scope (modules [A–C]) over a 60-year life cycle. BS EN 15978 provides a framework for appraising the environmental impacts of the built environment and is part of the BS EN 15643 family of standards for the sustainability assessment of buildings. It sets out the principles for whole life assessment of the environmental impacts from built projects based on life cycle assessment (LCA).

Upfront embodied carbon assessments are intended to estimate the actual carbon dioxide equivalent emissions due to construction. LETI's Embodied Carbon Primer (LETI, 2020b) across all building elements highlights the significant contribution of Modules A1–A3 to the total embodied carbon. Whereas, the whole life carbon assessment includes the operational carbon as well as the replacement related carbon emissions across the life cycle of the asset. In more detailed assessments, life-cycle phases are defined in the EPD standard BS EN 15804 (BSI, 2020), with operational energy/carbon use reported in Module B6.

The UK Net Zero Carbon Buildings Standard, when fully released following the pilot phase, will provide the benchmark metrics for embodied carbon targets, embedding this LCA process into more standardised processes. An increasing number of businesses report the carbon emissions from their properties and assets, in the form of Scope 1 and 2 carbon emissions (their direct and indirect emissions from fossil fuels, for example). Businesses are now increasingly attempting to quantify and report their Scope 3 emissions (those emissions for which they are indirectly responsible up and down the value chain). An example of this is when the company buys, uses and disposes of products from suppliers. Embodied carbon of buildings typically represents a significant share of a real estate company's overall carbon footprint, along with staff travel, and it is likely that this reporting will drive greater attention to embodied carbon in the future.

Naturally, embodied carbon in infrastructure construction is also significant. In response to the growing understanding of this issue, *PAS 2080: Carbon management in infrastructure* was developed by the British Standards Institution (BSI) in 2016, with an updated version since published by BSI as the national standards body sponsored by the Institution of Civil Engineers and the Green Construction Board (GCB) in 2023 (BSI, 2023). The revised PAS 2080 adopts an integrated approach to widening the scope of PAS 2080 beyond infrastructure to the whole built environment, taking a whole-life view to addressing the urgent need to retrofit existing stock and balance capital carbon investment with operational and user benefit.

Although embodied carbon reporting is not yet required by Building Regulations, the built environment industry has been driving a national effort, through the proposed 'Part Z' to integrate embodied carbon into UK regulatory frameworks. The RICS Professional Statement guidance (RICS, 2024) on Whole Life Carbon Assessment for the Built Environment has become the de facto standard against which all UK whole life carbon assessments (WLCAs) have been benchmarked since 2017. The updated edition, effective from July 2024, introduces comprehensive revisions that advance the standard and improve overall carbon accounting. This includes a requirement to account for emissions from demolition and pre-construction activities and a requirement for better measurement of operational carbon. The RICS standard incorporates elements from other existing standards, such PAS 2080, BS 8536 (BSI, 2022) and various European standards. This alignment has created a unified approach for assessing all types of built assets – including buildings, infrastructure development and external works, improving comparability across assessments.

Some countries in Europe already have policies in place that address whole-life embodied carbon emissions. In the Netherlands, all new buildings larger than 100 m^2 are required to undertake whole-life carbon calculations and carbon mitigation cost estimates using nationally approved methodologies, databases, and tools administered by an appropriate value-chain organisation.

In Sweden, embodied carbon calculation and reporting are required for new buildings to obtain building permit approval. The policy first mandates collecting data on current embodied carbon levels across the sector, followed by the introduction of limit values for future projects.

France's recently introduced *RE2020* policy has mandated whole-life cycle carbon calculations for all new houses built and refurbished since 2022. Within this policy, targets for reducing emissions will get progressively tighter each year and will apply to an increasingly wider scope of building projects. The ratcheting of targets and progressive increase in scope of RE2020 aim to enable the French construction sector to reach net zero by 2050.

The Carbon Leadership Forum (CLF) originated in the United States but now has a membership across 94 countries with a mission of eliminating embodied carbon in buildings, materials and infrastructure to create a just and thriving future. Since 2023, the topic of embodied carbon has become a central issue within the green building movement, prompting companies, organisations, policymakers and design professionals involved in this forum to confront the carbon challenge with a heightened urgency and attention. One particular challenge is embodied carbon associated with the building systems as the impacts of the systems installed can be significant. Mechanical equipment, delivery, refrigerants and refrigerant leaks are often excluded from life-cycle analysis making their full impacts difficult to quantify. MEP 2040 (Mechanical, Electrical and Plumbing) is a movement initiated by CLF members with many industry signatories aiming to tackle this challenge. It seeks to radically reduce total carbon emissions associated with building systems through collective action. This movement has gained substantial momentum and is shaping real estate practices and associated design approaches.

The MEP 2040 Commitment asks all members to

(*a*) establish a company plan
(*b*) request low-GWP (Global Warming Potential) refrigerants
(*c*) request EPDs (Environmental Product Declarations)
(*d*) participate in forums to share information and knowledge.

Focus area

Embodied carbon is a key theme within industry net zero targets, frameworks and the Government's future carbon budgets, even though there is no current government policy that legally requires the assessment or control of embodied carbon emissions from buildings. Some local governments are taking action through using their planning powers. For example, the Greater London Authority (GLA)'s 2021 London Plan sets out requirements for all developments over a certain size to calculate embodied impacts following the RICS methodology. Other local authorities including the West Midlands Combined Authority and Brighton and Hove City Council, have also incorporated embodied carbon and whole-life carbon assessments into their local plans. PAS 2080, revised in 2023, provides the standard for managing carbon across the built environment including infrastructure and buildings.

- Reporting on and reducing embodied carbon provides a clearer understanding of the whole life impacts and Scope 3 emissions associated with a building development.
- RICS Professional Statement guidance on Whole Life Carbon Assessment for the Built Environment, 2nd Edition incorporates elements from other established standards, including PAS2080, BS8536 and various European standards.
- Net Zero Carbon Buildings Standard will set out metrics for embodied carbon for most common building types.
- Reduces regulatory risks as embodied carbon becomes an increasing area of focus in emerging net zero policies.
- Datasets are becoming more consistent in developed economies, with an increasing number of Environmental Product Declarations (EPD) for construction products available now in accordance with International Standard ISO 14025 (ISO, 2006).

3.3.4 Ecological protection

The Government's 25 Year Environment Plan published in 2018, provides fresh impetus for the principles of sustainable development to be at the heart of decision-making (HM Government, 2018). It sets out clear objectives and common language for policy and decision makers to adopt policies and actions structured around six key areas.

- Using and managing land sustainably.
- Recovering nature and enhancing the beauty of landscapes.
- Connecting people with the environment to improve health and wellbeing.
- Increasing resource efficiency and reducing pollution and waste.
- Securing clean, healthy, productive and biologically diverse seas and oceans.
- Protecting and improving our global environment.

Environmental protection is much broader than simply carbon reduction. It includes improving biodiversity, air quality, water quality and reducing waste production. A failure to protect the wider environment will result in climate change, pollution, loss of biodiversity and depletion of natural resources, to name just a few examples.

3.4. Shift to climate resilience and adaptation

Despite global efforts to reduce the amount of carbon and other GHGs emissions, these gases continue to accumulate in the atmosphere, causing the planet to continue to warm.

With the advancing climate crisis come hazards such as increasing temperatures, changing patterns of rainfall, prolonged periods of drought and increased flooding, each poses major threats to the built environment, the people who inhabit it, and the natural systems on which communities depend. Decision makers and all those involved in the built environment are becoming acutely aware of the urgent need to respond to the increasing threat of climate risks by rapidly integrating resilience measures into buildings, communities and infrastructure. This is vital for reducing risks to businesses, protecting economic stability, and importantly, for safeguarding the health and quality of life of people and communities.

The UK Climate Change Risk Assessment (CCRA) is a 5-yearly evaluation of the major risks and opportunities from climate change for the UK (HM Government, 2022c). Six key risk areas were outlined in the latest version.

- Flooding and coastal change risks to communities, businesses and infrastructure.
- Risks to health, wellbeing and productivity from higher temperatures.
- Risks of water deficits in public water supply, and in agriculture, energy generation and industrial operations, with impacts on freshwater ecology.
- Risks to natural capital, including soils, coastal, marine and freshwater ecosystems, and biodiversity.
- Risks from climate-related impacts on domestic and international food production and trade.
- New and emerging pests and diseases, and non-native species, affecting people, plants and animals.

Risks to health and wellbeing (additional evidence-aligned facts)

- Heat-related deaths could triple by the 2050s, considering the effects of climate change and population growth.
- In 2020, an additional approximately 2500 deaths were recorded during the summer heatwave across England, which could not be explained by other factors.

Risks to infrastructure due to flooding (additional evidence-aligned facts)

- Data show that 178 power stations and 575 substations are currently at significant risk from surface water flooding and 67 power stations and 234 substations are at risk from river flooding across the UK.
- The risk increases significantly from surface water flooding in the future, potentially doubling the risk in a 4°C warming scenario.
- Present day expected annual damages to non-residential properties across the UK average around £670 million.
- Damages could increase in a 2°C warming scenario by around 27% by 2050 and 40% by 2080.
- Recent research conducted to support the CCRA has indicated that the vulnerability of interconnected systems may be significantly underestimated.
- The risk of network failures is already high and has the potential to affect hundreds of thousands or millions of people each year.

Risks to natural world – pests, pathogens and invasive species (additional evidence-aligned facts)

- Warmer winters have already had a clear influence on outbreaks of some pests and pathogens.
- The Environmental Audit Committee has indicated that these risks potentially cost the UK economy £1.7 billion per year.

The built environment has played a critical role in reducing vulnerability to climate risk. Cities and urban areas typically increase exposure to physical hazards – for example, built-up areas can intensify the urban heat island (UHI) effect, while extensive impermeable surfaces increase the likelihood of flooding by limiting natural water infiltration.

Well-designed buildings and natural environments reduce climate hazards, keeping users safe and comfortable. Disruptions to transport infrastructure from climate events can lead to major economic losses.

Built environment professionals have a critical role in climate adaptation. Their responsibilities include evaluating asset level risks, formulating practical adaptation strategies, designing climate-resilient buildings, and supporting investment in broader urban resilience initiatives.

Nevertheless, achieving meaningful progress demands a substantial shift toward systemic and institutionalised change across the industry, with an emphasis on delivering the scalable and widespread actions that enhance climate resilience throughout the built environment.

Key challenges and barriers to adaptation in the built environment – such as knowledge gaps, policy and regulatory uncertainty, as well as the limitations of traditional economic and financial appraisal methods, such as the use of discount rates, which make it difficult to justify investment. At present, the sector lacks clear definitions, targets and metrics to support sector-wide climate resilience.

The Climate Change Committee (CCC), an independent, statutory body established under the Climate Change Act 2008, has highlighted the absence of resilience standards across the UK. Both public and private investment will be essential, yet current barriers – especially in the private sector – are limiting progress in preparing for climate impacts (Climate Change Committee, 2023).

Adaptation Reporting Power (ARP)

The UK Climate Change Act 2008 enables the Government to require infrastructure providers and bodies with functions 'of a public nature' to provide reports on how they manage climate risk. ARP was introduced under the Act to help ensure reporting organisations are taking appropriate action to adapt to climate change. ARP supports this both directly, through engaging organisations in the reporting process and indirectly, by raising awareness, building capacity in organisations, and sharing good practice.

The ARP reports are aligned with the Government's climate change risk assessments and national adaptation programmes.

UKGBC's Climate Resilience Roadmap

UKGBC's Climate Resilience Roadmap is a collaborative effort with industry taskforce to help shape this pathway using cutting-edge insights at the forefront of global industry. The aim of this process is to advance climate resilience in the UK built environment through the following actions.

- Setting measurable metrics and science-based targets for climate resilience.
- Identifying key actions for stakeholders across the industry.
- Identifying key policies needed to drive forwards local and national policy.
- Creating a pathway for UK built environment stakeholders to follow in achieving a climate resilient built environment by 2050.

Globally, the effects of climate change are distributed unequally, with nations least responsible for emissions often the most vulnerable and least prepared. Within urban areas, extreme weather causes widespread disruption, yet certain communities consistently endure disproportionate levels of loss, damage, and long-term impacts, as shown in Figure 3.4, Climate Risk Index (Eckstein and Schafer, 2021). The Climate Risk Index is a tool that assesses extreme weather events' human and economic impacts on countries and regions, capturing the realised consequences of climate related risks over the preceding 30 years (Eckstein and Schafer, 2021).

Figure 3.4 Climate Risk Index (Eckstein and Schafer, 2021) (permission requested from Germanwatch)

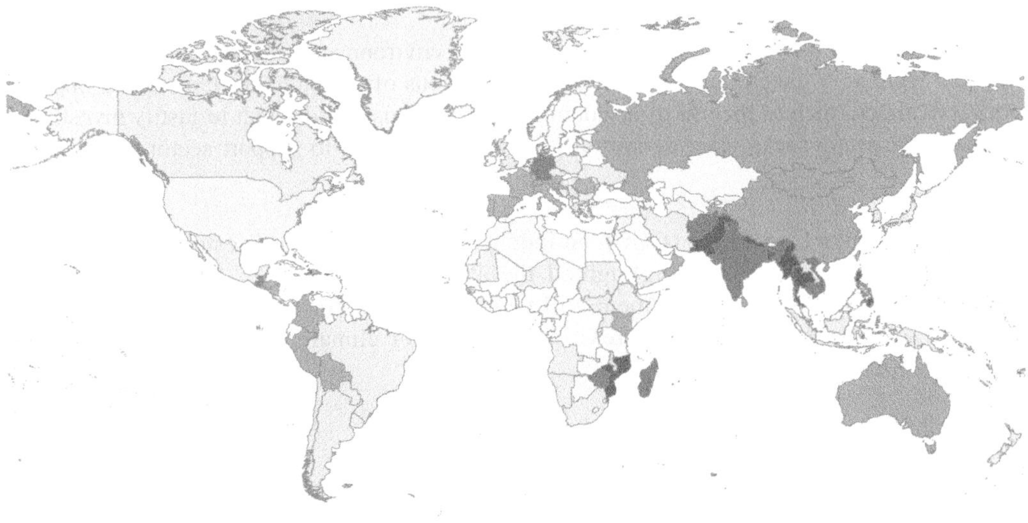

3.4.1 Methodology and key considerations

The World Bank provides a free climate and disaster risk screening tool to support the identification of short- and long-term potential climate and disaster risks in development projects, policies and programmes. By identifying risks early and integrating resilience measures at the project design stage, practitioners can improve the likelihood of achieving development objectives. The tool can be used by development professionals for high-level screening at an early stage of project design or in national level planning processes.

A systematic process is needed to identify climate-related threats, evaluate the risks they present, and implement rigorous adaptation and resilience measures. Figure 3.5 illustrates the methodology in a stepwise approach, beginning with an assessment of climate risks (Step 1) followed by an evaluation of climate vulnerabilities (Step 2).

Step 1 – Establish the key climate risks through desktop studies and engagement.

Step 2 – Identify the main climate related vulnerabilities across receiving environments and communities.

Climate risk and vulnerability are related but separate considerations.

> Climate risk assessments focus primarily on the projected changes in climatic conditions, inventory of potentially impacted assets, the likelihood of the impact happening and the resulting consequences.

Figure 3.5 Climate risk and adaptation methodology

Step 1 - Establish the key climate risks through desktop studies and engagement	• Hazard identification and assessment for water security, flood risk, high temperature, natural capital, pests and disease • Shock and stress identification and prioritisation • Determining existing preparedness
Step 2 - Identify the main climate-related vulnerabilities across receiving environments and communities	• Analysis of current and future climate risks using projections • Establish key impacts and vulnerabilities • Climate vulnerability assessment
Step 3 - Development of adaptation programmes	• Desktop assessment of longlist of climate resilience measures • Prioritise resilience and adaptation programmes • Cost analysis and finance plan • Climate adaptation and resilience strategy, to include governance and implementation
Step 4 - Delivery and project implementation	• Capital projects with resilient infrastructure, built and natural assets • Capacity building and raising awareness and post implementation evaluation

> Vulnerability assessments emphasise exposure, sensitivity and adaptive capacity of systems, assets and populations. (Climate-ADAPT, 2016b).

The process begins with an assessment of key climate risks, drawing on locally available data, stakeholder engagement and analysis of current and projected climate conditions, as outlined in Section 3.2.

A review of future climate projections will be required using either RCPs or SSPs projections to provide a high-level understanding of relevant climate variables. Depending on the location, the review may be limited by the potential coarse spatial resolution from the SSPs projections relative to specific masterplan or development area. UKCP Projections provide UKCP User Interface or CEDA Data Catalogue for the users to select the most relevant dataset. This review will inform the subsequent risk, hazard and vulnerability assessment. An example of climate risk appraisal for One Angel Square building is indicated in Table 3.3.

Impact of climate risks is measured considering the impact of both shocks and stresses.

> Shocks can be slow-onset like drought, or relatively rapid onset like flooding, disease outbreak, or market fluctuations. Stresses are 'long-term trends or pressures that undermine the stability of a system and increase vulnerability within it' (USAID Center for Resilience, 2018).

One Angel Square, Manchester

One Angel Square, as the Co-operative Head Office offers a 30 000 m^2 high-quality office space in Manchester city centre. The head office building aspires to achieve the highest energy standard that is economically viable. The environmental targets such as BREEAM Outstanding, EPC A, DEC A as well as stringent energy benchmarks have been achieved throughout the detailed design and construction stages. A climate change adaptation strategy was developed to assess how and when the building and its operation may need to be adapted in the future climate conditions. Table 3.3 shows the climate risk assessment for this commercial development in Manchester.

Table 3.3 Climate risk assessment for One Angel Square, Manchester

Climate variable	Impact/consequence	Threats	Risk
Hot Summer Dry bulb termperature (average temperature in the summer)	• Overheating • Occupant complaints • Increased electricity bills • Risk of power backup	• Function of the building and equipment • Insufficient budgets for incorporating adaptation measures • Inadequate indoor environment resulting indirect losses	High
Dry summer Reduced summer average rainfall	• More irrigation for green roof/landscape • Higher potable water use	• Building foundation • Sub-structure design	Low
Mild Winter Minimum dry bulb temperature	• Building users • Heating system	• Heating system inefficiency if no controls in place	Low
Wet Winter Increased average winter rainfall	• Increased pressure on drains • Increased water management costs	• Road safety • Increased cost	High
Heat Wave Summer Peak temperature	• Overheating • Reduced productivity • Increased cost • Occupant complaints	• Function of the buildings • Risks to ICT, energy and utilities assets • Cost to the asset owners	High
Very Wet Winter Sudden increased rainfall	• Risk to equipment damage • Increased pressure on drains • Road safety	• Damage and loss of functionality • Increased cost • Access to buildings	High

For the built environment, Gething's book provides a useful guide on assessing the vulnerability of a built environment asset to the identified risks (Gething, 2010). These vulnerabilities can be broadly grouped to three categories.

- Vulnerability to extreme temperatures: Keeping cool, for internal spaces and for external spaces, often referred to as 'heat stress'. Keeping warm is also a key built environment vulnerability, especially as extreme temperatures and cold spells become more frequent.
- Vulnerability to building or infrastructure failure: A systematic risk assessment, focused on structural and fabric resilience, is required, to identify and evaluate how expected extreme weather conditions arising from climate change may affect a building over its projected life cycle, and to determine where mitigation measures are feasible. Although most of the environmental factors and materials degradation shown in Table 3.4 are already covered in international and relevant national design codes, the changing climate may exacerbate the impact. The aims are to protect vulnerable components of the building from damage and prevent degradation of exposed materials.
- Water vulnerability: the rising occurrence of extreme climate conditions, such as drought and high-intensity rainfall, necessitates robust water conservation strategies and drainage systems specifically designed to accommodate flood events.

Table 3.4 illustrates the BREEAM recommendations (based on *BREEAM New Construction* (BRE, 2025)) for considering the building material choice and design parameters to improve climate resilience.

While methodologies vary across sectors, the built environment has seen targeted innovation funding to advance assessment tools. Informed by UK Climate Projections (UKCP, n.d.), a holistic methodology is recommended, employing tools such as the UKCIP Risk Framework (UKCIP, 2013), Adaptation Wizard (UKCIP, 2003), and Adaptation Support Tool (Climate-ADAPT, 2016a) to evaluate the climatic risks, including flooding risks, heat stress, water stress and extreme wind and their consequential impact. Climate vulnerability assessment is primarily conducted by expert built-environment professionals in consultation with the client team.

Table 3.4 Considerations of building materials design to address climate resilience (created using information from BRE (2025))

Applicable building elements	Environmental factors	Material degradation
Substructure/retaining walls	Solar radiation	Corrosion
External walls	Temperature variation	Dimensional change
Roof/balconies	Water/moisture/precipitation	Fading/discolouration
Glazing (windows / rooflights)	Wind	Rotting
External doors	Extreme weather	Leaching
Cladding	Vegetation	Blistering and melting
External railings/balustrades/ external staircases	Pollutants (air, water and ground)	Salt crystallisations
Hard landscaping	Pests/insects	Abrasion

Once the risks and vulnerabilities have been identified, the next steps in the process involve developing potential resilience measures (Step 3) and delivering the adapted project (Step 4).

Step 3 – Development of adaptation programmes

Step 4 – Delivery of projects that are resilient to climate change

As part of the development of appropriate measures, a long list of potential adaptation opportunities can be devised based on the identified vulnerability risk area. It is essential to appraise these measures to avoid 'over-adaptation', which may lead to unintended negative consequences in other areas or become cost prohibitive. The extent to which a measure constitutes 'over-adaptation' is case specific, depending on the impact of the risk and the likelihood of the risk being realised. Table 3.5 gives an example of the appraisal of a long list of adaptation measures for a higher education building project in the UK. Concepts such as no-regret adaptation, win-win measures and over-adaptation are used as criteria for shortlisting.

From the initial appraisal, a short list of the adaptation measures can then be assessed using appropriate professional design tools for cost and benefit analysis.

In some cases, grouping adaptation measures into packages can help communicate options more effectively to the design team and support the assessment of their feasibility. Table 3.6 shows an example of measures designed to address thermal comfort risk in the future warmer climate for a housing development in the southeast of the UK, where air conditioning or comfort cooling in a house is not commonly used.

Table 3.5 Adaptation measures appraisal

Adaptation measure(s)	Appraisal	Comments
Building measures to reduce overheating risk	**No regret**	A 'best outcome' regardless of any difference in climate change scenarios
Energy efficient design to reduce reliance on fossil fuels	**Win-win**	Decisions taken for reasons not directly motivated by need to adapt to climate change
Management and operational measures to improve efficiency	**No regret**	Decisions considered to have very low cost implications taking into account uncertainties of scale of future climate change
Improved building users awareness and preparedness	**No regret**	Decisions considered to have very low cost implications taking into account uncertainties of scale of future climate change
Mechanical cooling throughout	**Over adaptation**	Decisions with high investment cost but limited benefit
Rainwater collection/green roof	**Win-win**	Decisions taken for reasons not directly motivated by need to adapt to climate change
Increased flood attenuation tank/ improved surface water distribution network	**Low regret**	Decisions considered to have very low cost implications taking into account uncertainties of scale of future climate change

Table 3.6 Packages of adaptation measures to adapt thermal comfort in a residential building

Package 1: Passive option	Solar control glazing Stack ventilation Supply air pre-cooling (e.g. earth tube)
Package 2: Mechanical option	Solar control glazing External shutters Mechanical ventilation rate increase Evaporative cooling
Package 3: Thermal mass option	Solar control glazing Thermal mass (with night cooling) Reflective internal blinds Evaporative cooling

In the case of flooding risks, civil engineering design is typically informed by the probabilistic assessment of flood events, expressed as a return period in accordance with relevant local codes and the established risk profile. In this context, a climate consultant is required to work in close collaboration with the design team to identify the implications of projected changes in storm event frequency and to recommend supplementary testing of proposed adaptation measures in response to anticipated increases in rainfall intensity and duration. This assessment should be undertaken jointly with the design team to ensure that the influence of evolving climatic variables on the baseline design parameters is fully evaluated and appropriately integrated into the overall engineering approach. The work will need to be carried out in collaboration with the design team in assessing the impact of changing climatic variables on the 'baseline' design parameters adopted by the design team.

In many countries, the consideration of higher rainfall events has been built into planning policy or equivalent. In the UK, planning policy and Environment Agency (the Government regulator) currently require the volume of attenuation storage to mitigate a 1-in-100-year storm event with an additional 45% allowance to account for projected climate change impacts. In parallel, surface water discharge rates to public sewers are required to be restricted in accordance with local regulatory limits. These requirements may vary geographically and are influenced by the level of exposure and associated risk profile of the proposed development, providing a structured means of communicating the scale of risk and the implications of different mitigation options to inform decision making, while recognising that not all benefits can be readily quantified.

At the macro level, tools such as The World Bank Climate and Disaster Risk Screening Tools by World Bank Climate and Disaster Risk Screening (n.d.) and the University of Reading's (2023) Climate Risk Indicators for local authorities and nations in the UK, provide valuable early-stage screening to inform initial assessments. Once the risks and vulnerability are identified, more detailed assessment of resilience and adaption measures are required. City scale analysis will often require geospatial analysis of social and infrastructure vulnerabilities, this could include public land ownership, households poverty levels, and distribution of older adults to support social sensitivity analysis.

The success of an adaptative programme is underpinned by how a city and a local community build their adaptive capacity. Broadly speaking, adaptive capacity denotes the ability of a system to adjust, modify or change its characteristics or actions to reduce potential damage, leverage opportunities or cope with the consequences of shock or stress. A key component of this is ensuring that individuals, communities and societies are actively engaged in processes of change. Importantly, this involves behavioural adaptation as well as adjustments in resources and technologies.

As climate science continues to evolve the built environment sector must integrate emerging knowledge to address the risk. Challenges will be inevitably encountered, such as the following.

- Lack of guidance on the application of climate projections in the construction industry. As climate projections, including UKCP18's RCPs and SSPs, are provided on a probabilistic basis, the associated climate variable risks largely depend on the client's requirements and the professional advice of the project team.
- Lack of data compatibility that is suitable for built environment professionals to assess the impact. RCPs and SSPs provide climatic projections for many global locations, however, the datasets are not available on an hourly basis, which is essential for building performance simulations. In the UK, The Chartered Institute of Building Services Engineers (CIBSE) used various techniques including morphed methods to generate the hourly dataset of the future weather files for numerous UK cities using monthly and daily projection data to support building design (CIBSE, 2014). Energy and Climate Change Division at the University of Southampton also constructed design weather data for future climates for worldwide locations based on UKCIP02 projections (University of Southampton, 2022).
- In many cases, the best and most effective adaptation measures must be implemented at infrastructure level, often extending beyond the redline boundary of a masterplan or individual development site. Evaluating the vulnerability of buildings to failures in infrastructure, and identifying appropriate adaptation measures require collaboration with adjacent land, often with public sector bodies. Such partnerships enable wider interventions, including the use of nature based solutions, or landscape scale flood management strategies to create buffer zones in high risk areas.
- An adaptation programme that engages all stakeholders is the most effective way in achieving climate resilience. This includes capital programme managers and clients, governance and HR policy, funding programme managers and building occupants or affected communities.

Focus area

Shifting to climate adaptation is inevitable, as we are living with increasingly extreme climate conditions in the years ahead.

- There are four essential steps to follow in order to identify an appropriate adaptation strategy, from assessing climate risks, vulnerability, to identifying adaptation measures through using cost and benefit analysis and implementing selected actions. Stakeholder engagement is required throughout the process.
- Concepts such as no-regret adaptation, win-win measures and the avoidance of over-adaptation can be used as criteria for early appraisal of potential adaptation measures.
- The level of climate resilience and decision on adaptation remain largely voluntary and are often driven by the client's specific requirement – for example, a client may need to decide whether to use the 90th-percentile RCP8.5 dataset for the 2050s as a representation of a 'near-extreme' scenario.

- UKGBC is working with the sector-wide organisations to co-create science-based, industry-wide targets through its Resilience Roadmap project.
- City scale analysis typically requires geo-spatial analysis of social and infrastructure vulnerabilities.
- Building a community adaptive capacity is essential to the success of an adaptative programme as it ensures that individuals, communities and societies are actively engaged in processes of change.

3.4.2 Urban heat island

Climate change is driving increases in temperature and more frequent extreme heat events, causing multiple risks for society, particularly for the most vulnerable. Impacts include increases in heat-related deaths and decreased workplace productivity. The health impacts of heatwaves are an emerging environmental health concern across the globe. Recent heatwave events – most notably, the 2003 European event with up to 80 000 victims (Robine *et al.*, 2007) and the 2010 Russian event with an estimated 54 000 fatalities (Revich, 2011) – have been well documented and their impacts quantified. Events such as the 2003 heatwave in Europe, which caused at least 30 000 deaths (UNEP, 2004), were followed by an even more severe heat event in the summer of 2022, when over 70 000 fatalities were recorded across Europe according to a recent study (Ballester *et al.*, 2023). These heatwaves occurring in all parts of the world are projected to become more frequent and severe in the next decades. People who already live in hot climates will face longer periods of extreme heat, with some events reaching 'unliveable' conditions - defined by wet-bulb temperatures exceeding 35°C for sustained periods - forcing the closure of schools and businesses and creating significant economic and health impacts. Even in cooler cities, residents will be exposed to levels of extreme heat to which they are unaccustomed, and many buildings and homes are not equipped with adequate cooling technologies, leading to rapid overheating and health impacts.

Approaches to reducing these impacts include adaptation of buildings, reducing Urban Heat Island (UHI) through increased green space and better medical planning for heatwaves. Building resilience requires a detailed understanding of the underlying drivers of heat risk and how these risks vary spatially.

UHI effect occurs when urbanised areas experience higher temperatures than outlying areas. Structures such as buildings, roads, and other infrastructure absorb and re-emit the sun's heat more than natural landscapes such as forests and water bodies. These surfaces have a low albedo, which increases the amount of energy from solar radiation they absorb, hence trapping the heat causing an UHI effect. More design guidance associated with UHI is discussed in Chapter 5.

3.4.3 Nature-based solutions

Nature-based solutions can provide multiple holistic benefits, not only for the natural environment, but also for people. They can enhance air quality, support health and wellbeing, and provide significant biodiversity gains.

The Building with Nature Standards (BwNS) and Building with Nature Accreditation scheme were created by Building with Nature (Building with Nature, n.d.) to provide best-practice guidance for planning authorities, professionals and developers for delivering high-quality green infrastructure. The BwNS outline good practice at each stage of the development process, providing valuable guidance that enables planning authorities to test emerging policy and supports designers in engaging with clients and enhancing the resilience of new development.

The International Union for Conservation of Nature (IUCN) provides eight design principles (IUCN, 2020) for nature-based solutions.

(*a*) Embrace nature conservation norms and principles.
(*b*) Can be implemented alone or in an integrated manner with other solutions to societal challenges.
(*c*) Determined by site-specific natural and cultural contexts including traditional, local and scientific knowledge.
(*d*) Produce societal benefits in a fair and equitable way in a manner that promotes transparency and broad participation.
(*e*) Maintain biological and cultural diversity and the ability of ecosystems to evolve over time.
(*f*) Are applied at a landscape scale.
(*g*) Recognise and address the trade-offs between the production of a few economic benefits for development and future options for the production of the full range of ecosystem services.
(*h*) An integral part of the overall design of policies, and measures or actions to address a specific challenge.

It is useful to understand the inter-relations and concurrence between nature-based solutions and the eight principles in five frameworks: Protected Areas, Ecological Restoration, Ecosystem-based Adaptation, Forest Landscape Restoration and Ecosystem Approach. The eight principles and five frameworks are wide ranging and cover a large variety of green and blue infrastructure for rural and urban development scales.

Nature-based design solutions deliver an extensive range of potential co-benefits, including direct and indirect contributions to climate resilience with a sample illustrated below.

- Flood risk mitigation.
- Increased extent of and access to green public open space including increased play space.
- Increased health and wellbeing leading to reduced pressure on health and locally delivered social services and thus greater social equity.
- Avoided costs associated with poor physical and mental health.
- Improved air quality, microclimate and comfort.
- Increased climate change resilience and adaptation.
- Habitat restoration and creation including improved connectivity.
- More resilient natural capital.
- Increased recreation values.
- Increased asset values due to the amenity value of living close to green spaces.

Nature-based design essentially considers ecosystem of wider development area and delivers cumulative, compounded benefits to both biodiversity and human wellbeing.

Biodiversity Net Gain

The UK Government published a new Biodiversity Net Gain (BNG) framework, which became a legal requirement in England for most new major developments under the Town and Country

Planning Act (TCPA) from January 2024 and for small sites from April 2024. Biodiversity Net Gain (BNG) requires developers to create or enhance natural habitats so that new development results in a measurable positive impact ('net gain') on biodiversity compared with the pre-development baseline. This legal requirement for new development to deliver at least a 10% net gain represents a significant shift in environmental policy. By contrast, in many other countries, the focus remains on minimising ecological harm and proposing measures to enhance existing biodiversity, with no mandatory requirement to exceed a defined net-gain threshold.

This policy drives a significant shift in business decision-making, as impacts on nature – wildlife and plant species and ecosystems – must be fully built into business decisions and costs.

BNG is expected to stimulate the development of creating markets for green infrastructure and biodiversity investment. This facilitates the markets for adaptation outcomes across relevant legislation and policy programmes, including initiatives on carbon market integrity and the Environmental Land Management Schemes.

By applying existing standards and supported by best-practice guidance, planning authorities, professional experts and developers are delivering high quality green infrastructure.

BwNS shows potential interventions at each stage of the development process. Planning authorities are beginning to use these standards to develop and test new planning policy, benefitting from a clear picture of what good looks like, a shared framework of principles, and the ability to draw on supporting specialist knowledge that might not exist in-house.

Wellbeing

Humans have an inherent affinity for nature, a deeply ingrained instinct shown to provide significant mental and physical health benefits. The term *biophilia* originates from the Greek – it literally means a love of life or living things. Biophilic design is an approach to architecture of built environment that seeks to strengthen the connection between building occupants and nature. Biophilic designs respond to the human ability to experience and 'feel' nature through five senses: sight, smell, touch, taste and hearing. Accordingly, biophilic designed buildings incorporate features that amplify and enhance the five senses, such as natural lighting and ventilation, and landscape elements. These measures, help create healthier, more comfortable, and more productive environments for occupants. Biophilia design principles apply to both indoor and outdoor spaces.

A recent good example is the Jewel, Changi Airport in Singapore (Figure 3.6). The development was designed to bring the outdoors to the indoors, creating an interior environment that integrates substantial natural elements. This immersive green environment enhances visitor wellbeing and, in turn, supports the airport's commercial performance.

3.4.4 Resilient infrastructure

Resilient infrastructure provides critical services to communities, supporting economic functions, and serving as the first line of defence against shocks and disasters. Such services cover utility supplies (water, power, communication), emergency services such as flooding and natural disaster relief, as well as social services such as healthcare.

Figure 3.6 Jewel, Changi Airport, Singapore (photo credit to Mei Ren)

Power

In the UK and other highly urbanised nations, daily life relies on robust provision of electrical supplies. These supplies are particularly vulnerable to interruptions from extreme weather and cyber attack. The risk of cyber attack of the control systems that underpin electrical transmission networks is a significant and growing concern. However, despite this, extreme weather remains the primary cause of recent energy supply disruptions. Unpredictable weather and more frequent extreme conditions due to climate change have heightened the risk of power outages and this risk will only increase as weather conditions become less predictable.

Depending on the extent and severity of the power outage, its impact can range from minor disruption to daily activities to potentially life-threatening consequences. Within a hospital environment, for example, power disruption can profoundly affect the quality of healthcare, especially if it occurs during a major operation.

Weather related threats have traditionally posed major challenges to energy systems, particularly through short-term extreme storms, which in some locations are increasing in intensity and frequency due to climate change (Jasiūnas *et al.*, 2021). However, climate change is also relevant in terms of long-term shifts in typical weather patterns, which can leave capital intensive infrastructure operating under conditions for which it was not originally designed – for example, some electrical sub-stations are located in areas vulnerable to flooding. This presents significant risks as energy systems – whether based on fossil fuel or renewable energy dominated, are among the infrastructure types most sensitive to environmental hazards.

More significant for the near future are the transformational changes needed to the electricity system as a consequence of decarbonisation. These include the introduction of large amounts of self-dispatching renewable generation, the potential electrification of much of transport and space heating, and the rise of the smart consumers and smart homes. Combined with the need to upgrade power networks for a decarbonised system, these shifts greatly increase system complexity and demand a level of engineering coordination and integration that current industry structures and market arrangements are not yet equipped to deliver. Governments and system operators recognise the urgent need to upgrade power infrastructure and deliver transformative change to manage emerging demands, although such transitions take time and can be disruptive in their own right.

Design solutions and management approaches can be deployed to enhance the resilience of energy systems. Some solutions are relatively simple measures such as moving building's substation to be above potential flooding level to protect the local power supply to a building. Investing in innovative solutions and exploring new ways to optimise existing infrastructure can help safeguard businesses against the operational and financial risks associated with power instability. Two particularly relevant technology categories are distributed energy resources such as solar photovoltaic panels, combined heat and power, battery storage, and software solutions including Supervisory Control and Data Acquisition (SCADA), active network management. This increasing digitalisation of energy systems – for example, through smart grids that support renewable electricity integration, can strengthen system resilience to weather and technical failure. However, the deployment of more sophisticated and interconnected control systems may also introduce new vulnerabilities by creating additional points of exposure to cyberattacks.

Microgrids can minimise the impact of disruptions on the wider network by enabling the power to flow independently through local generation. In doing so, they enable organisations to enhance power resilience and mitigate risks in the face of environmental distress. An example of this was seen during the 2018 California wildfires and Hurricane Sandy in New York in 2012. Both events caused widespread disruption and destruction. During both of these events, microgrids were used in an 'islanded mode' to allow power to continue autonomously over a 10-day period. These systems provided continuous electricity to emergency and critical services without relying on the national grid, highlighting their value in sustaining essential operations during severe environmental events.

Flooding

Flooding is one of the most prevalent natural hazards, affecting livelihoods and development opportunities globally. Climate change has already shown strong evidence of increasing both the severity and frequency of flood risks, largely through more intense rainfall events and raising sea levels (Gunn and Rye, 2023). Furthermore, as populations continue to grow, more people are living in flood prone areas, increasing the exposure of vulnerable communities. Recent research from the World Bank has found that 1.8 billion people (23% of the world's population) live in areas that are directly exposed to 1-in-100 year flooding. Most of these people (89%) are in low and middle income nations (IPCC, 2019; Rentschler *et al.*, 2022).

Given that floods affect more people than any other climate-related natural hazards, it is imperative for governments, organisations and communities to proactively invest in resilience measures. These steps include implementing sustainable infrastructure, improving early warning systems and developing robust disaster response strategies.

In the first six months of 2023, nearly 100 major flood events recorded across all populated continents. Most high impact floods were driven by heavy rainfall that overwhelmed rivers and urban drainage systems, or by the combined effects of inland and coastal flooding associated with tropical cyclones. These events occurred in locations as geographically dispersed as New Zealand, Italy, and California, demonstrating that no region of the world is insulated from flood risk.

Floods are an increasing danger in the UK. The Environment Agency estimates that around 5.7 million homes and businesses in England are at risk of flooding, a figure that could double by 2055, driven in part by the impacts of climate change.

River flooding poses a major risk. Over time, a large majority of UK rivers have been modified to support human development, whether through canalised for transport, culverting to enable urban expansion, or the drainage of open land to create agricultural fields. These artificially straightened and engineered watercourses accelerate the movement of rainwater downstream, making channels overflow more quickly during heavy rainfall. Without natural floodplains to slow the flow, downstream communities face heightened flood risk.

Surface water flooding occurs when the volume of rainfall exceeds the capacity of drains and surface water sewers preventing water from entering drainage systems or infiltrating into the ground, causing it to flow across the land instead. Capacity of sewers and drains was calculated based on historic average weather patterns, which is now rapidly changing. As a result, the existing systems often cannot cope with the higher volumes of surface water produced by more intense rainfall.

Implementation of sustainable infrastructure often involves the use of sustainable drainage systems (SuDS). The most effective SuDS manage rainwater close to where it falls, are located on (or close to) the surface and often include vegetation. Examples include rain gardens, tree pits, swales and wetlands. All these green features act as a sponge to absorb rainwater. Where such green features are not feasible, permeable materials for surfaces are alternative by allowing water to infiltrate into the ground, filtering pollutants and recharging groundwater. Examples of permeable paving include porous asphalt, permeable concrete, gravel systems, and permeable pavers.

Adapting to increasingly extreme conditions requires forward-thinking design that finds proactive solutions to improve a development's flood resilience. Rather than treating floods simply as a hazard, Watson and Adams (2010) suggested a series of concepts of design for flooding to create more sustainable, regenerative and resilient places. These principles include integrating flood-mitigation and defence features tailored to different levels of severity directly into development planning. An example is the Mayfield Park development (Figure 3.7), which incorporates upstream flood resilience measures designed to protect the downstream area in Manchester.

Mayfield Park

The 6.5 acre park provides a sequence of character spaces connected by a meandering River Medlock; the design strikes a careful balance between active useable green spaces for people with natural habitat creation for wildlife. The park has been delivered from a nature-based solution with naturalistic planting design that can adapt to changing climate conditions. In drier areas along the path, the plant selection takes into account their resilience based on topography and sun exposure. In areas designed to accommodate flood water and high river flows during the flooding periods, nature-based systems intercept and slow water discharge and encourage evapotranspiration – for example, rain gardens feature moisture-tolerant planting that absorbs excess surface runoff, instead of allowing rain to drain into the river and contribute to flood risks downstream.

Figure 3.7 Mayfield Park, Manchester, UK (photo credit to yellow book licensed under CC BY 2.0)

Heat vulnerability

To successfully build resilience to heat, a detailed understanding of the causes and the spatial variation of associated risks is required. This can be achieved by the development of a Heat Vulnerability Index (HVI) that illustrates where and why people and infrastructure are most at risk. The HVI shows neighbourhood areas whose residents are more at risk during and immediately following extreme heat. It uses a statistical model to summarise the most important social and environmental factors that contribute to neighbourhood heat risk. The factors included in the HVI are surface temperature, green space, access to home air conditioning, and the percentage of residents who are low-income (Wolf and McGregor, 2013). Heat increases mortality for older people and those on low incomes more than any other extreme weather event, this pattern is evident worldwide, including New York, London, Bristol and Manchester.

Completion of a HVI can help city governments to develop an appropriate Heat Resilience Plan, such as the mapping project developed by Bristol City Council (2022). Part of the Heat Resilience Plan is to consider urban development to reduce the UHI as much as possible.

For warmer climate conditions, this is often referred as heat stress mitigation. In an effort to share knowledge and raise awareness, Climate-ADAPT published a visualisation tool for a series of regional Universal Thermal Climate Index (UTCI) maps. UTCI is a human biometeorology parameter to assess how the outdoor environments are experienced by the human body. The index helps governments and healthcare professionals to access to the UTCI predictions and the underlying data for use in policy-making and emergency planning, ultimately benefitting the communities that they seek to protect. The European Climate Adaptation Platform Climate-ADAPT is a partnership between the European Commission and the European Environment Agency (EEA). Climate-ADAPT is maintained by the EEA with the support of the European Topic Centre on Climate Change Impacts, Vulnerability and Adaptation (ETC/CCA). Climate-ADAPT aims to support Europe in adapting to climate change, helping users to access and share data and information (European Commission and European Environment Agency, 2024).

Climate related emergency planning

As global weather patterns become increasingly unstable and less predictable, determining how best to respond to emerging risks – and how to plan for effective preparedness, response, and recovery – has become a critical priority. Numerous international frameworks and policy and guidance documents are available to address resilience planning. The UK Government Resilience Framework (HM Government, 2022b) builds on UK's established civil contingencies frameworks, the Civil Contingencies Act 2004, and aims to strengthen national resilience by improving the ability to prevent, mitigate, respond to and recover from the risks facing the nation. The Framework underscores the UK's 'whole of society' approach to resilience and is a multi-year framework with various milestones and alignment to other UK resilience-focused strategies and guidance. The Cabinet Office also conducts a biennial national survey of the capabilities available to respond to emergencies. British Standard for City Resilience BS67000 (BSI, 2019), is intended for use by all stakeholders involved in strengthening city resilience - including those in the public, private, and third sectors across communities, governance, the environment and infrastructure.

Many local resilience forums (managed at local government level) and their first responder organisations work collaboratively to maintain preparedness, strengthen capabilities, and drive continuous improvement in resilience practices. They hold responsibility for coordinating and overseeing emergency planning to ensure effective responses to both localised incidents and large-scale emergencies. They work to identify potential risks and produce emergency plans to either prevent or mitigate the impact of any incident on their local communities including shocks associated with natural hazards such as flooding and heatwaves, infectious diseases such as Covid and pandemic influenza and infrastructure or technology failures or those resulting from human activity such as terrorism.

Several tools have been developed through the Making Cities Resilient programme and 100 Resilient Cities network to measure a city's resilience, establish baselines for comparison, and help cities prioritise their resilience building efforts. These include the Disaster Resilience Scorecard for Cities (UNDRR) and City Resilience Index (100 Resilient Cities) and the Smart Mature Resilience toolkit developed by a European funded project to assist cities in building their resilience. Tools to explore specific resilience themes include the City Water Resilience Approach (CWRA), which helps cities plan and implement actions to build resilient urban water systems.

As global environmental weather patterns change and become less predictable, the approach of those who construct and manage the built environment may need to alter. In the face of rapid change, the question of how best to respond to risks and how to plan for effective preparedness, response and recovery from emergencies is becoming a high priority.

Integrated emergency management (IEM) has six steps

- anticipation
- assessment
- prevention
- preparation
- response
- recovery.

Emergency preparedness, response, and recovery focus on the consequences and wider impacts of events rather than solely on their underlying causes.

In addition, accessible and comprehensive insurance solutions are vital to close the protection gap and ensure that communities have the necessary financial support to recover and rebuild after natural catastrophes.

In Greater Manchester, to enhance preparedness to emergency scenarios, a local resilience forum ran a comprehensive training and exercising programme in 2016 to help emergency responders to practise their skills and to validate emergency plans: Exercise Triton II (Greater Manchester Resilience Forum, 2021). The complex scenario used in the exercise was designed to test how emergency services and other partners including the military, local and central government, health services, utility companies and voluntary agencies were equipped to tackle a major emergency characterised by high rainfall, river flooding and the breaching of a reservoir dam.

Although large-scale incidents such as the scenario tested in Exercise Triton II are unlikely, the training of emergency planning proved to be of benefit. In 2019, intense rainfall caused several panels on the Toddbrook Dam spillway in Derbyshire to collapse, prompting the evacuation of around 1500 Whaley Bridge residents for up to six nights. The incident underscored the importance of preparing for low probability, high impact events and demonstrated the value of effective emergency preparedness measures.

To respond to the challenge of climate change it is essential not only to achieve rapid reductions in carbon and GHG emissions to reach a net-zero position, but also to prepare for increasingly variable and extreme weather patterns that existing built environment systems are not designed to withstand. Assessing risks, sharing solutions and learning lessons will be crucial to ensure that communities and economies are adequately protected.

REFERENCES

Arrhenius S (1896) On the influence of carbonic acid in the air upon the temperature of the ground. *Philosophical Magazine and Journal of Science* **41**: 237–276. https://www.rsc.org/images/arrhenius1896_tcm18-173546.pdf. Reproduced by Royal Society of Chemistry.

Ballester J *et al.* (2023) The effect of temporal data aggregation to assess the impact of changing temperatures in Europe: an epidemiological modelling study. *The Lancet Regional Health – Europe* **36**: 100779, 10.1016/j.lanepe.2023.100779.

BRE (Building Research Establishment) (2025) *BREEAM New Construction UK Version 7: Technical Manual*. SD6074. BRE Global. https://breeam.com/en/breeam-newconstructionV7/downloads (accessed 17/01/2026).

Bristol City Council (2022) *The Keep Bristol Cool Mapping Tool*. https://www.arcgis.com/apps/instant/portfolio/index.html?appid=9bc417a1906545f38ab9f107cbd8051e (accessed 17/01/2026).

BSI (2011) BS EN 15978: Sustainability of construction works – Assessment of environmental performance of buildings. BSI, London, UK. https://standardsdevelopment.bsigroup.com/projects/2023-01815 (accessed 17/01/2026).

BSI (2019) BS 67000:2019: City resilience guide. BSI, London, UK. https://knowledge.bsigroup.com/products/city-resilience-guide (accessed 17/01/2026).

BSI (2020) BS EN 15804:2012+A2:2019. Sustainability of construction works. Environmental product declarations. Core Rules for the product category of construction products. BSI, London, UK.

BSI (2022) BS 8536:2022 Design, manufacture and construction for operability. Code of practice. BSI, London, UK.

BSI (2023) PAS 2080:2023: Carbon management in buildings and infrastructure. BSI, London, UK. https://knowledge.bsigroup.com/products/carbon-management-in-buildings-and-infrastructure (accessed 17/01/2026).

Building with Nature (n.d.) https://www.buildingwithnature.org.uk/ (accessed 17/02/2026)

Caesar J. (2018) *UKCP18 Guidance: Representative Concentration Pathways*, p. 2. https://www.metoffice.gov.uk/binaries/content/assets/metofficegovuk/pdf/research/ukcp/ukcp18-guidance--representative-concentration-pathways.pdf (accessed 17/01/2026).

CIBSE (The Chartered Institution of Building Services Engineers) (2014) *TM55 Design for Future Climate: Case Studies*. https://www.cibse.org/knowledge-research/knowledge-portal/technical-memorandum-55-design-for-future-climate-case-studies-2014-pdf?id=a0q20000008I76aAAC (accessed 17/01/2026).

CIBSE (2022) *TM54 Evaluating Operational Energy Use at the Design Stage*. https://www.cibse.org/knowledge-research/knowledge-portal/tm54-evaluating-operational-energy-use-at-the-design-stage-2022 (accessed 17/01/2026).

Climate-ADAPT (2016a) *Urban Adaptation Support Tool: 2. Assessing Climate change risks and vulnerabilities*. https://climate-adapt.eea.europa.eu/knowledge/tools/adaptation-support-tool/step-2-0 (accessed 17/01/2026).

Climate-ADAPT (2016b) *Urban Adaptation Support Tool: 2.4 Conducting risk and vulnerability assessments*. https://climate-adapt.eea.europa.eu/en/knowledge/tools/urban-ast/step-2-4 (accessed 17/01/2026).

Climate Change Act 2008 (2050 Target Amendment) Order 2019 (2019) https://www.legislation.gov.uk/ukdsi/2019/9780111187654/contents (accessed 17/01/2026).

Climate Change Committee (2023) *Investment for a Well-Adapted UK*. https://www.theccc.org.uk/wp-content/uploads/2023/01/Investment-for-a-well-adapted-UK-CCC.pdf (accessed 17/01/2026).

Eckstein K and Schafer (2021) *Global Climate Risk Index 2021, Who suffers most from extreme weather events? Weather-related loss events in 2019 and 2000–2019, Germanwatch*. https://www.germanwatch.org/en/19777 (accessed 17/01/2026).

Edwards PN (2010) History of climate modeling. *Wiley Interdisciplinary Reviews Climate Change* **2(1)**: 128–139, 10.1002/wcc.95.

Energy Act 2023 (2023) https://www.legislation.gov.uk/ukpga/2023/52 (accessed 17/01/2026).

European Commission and European Environment Agency (2024) *Key Services, Thematic Features and Tools of Climate-ADAPT*. https://climate-adapt.eea.europa.eu/en (accessed 17/01/2026).

Fung *et al.* (2020) *UKCP Global (60km) – Low emissions scenario (RCP 2.6), UKCP Global*. Met Office, Exeter, UK, pp. 2–6. https://www.metoffice.gov.uk/binaries/content/assets/metofficegovuk/pdf/research/ukcp/ukcp18_factsheet_global_rcp_2_6.pdf (accessed 17/01/2026).

Gething B (2010) *Design for Future Climate: Opportunities for Adaptation in the Built Environment*. Technology Strategy Board. https://www.ukgbc.org/wp-content/uploads/2018/09/Design-for-future-climate-full-report.pdf (accessed 17/01/2026).

Greater Manchester Resilience Forum (2021) *Greater Manchester Resilience Strategy 2020–2030*. https://www.greatermanchester-ca.gov.uk/media/4541/greater-manchester-resilience-strategy-2020-2030-accessible.pdf (accessed 17/01/2026).

Gunn N and Rye C (2023) A view of catastrophic flooding from across the world. *WTW*, 20 July. https://www.wtwco.com/en-gb/insights/2023/08/a-view-of-catastrophic-flooding-from-across-the-world (accessed 17/01/2026).

HM Government (2017) *The Clean Growth Strategy: Leading the Way to a Low Carbon Future*. https://assets.publishing.service.gov.uk/media/5ad5f11ded915d32a3a70c03/clean-growth-strategy-correction-april-2018.pdf (accessed 17/01/2026).

HM Government (2018) A Green Future: Our 25 Year Plan to Improve the Environment. https://assets.publishing.service.gov.uk/media/5ab3a67840f0b65bb584297e/25-year-environment-plan.pdf (accessed 17/01/26).

HM Government (2022a) *The Building Regulations 2010 Conservation of Fuel and Power Approved Document L (as amended 2023)*. https://assets.publishing.service.gov.uk/media/662a2e3e55e1582b6ca7e592/Approved_Document_L__Conservation_of_fuel_and_power__Volume_1_Dwellings__2021_edition_incorporating_2023_amendments.pdf (accessed 17/01/2026).

HM Government (2022b) *The UK Government Resilience Framework*. https://assets.publishing.service.gov.uk/media/63cff056e90e071ba7b41d54/UKG_Resilience_Framework_FINAL_v2.pdf (accessed 17/01/2026).

HM Government (2022c) *UK Climate Change Risk Assessment 2022*. https://assets.publishing.service.gov.uk/media/61e54d8f8fa8f505985ef3c7/climate-change-risk-assessment-2022.pdf (accessed 17/01/2026).

HM Government (2024) *UK Third National Adaptation Programme (NAP3)*. https://www.gov.uk/government/publications/third-national-adaptation-programme-nap3 (accessed 27/10/2024).

IPCC (The Intergovernmental Panel on Climate Change) (2013) *Climate Change 2013: The Physical Science Basis. Contribution of Working Group I to the Fifth Assessment Report of the Intergovernmental Panel on Climate Change*. Cambridge University Press, Cambridge, UK and New York, USA, 10.1017/CBO9781107415324.

IPCC (2014) *Climate Change 2014: Impacts, Adaptation, and Vulnerability. Part A: Global and Sectoral Aspects. Contribution of Working Group II to the Fifth Assessment Report of the Intergovernmental Panel on Climate Change* (Field CB *et al.* (eds)). IPCC, Geneva, Switzerland.

IPCC (2019) *IPCC Special Report on the Ocean and Cryosphere in a Changing Climate*. https://www.ipcc.ch/srocc/ (accessed 17/01/2026).

IPCC (2023) *Climate Change 2023: Synthesis Report. Contribution of Working Groups I, II and III to the Sixth Assessment Report of the Intergovernmental Panel on Climate Change*. IPCC. https://www.ipcc.ch/report/ar6/syr/downloads/report/IPCC_AR6_SYR_LongerReport.pdf (accessed 17/01/2026).

IUCN (International Union for Conservation of Nature) (2020) *Guidance for Using the IUCN Global Standard for Nature-based Solutions: First Editions*, 10.2305/iucn.ch.2020.09.en.

ISO (International Organization for Standardization) (2006) ISO 14025:2006: Environmental labels and declarations — Type III environmental declarations — Principles and procedures. ISO, Geneva, Switzerland.

Jasiūnas J, Lund PD and Mikkola J (2021) Energy system resilience – a review. *Renewable and Sustainable Energy Reviews* **150**: 111476, 10.1016/j.rser.2021.111476.

LETI (2020a) *LETI Climate Emergency Design Guide*. https://www.leti.uk/_files/ugd/252d09_3b0f2acf2bb24c019f5ed9173fc5d9f4.pdf (accessed 17/01/2026).

LETI (2020b) *LETI Embodied Carbon Primer*. https://www.leti.uk/_files/ugd/252d09_8ceffcbcafdb43cf8a19ab9af5073b92.pdf (accessed 17/01/2026).

Met Office *et al.* (2019) *UKCP Factsheet: UKCP Global (60km) – Low emissions scenario (RCP 2.6)*. https://www.metoffice.gov.uk/binaries/content/assets/metofficegovuk/pdf/research/ukcp/ukcp18_factsheet_global_rcp_2_6.pdf (accessed 17/01/2026).

Met Office (2022) *UKCP Summaries and Headline Findings*. https://www.metoffice.gov.uk/research/approach/collaboration/ukcp/summaries/index (accessed 27/10/2024).

MHCLG (Ministry of Housing Communities and Local Government) (2024) *National Planning Policy Framework*. https://assets.publishing.service.gov.uk/media/67aafe8f3b41f783cca46251/NPPF_December_2024.pdf (accessed 17/01/2026).

Murphy JM *et al.* (2023) *Comparison Between Climate Change Projections from the UKCP Land Scenarios and CMIP6 models, Hadley Centre Technical Note 109*. Met Office, UK. https://www.metoffice.gov.uk/binaries/content/assets/metofficegovuk/pdf/research/ukcp/hctn_109_comparison_ukcp_cmip6.pdf (accessed 17/01/2026).

Myers KF *et al.* (2021) Consensus revisited: quantifying scientific agreement on climate change and climate expertise among Earth scientists 10 years later. *Environmental Research Letters* **16(10)**: 104030, 10.1088/1748-9326/ac2774.

Rentschler J, Salhab M and Jafino BA (2022) Flood exposure and poverty in 188 countries. *Nature Communications* **13(1)**, 10.1038/s41467-022-30727-4.

Revich B (2011) Heat-wave, air quality and mortality in European Russia in summer 2010: preliminary assessment. *Yekologiya Cheloveka/Human Ecology* **7(3–9)**. cabidigitallibrary.org/doi/full/10.5555/20113273738 (accessed 17/01/2026).

RIBA (Royal Institute of Chartered Surveyors) (2021) *2030 Climate Challenge*. https://www.architecture.com/about/policy/climate-action/2030-climate-challenge?srsltid=AfmBOoryVwoat1G2Ir3FO81I1utrThjygdaAJxip0Mf7tmqessjDznSs (accessed 17/01/2026).

RICS (Royal Institution of Chartered Surveyors) (2017) *Whole Life Carbon Assessment for the Built Environment, RICS Professional Statement*. https://www.rics.org/profession-standards/rics-standards-and-guidance/sector-standards/construction-standards/whole-life-carbon-assessment (accessed 17/01/2026).

RICS (2024) *Whole Life Carbon Assessment for the Built Environment Version 3*. https://www.rics.org/content/dam/ricsglobal/documents/standards/Whole_life_carbon_assessment_PS_Sept23.pdf (accessed 17/01/2026).

Robine J-M *et al.* (2007) Death toll exceeded 70,000 in Europe during the summer of 2003. *Comptes Rendus Biologies* **331(2)**: 171–178, 10.1016/j.crvi.2007.12.001.

Slingo J (2016) The evolution of climate science: a personal view from Julia Slingo. *WMO Bulletin* **66(1)**. https://library.wmo.int/viewer/46039?viewer=picture#page=18&viewer=picture&o=bookmarks&n=0&q = (accessed 17/01/2026).

Target Zero (2012) *Guidance on the Design and Construction of Sustainable, Low Carbon Mixed-Use Buildings*. Report V 2.0. British Constructional Steelwork Association (BCSA),

UK. https://steelconstruction.info/images/3/33/Mixed-Use_guidance_doc_v2.pdf (accessed 17/01/2026).

UK Climate Projections (UKCP) (n.d.) https://www.metoffice.gov.uk/research/approach/collaboration/ukcp (accessed 27/10/2024).

UK Net Zero Buildings (2024) UK Net Zero Carbon Buildings Standard. https://www.nzcbuildings.co.uk/ (accessed 17/01/2026).

UKGBC (UK Green Building Council) (2019) *Net Zero Carbon Buildings Framework*. https://ukgbc.org/resources/net-zero-carbon-buildings-framework/ (accessed 17/01/2026).

UKCIP (2003) *UKCIP Adaptation Wizard Stage 1: Identify problem and objectives*. https://www.ukcip.org.uk/wp-content/Wizard/UKCIP-risk-stages-1and2.pdf (accessed 17/01/2026).

UKCIP (2013) *UKCIP Risk Framework*. https://www.ukcip.org.uk/wizard/about-the-wizard/ukcip-risk-framework/ (accessed 17/01/2026).

UNEP (United Nations Environment Programme) (2004) *Impacts of Summer 2003 Heat Wave in Europe: Environmental Alert Bulletin*. https://www.unisdr.org/files/1145_ewheatwave.en.pdf (accessed 17/01/2026).

UNEP (2022) *CO_2 Emissions from Buildings and Construction Hit New High, Leaving Sector Off Track to Decarbonize by 2050: UN*. https://www.unep.org/news-and-stories/press-release/co2-emissions-buildings-and-construction-hit-new-high-leaving-sector#:~:text=Released%20at%20the%20latest%20round,related%20CO2%20emissions%20in%202021 (accessed 17/01/2026).

United Nations (2015) *Paris Agreement*. https://unfccc.int/files/essential_background/convention/application/pdf/english_paris_agreement.pdf (accessed 17/01/2026).

United Nations Environment Programme and Yale Center for Ecosystems + Architecture (2023) *Building Materials and the Climate: Constructing a New Future*. https://wedocs.unep.org/20.500.11822/43293 (accessed 17/01/2026).

University of Reading (2023) *Climate Risk Indicator Explorer*. https://uk-cri.org/ (accessed 27/10/2024).

University of Southampton (2022) *Climate Change World Weather File Generator for World-Wide Weather Data – CCWorldWeatherGen – University of Southampton Blogs*. https://energy.soton.ac.uk/ccworldweathergen/ (accessed 27/10/2024).

USAID Center for Resilience (2018) *Resilience Measurement Practical Guidance Note Series 2: Measuring Shocks and Stresses*. https://www.fsnnetwork.org/sites/default/files/GN02_ShocksandStressesMsmt_Final4-11508_0.pdf (accessed 17/01/2026).

Watson D and Adams M (2010) *Design for Flooding: Architecture, Landscape, and Urban Design for Resilience to Climate Change*. John Wiley and Sons, Hoboken, NJ, USA.

Wolf T and McGregor G (2013) The development of a heat wave vulnerability index for London, United Kingdom. *Weather and Climate Extremes* **1**: 59–68, 10.1016/j.wace.2013.07.004.

World Bank Climate and Disaster Risk Screening (n.d.) The World Bank Climate and Disaster Risk Screening Tools (n.d.) https://climatescreeningtools.worldbank.org/ (accessed 17/01/2026).

Navigating an Evolving Sustainable Built Environment: Delivering Net Zero Carbon and Climate Resilience

Mei Ren and Rachel Waggett
ISBN 978-1-83662-853-8
https://doi.org/10.1108/978-1-83662-850-720261007

Chapter 4
The sustainable built environment: responses to the challenge

4.1. From understanding the challenge to shaping the response

Previous chapters have set out the scientific, policy and practical foundations that underpin today's climate and sustainability agenda for the built environment. From earliest indications that environmental action was necessary, through to the establishment of an overwhelming scientific consensus on human-driven climate change, the scale and immediacy of the challenge is now clear. Despite major progress in decarbonisation policy, operational energy standards and emerging approaches to embodied carbon, the built environment remains a significant contributor to global emissions and is increasingly exposed to climate-related risks such as extreme heat, water scarcity and flooding.

The unprecedented clarity of the need for action, supported by detailed climate projections, whole-life carbon methodologies and expanding regulatory frameworks is certainly to be celebrated; for the first time there is full scientific consensus on the need for action and the outcomes to be achieved. However, built environment practitioners face a rapidly evolving landscape of guidance, standards and expectations, ranging from net zero definitions and embodied carbon frameworks to nature-based solutions and community resilience planning. Lack of information is no longer a barrier to action, but rather how to navigate it effectively, make sound decisions based on the unique needs and opportunities of each project, and embed sustainability into every stage of project development as well as for corporate environmental, social and governance (ESG) reporting.

This chapter explores how the built environment sector is responding to the net zero imperatives and examines the growing suite of frameworks, standards and tools intended to guide that response. Acknowledging that technical solutions alone cannot deliver the systemic change required, the need for clear decision-making processes, alignment across multi-disciplinary teams and suitable governance structures is a crucial part of this picture. This understanding helps to shape the essential shift from understanding what must be done, to determining how to set about doing it effectively.

4.2. Responding to net zero challenge

In response to an increasingly dire climate change outlook, countries around the globe are already experiencing the consequences of approaching 1.5°C in global average temperature increase (Adil *et al.*, 2025). Figure 4.1 shows that this is now a crucial point in time. With global warming rates at their highest historical levels, this is the time to seize the narrow window of opportunity and accelerate the action that has already started (Mann, 2023).

The UK Green Building Council (UKGBC) launched an Advancing Net Zero programme in 2018 as part of a World Green Building Council initiative globally, with a mission to lead the UK's transition to a net zero built environment. It has produced leading guidance for net zero

Figure 4.1 An illustration of global climate change to date and possible future scenario (SSP3) compared with global target (created using data from Forster *et al.*, 2025; IPCC, 2023b)

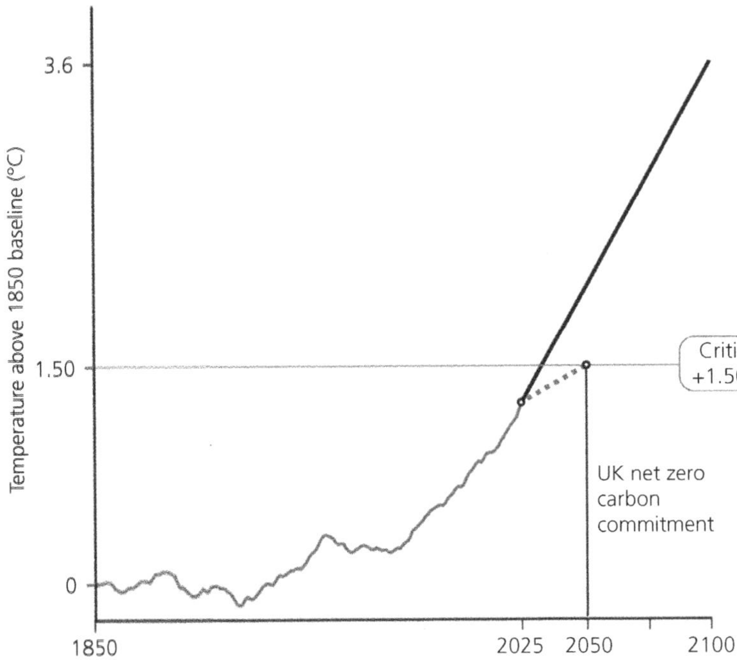

such as the Net Zero Carbon Buildings Framework (UKGBC, 2019) and the Net Zero Whole Life Carbon Roadmap (UKGBC, 2021b). One of the key outcomes of this process has been the recognition that the industry must take meaningful action to measure and reduce embodied carbon emissions alongside its operational emissions. Momentum around net zero development has accelerated significantly within the built environment. Although many Net Zero Carbon frameworks and guidance documents focus primarily on buildings, they are increasingly being adopted by corporates to support reporting on environmental, social and governance (ESG) performance.

4.2.1 UKGBC net zero carbon definition

The Net Zero Carbon Buildings Framework defined by the UKGBC (2019) provides guidance on the definition of net zero carbon buildings – both homes and non-domestic – and a way to demonstrate how a building has achieved net zero carbon status. It focuses on carbon impacts that can be readily measured and mitigated – operational energy and embodied impacts of construction. The framework sets out the key principles that must be followed to demonstrate alignment with net zero carbon for construction and for operational energy. The Whole Life Carbon Management Framework, introduced in PAS 2080:2023 (BSI, 2023) provides a structured approach to support decision making, and is the world's first framework to bring together both buildings and infrastructure in the decarbonisation of the built environment. The revised PAS demands that every part of the value chain works collaboratively to consider the whole life carbon of projects. Figure 4.2 outlines UKGBC net zero carbon framework (UKGBC, 2019).

Net zero carbon – construction	Net zero carbon – operational energy
UKGBC defines net zero carbon as: "When the amount of carbon emissions associated with a building's product and construction stages up to practical completion is zero or negative, through the use of offsets or the net export of on-site renewable energy".	UKGBC defines net zero carbon as: "When the amount of carbon emissions associated with the building's operational energy on an annual basis is zero or negative. A net zero carbon building is highly energy efficient and powered from on-site and/or off-site renewable energy sources, with any remaining carbon balance offset".

Since publication, both national and local policy makers have begun to integrate this approach into local policies to help drive the transition to a net zero carbon built environment. The principles have increasingly been used to help inform the direction of policies and public procurement, with a particular focus on in-use energy performance and addressing embodied carbon at both national and local level.

Further net zero guidance from UKGBC include Renewable Energy Procurement and Carbon Offsetting Guidance for Net Zero Carbon Buildings; this suite of net zero guidance provides greater transparency and robustness to the claim of net zero buildings.

The term 'zero carbon' and carbon neutrality have been increasingly used in national climate policies (see Figure 4.3 for definitions). In line with the UKGBC Net Zero Carbon Buildings Framework and other industry initiatives, net zero carbon buildings must maximise building efficiency before turning to carbon offsetting.

Net Zero Whole Life Carbon Roadmap (UKGBC, 2021b) aims to develop the carbon footprint for the UK built environment – the Net Zero Carbon trajectory to 2050 – and policy recommendations with industry action plans to deliver the 2050 trajectory.

- Identify and set out a forward emissions budget to 2050 for the built environment, which is consistent with the wider UK carbon budget, and with a suitable system boundary.
- Provide a whole life carbon pathway or emission trajectory for the UK built environment, and the timeframes in which they need to be taken based on economic or technical considerations.

In 2021, momentum toward global climate action increased substantially. Informed by the latest IPCC assessment reports, a growing number of companies, countries, cities and regions adopted immediate strategies to achieve net zero emissions, with many formalising their commitments through participation in the UN Race to Zero. This trend has been particularly pronounced within the built environment sector. As the built environment contributes to 38% of global emissions, coordinated and credible pathways to net zero are essential for individual markets.

One of the principal findings from the roadmap concerns the establishment of an overarching consumption-based carbon budget for the built environment. This budget provides a top-down dataset that can inform and guide the development of asset level and project level targets. Figure 4.4 reproduces the UK Greenhouse emissions for built environment based on 2018 UK carbon budget (Climate Change Committee, 2020). This shows that 42% of the UK total

Figure 4.2 UKGBC steps to achieving a net zero carbon building (reproduced with kind permission of the UKGBC (2019))

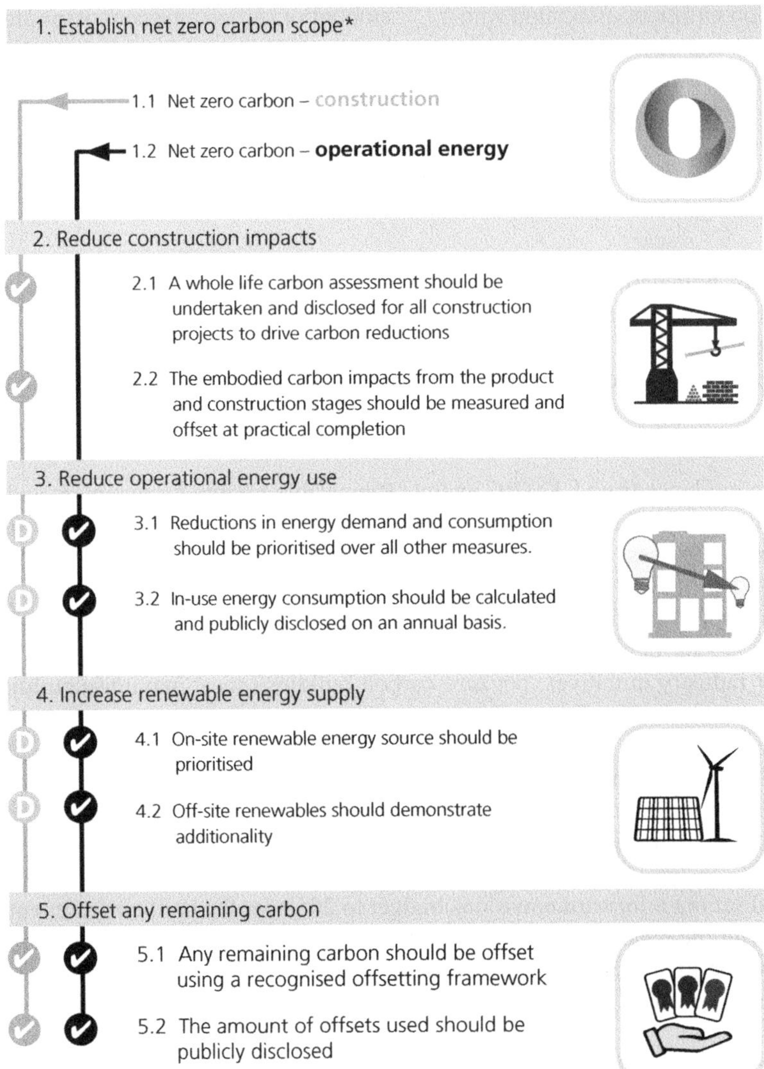

D New buildings and major refurbishments targeting net zero carbon for construction should be designed to achieve net zero carbon for operational energy by considering these principles.

* Please also note, a further scope for net zero whole life carbon (1.3) will be developed in the future.

greenhouse gas (GHG) emissions calculated on consumption basis, including electricity and fuel supply, transport, waste, aviation and agriculture, come from the built environment and associated sectors. Within the built environment sector, the largest sources of emissions come from domestic and non-domestic buildings, surface transport and infrastructure related embodied carbon.

Figure 4.3 Carbon definition explained

Carbon
The term "carbon" is widely referred to multiple greenhouse gases. "Carbon" is used as a proxy for "carbon dioxide equivalent" (CO_2e) and describes the aggregated Global Warming Potential (GWP) of multiple greenhouse gases in a common unit

CO_2e

GWP Equivalent
$CO_2 = 1$
$CH_4 = 28$
$N_2O = 298$
$SF_6 = 22,800$

Net zero carbon
The development must make significant reductions in energy and CO_2 emissions and maximise onsite renewable generation before offsetting any residual as a last resort through atmospheric removal

CO2 Emissions + $-CO_2$ Offsets
Offset RESIDUAL emissions

Absolute zero
The development must eliminate all carbon emissions without the use of offsets

CO2 Emissions + NO RESIDUAL emissions to offset

Figure 4.4 Total UK GHG emissions (utilising data from Climate Change Committee, 2020) showing proportion of Built Environment emissions (reproduced with kind permission of UKGBC (UKGBC, 2021b))

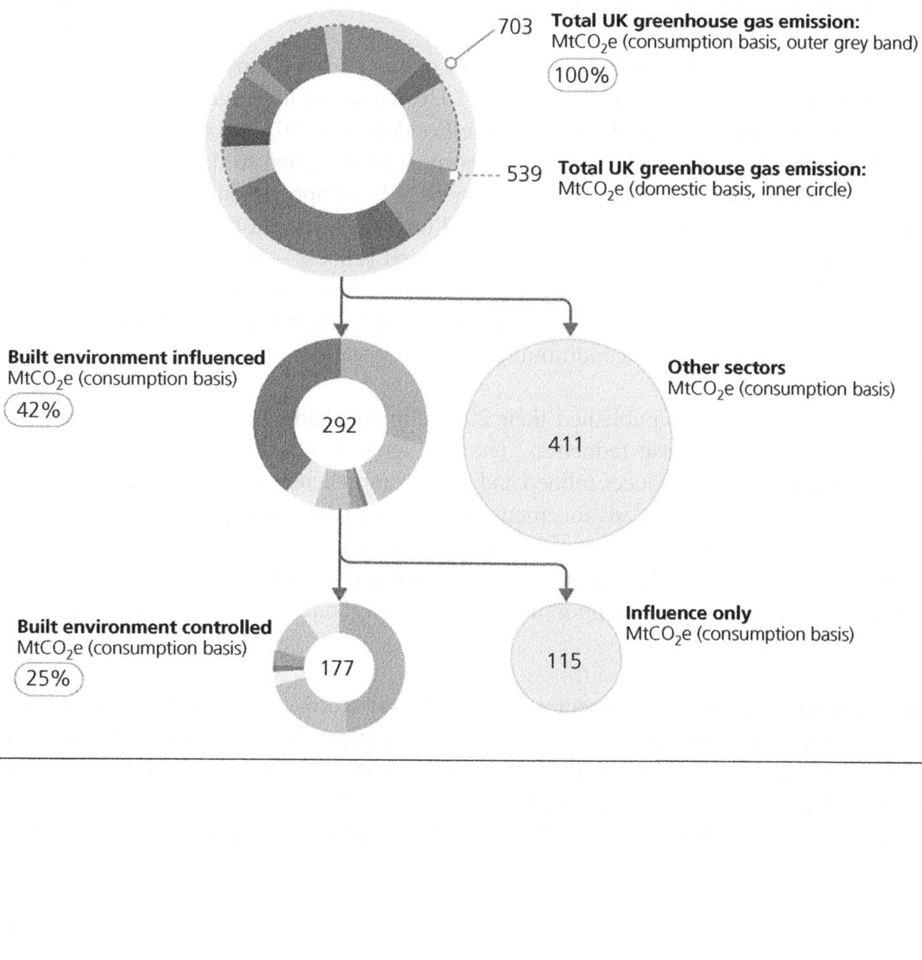

The Net Zero Whole Life Carbon Roadmap (UKGBC, 2021a) sets out the anticipated design standards and recommendations for both policy makers and industry stakeholders. For clarity, a selection of key recommendations are summarised in Table 4.1. These recommendations offer valuable insight and direction for shaping future legislation, policy development, and strategic pathways toward achieving net zero.

Embodied carbon from the construction and refurbishment of buildings makes up 20% of built environment emissions. As operational emissions from buildings decrease, the trajectory results indicate that embodied carbon will form over half of built environment emissions by 2035.

The Net Zero Whole Life Carbon Roadmap also provides recommendations aimed at addressing embodied and operational carbon in infrastructure.

> Integrated Decarbonisation of Infrastructure Systems:
>
> - Introduce the role of a National Infrastructure Integrator to enable holistic decision-making across UK infrastructure investment with full visibility of all carbon impacts.
> - Demonstrate leadership within public procurement by way of Infrastructure and Projects Authority (IPA) commitment to the CSIC Carbon Reduction Code (which includes integrating carbon reduction targets and reporting commitments explicitly in all procurement documents from 2021).
> - Set a requirement for all regulators to develop an explicit first-order objective to support the transition to net zero by 2050.
> - Mandate for PAS 2080 (or equivalent standard) to be fully implemented across all Infrastructure projects by 2025. (UKGBC, 2021b)

Other influential net zero carbon guidance includes London Energy Transformation Initiative (LETI) and Royal Institute of British Architects (RIBA) 2030 Climate Challenge as well as those leading practices required by The London Plan 2021 (The London Plan, 2021). LETI believes that to meet national 2050 reduction targets, all new buildings must be designed to deliver net zero carbon by 2025. This will require the construction industry to be fully equipped with the knowledge and skills necessary. With regards to operational energy, LETI outlines a series of indicative design measures for offices, residential properties and naturally ventilated schools – For example, in order to meet the operational energy consumption target of 55 kWh/m²/yr for commercial offices, recommended measures include enhanced building fabric design standards, glazing ratio, natural ventilation or low carbon heating, ventilation and air conditioning (HVAC) systems.

RIBA 2030 Climate challenge published their 2030 climate change targets for operational energy, embodied carbon and water use reduction. The Challenge has since been updated and reissued as Version 2 (2021), which introduces refined and more comprehensive targets for 2025 and 2030 (see Figure 4.5). These updates reflect advancements in embodied carbon assessment practices and align the framework with LETI and the latest jointly authored guidance. Version 2 of the Climate Challenge presents ambitious yet achievable forward-facing performance outcomes, that move beyond conventional compliance approaches and support a forward-looking trajectory toward net-zero carbon.

4.2.1.1 UK Net Zero Carbon Buildings Standard

Net Zero Carbon Buildings Standard represents the UK's first cross-industry initiative to consolidate net-zero carbon requirements for all major building typologies, aligned with a 1.5°C trajectory. The primary objective is to establish a unified Net Zero Carbon Buildings Standard for the UK built environment (UKNZCBS, 2025). Leading organisations including Better Building Partnership (BBP) (https://www.betterbuildingspartnership.co.uk/), Building Research Establishment (BRE)

Table 4.1 Key recommendations of the net zero whole life carbon roadmap, reproduced with kind permission of UKGBC (UKGBC, 2021b)

Operational carbon		Embodied carbon
New buildings	Existing	All buildings
Update the regulatory and policy framework for new homes/non-domestic buildings to include the following. ■ Update the National Calculation Methodology and the EPC methodology to create a fit-for-purpose predictive methodology for energy performance of dwellings that better reflects in-use energy performance. ■ 2025 Future Homes Standard and Building Regulations Part L 2025 update for new non-domestic buildings to align the introduction of the EUI compliance approach per sector with the timings of a confirmed mandatory energy disclosure timetable to introduce the following. – energy Usage Intensity (EUI) targets including regulated and unregulated loads (kWh/m2/yr) – thermal energy demand limits (kWh/m^2/yr) – low carbon heating for all new buildings (no fossil fuel) – measures to limit peak demand and enable load shifting (with limits on peak demand from 2030) – minimum standards for currently unregulated key appliances or equipment.	Introduce and clearly signpost a cut-off date of 2030 for sales of gas and oil boilers. ■ Adopt a national retrofit strategy by 2022, setting out a clear national homes upgrade programme, fully coordinated with local government, industry and relevant stakeholders. ■ A clear trajectory and regulatory framework to introduce mandatory minimum EPC rating of C (or equivalent under updated EPC methodology) for owner-occupied homes at the point of sale by 2028, for improving the MEES for domestic rented sector by 2028. ■ Tightening of building regulations requirements for works to existing buildings. ■ Introduce the planned performance-based rating system for large office buildings (>1000 m^2), including mandatory energy performance disclosure. ■ Introduce minimum standards and fiscal incentives for large existing office buildings by 2025.	Implement a regulatory policy framework for embodied carbon in new buildings, with clear signposting of a phased pathway. ■ Mandatory measurement and reporting of whole life Carbon by 2023 for large buildings (>1000 m^2) and residential developments (>10 dwellings). ■ Minimum standards (limits) for upfront embodied Carbon by 2025 for large buildings (>1000 m^2) in more mature sectors (i.e. those with sufficient asset level benchmark data), with associated fiscal incentives and penalties. ■ Minimum standards (limits) for upfront embodied Carbon by 2027 in all sectors. ■ Final phase to introduce minimum standards for all size buildings (with a suitable minimum threshold) in all sectors by 2030.

Figure 4.5 RIBA 2030 Climate Challenge target metrics for non-domestic (reproduced with kind permission of the Royal Institute of British Architects (RIBA, 2021))

RIBA sustainable outcome metrics	Business as usual (New build, compliance approach)	2025 Targets	2030 Targets	
Operational energy $kWh/m^2/y$	130 $kWh/m^2/y$ DEC D (90)	<75 $kWh/m^2/y$ DEC B (50) and/or NABERS base build 5	<55 $kWh/m^2/y$ DEC B (40) and/or NABERS base build 6	Targets based on GIA. Figures include regulated and unregulated energy consumption irrespective of source (grid/renewables) *(a)* Use a 'Fabric First' approach *(b)* Minimise energy demand. Use efficient services and low carbon heat *(c)* Maximise onsite renewables
Embodied carbon $kgCO_2e/m^2$	1400 $kgCO_2e/m^2$	<970 $kgCO_2e/m^2$	<750 $kgCO_2e/m^2$	Use RICS whole life carbon (modules A1-A5, B1-B5, C1-C4 incl sequestration). Analysis should include minimum of 95% of cost, include substructure, superstructure, finishes, fixed FF&E, building services and associated refrigerant leakage. *(a)* Whole life carbon analysis *(b)* Use circular economy strategies *(c)* Minimise offsetting and use as last resort. Use accredited, verifiable schemes (see checklist). BAU aligned with LETI band E; 2025 target aligned with LETI band C and 2030 target aligned with LETI band B.
Potable water use Litres/person/day	16 l/p/day (CIRA W11 benchmark)	13 l/p/day	10 l/p/day	CIBSE Guide G

(https://bregroup.com/), the Carbon Trust (https://www.carbontrust.com/), CIBSE (https://www.cibse.org/), Institution of Structural Engineers (IStructE) (https://www.istructe.org/), LETI (https://www.leti.uk/), Royal Institute of British Architects (RIBA), Royal Institution of Chartered Surveyors (RICS) and UKGBC have joined forces to champion this initiative. Once published in full, the new standards would supersede the existing guidelines from LETI and RIBA 2030. It would do so with the support of a far more extensive body of evidence-based performance data available and the rigour to inform the setting of performance targets.

Performance targets will align with science-based trajectories needed to achieve net zero by 2050 and a 78% reduction by 2035 in the UK – thresholds recognised as necessary to maintain a reasonable chance of mitigating global warming to 1.5°C. It will also align with the energy demand reductions projected to be required to enable a net zero carbon energy supply sector. Extensive consultations undertaken between 2022 and 2024, culminating in the launch of the pilot standard in September 2024. The approach adopted is the first of its kind globally, combining both top-down and bottom-up evidence-based assessment methodologies to define robust, science-aligned performance targets for buildings.

The standard defines the metrics used to assess net zero carbon performance, along with performance targets, or limits that must be achieved. These are likely to include energy use, upfront embodied carbon (Modules A1 to A5, see Section 4.2.3 for more details) and life cycle embodied carbon. Additional indicators – such as space heating/cooling demand and peak load – may also be considered. The Standard will also define the methodology to carbon accounting, procuring renewable energy and the treatment of residual emissions, including carbon 'offsetting'. It will cover the most common building typologies for both existing and new buildings, including homes, offices, education, industry, retail, hotels, healthcare. However, the standard will not apply to infrastructure assets.

The Net Zero Carbon Buildings Standard is expected to have a significant impact on new developments across the built environment, affecting building owners, investors, designers, contractors to occupiers. It provides a clear consistent net zero benchmark against which buildings and assets can be verified to help prevent duplication, to strengthen integrity and reduce the risk of greenwashing. For investors, the Standard provides a trusted framework to support investment decision making, including sustainable finance, lending and debt, and for which policy-makers at national, city and local level may choose to embed within planning, building regulations, procurement and leasing. At time of writing, it is anticipated that carbon offsetting will be voluntary, a whole building approach will be adopted, third-party verification will be required to ensure robust verification to a net zero building claim.

Focus area

Buildings represent an approximately 40% global carbon emissions, including building operations and building materials and construction, therefore carbon reduction of building assets remains the priority for policy makers and industry leading partners.

- Set out a clearly defined Net Zero Buildings Standard for most of building sectors with a consistent approach agreed with built environment professionals across the majority of institutions and organisations.
- The Standard has been developed using top-down and bottom-up approach to ensure its deliverability in practice, while maintaining alignment with the science-based target and the UK Carbon Budget and Net Zero Whole Life Carbon Roadmap.
- The standard will consolidate all the relevant metrics and performance limits under a single unified framework, including Energy Use Intensity (EUI), upfront embodied carbon, life cycle embodied carbon, demand management, on-site renewables, renewable procurement offsetting and refrigerant leakage.

4.2.2 Zero carbon verification

The UKGBC's Net Zero Buildings Framework provides a consistent approach to defining, measuring and reporting a building's carbon performance, calling for more robust verification of assets or portfolios to ensure the credibility and consistency of net zero claims. At time of writing, the built environment has not established a 'one scheme fits all' solution.

At time of writing, the UK Net Zero Carbon Buildings Standard is considering options for a verification approach, for example, through a third party – that is, those wanting to describe their building as Net Zero Carbon will require verification of the building, in order to determine conformity with the Standard's requirements to confirm the reliability of information declared in claims.

Three potential options to making the process of creating a net zero carbon 'badge' more consistent and robust, have been evaluated by the UKGBC.

- An independent third-party verifier: the use of accredited verifiers who have undertaken relevant training to undertake verification; this accreditation could be offered through professional institutes and integrated into existing training modules.
- An independent third-party verifier as an extension to the existing rating tools, such as BREEAM In-Use or NABERS UK (NABERS stands for the National Australian Built Environment Rating System; this option could be relatively simple to adopt, enabling

the existing assessor to verify a certain portion of the net zero carbon claim, typically the onsite requirements for energy efficiency and renewables and a third-party verification of the remaining elements of a net zero claim – for example, the use of renewable energy procurement offsite or procurement offset.

- A new certification of net zero carbon buildings: move away from an open third-party verification towards establishing a new net zero carbon certification as the primary basis for verification. This approach would be similar to the net zero certification schemes already on offer in the UK market, including under the Living Building Challenge and LEED Net Zero, however, these existing schemes are for new buildings only. It is therefore likely that a new system will be required, enabling ongoing verification of in-use energy performance and over time, expanding to address whole life carbon.

4.2.3 Whole life cycle carbon

The definition and calculation of Whole Life Cycle carbon (WLC) in the UK Net Zero Carbon Buildings Standard follow the established modular structure of BS EN 15978 (BSI, 2012a) and they account for all components making up the finished building covering a cradle to grave scope (modules [A–C]) over a 60-year life cycle. BS EN 15978 provides a framework for appraising the environmental impacts of the built environment and is part of the BS EN 15643 family of standards for the sustainability assessment of buildings. It sets out the principles for whole life assessment of the environmental impacts from built projects based on life cycle assessment (LCA).

In recent years, harmonised approaches and coordinated efforts have led to more consistent methodology reducing differences in varying interpretations by practitioners and clients. To help describe the environmental impact of an asset – measured in carbon dioxide equivalent emissions ($kgCO_2e$) – its life cycle is split into stages and modules as defined by BS EN 15978 for buildings (BSI, 2012a) and PAS 2080:2023 for infrastructure (BSI, 2023). Figure 4.6 illustrates this framework showing the importance of the use, end-of-life and beyond-life-cycle stages, alongside the design and construction stage.

Figure 4.6 Whole life carbon life cycle stages showing modules A-D (created using data from BS EN 15978 (BSI, 2012a) and PAS 2080:2023 for infrastructure (BSI, 2023))

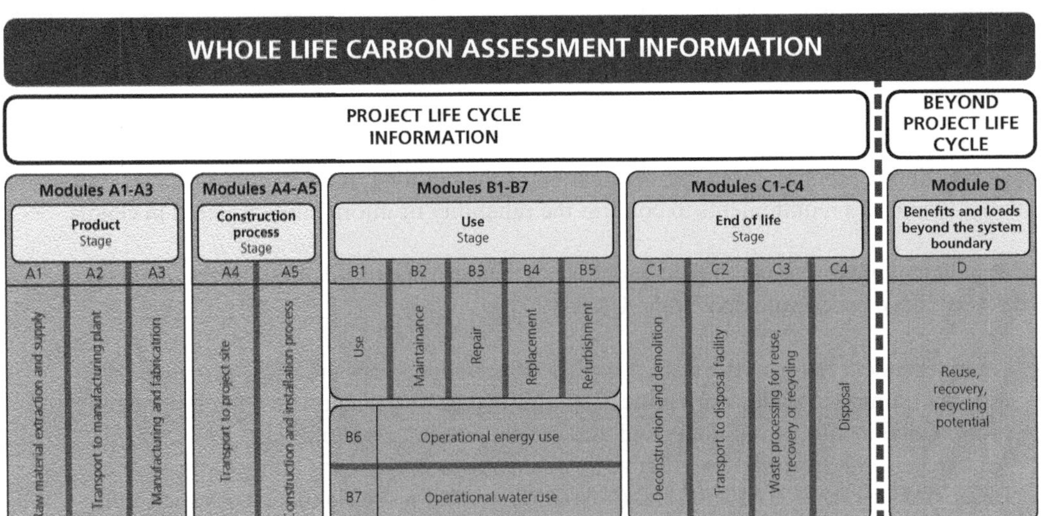

Upfront embodied carbon assessments are intended for estimation of actual carbon dioxide equivalent emissions due to construction. *LETI's Embodied Carbon Primer* (LETI, 2020) across all building elements highlights the significant contribution of Modules A1–A3 to the total embodied carbon. In contrast, the whole life carbon includes the operation carbon as well as the replacement related carbon emissions. Life cycle phases defined in the Environmental Product Declarations (EPDs) standard BS EN 15804 (BSI, 2012b), operational energy/carbon use is reported in Module B6. BS EN 15804 (BSI, 2012b) has been replaced by BS EN 15804+A2:2019 (BSI, 2020), the major European standard for EPDs for construction products, becoming mandatory in 2022.

Product stage, also known as 'cradle to gate' – includes Modules A1–A3. kgCO_2e released during extraction, processing, manufacture and transportation of materials between these processes until the product leaves the factory gates to be taken to site.

Construction process stage – includes Modules A4 and A5. kgCO_2e released during transport of materials/products to site, energy, usage due to activities on site (machinery use etc.).

Use stage – includes Modules B1–B7. kgCO_2e released due to use, maintenance, repair, replacement, refurbishment and operational energy and water while the building is in use. Module B4 (replacement) is often the focus of the use stage when embodied carbon is being considered.

End of life stage – Modules C1–C4. kgCO_2e released during decommissioning, stripping out, demolition deconstruction and transportation of materials away from the site, waste processing and disposal of materials.

Beyond life cycle – Module D. This estimates any net kgCO_2e benefits or loads beyond the project's life cycles associated with recycling of materials, and energy recovered from materials, or full reuse of materials/products when compared with the standard practice/standard product it would be replacing.

The upfront embodied carbon includes Modules A1 to A5 of the life cycle assessment, covering all emissions from raw material extraction to practical completion. The upfront carbon for new construction of office buildings in the UK typically ranges from 500–900 kgCO_2e/m^2 of GIA. This figure represents GHG emissions associated with materials and construction processes up to practical completion (Modules A1–A5). Upfront carbon excludes the biogenic carbon sequestered in the installed products at practical completion. This is equivalent to 5–10 years of the carbon dioxide equivalent emissions due to the energy consumption.

Whole life carbon assessment (WLCA) standard published by RICS, 2nd edition version 3 (RICS, 2024) launched in July 2024, has incorporated wide consultation and input from the global networks of professional bodies and industries. The key changes reflect the advancement of the standard to improving overall carbon accounting, an important step that supports the move towards national regulation of embodied carbon in the form of 'Part Z', which aims to integrate embodied carbon requirement into UK Building Regulations (RICS, 2024). These important developments to the standard include the requirement for demolition emissions, pre-construction emissions, and a greater level of detail required for reporting operational energy emissions, now requiring operational energy modelling to be assessed to account for both regulated and unregulated energy (for more information on regulated and unregulated energy, refer to Chapter 3).

Assessment of WLCA depends on robust embodied carbon from each material or product. Driven by initiatives and policy requirements, an increasing number of suppliers now use EPDs for construction products. When manufacturers report the embodied carbon emissions of their products, the data calculated and measured must follow International Standard ISO 14025: Environmental Labels and Declarations (ISO, 2006). Embodied carbon data of materials and products now include the major construction materials such as concrete, steel, aluminium, timber and glass, and have recently been extended to major HVAC equipment like heat pumps or chillers. However, a large gap in dataset of construction materials and products still exist, requiring collective effort from built environment professionals to promote, establish and enforce minimum standard and requirement for the industry. The complexity of these tools and the lack of accessibility to all disciplines can be a barrier to implementation. This is a rapidly evolving field and tools to assist implementation are still evolving and not yet widespread in use. Nevertheless, sufficient tools are now available to enable built environment professionals to embed these concepts more rapidly and integrate sustainable principles into all projects.

4.3. Framing purposeful sustainability

Previous chapters have set out the solid foundation that exists in applying sustainable development to the built environment. The science behind a changing global climate and what this could imply for all countries and habitations is now established and over the past few decades, a firm basis of knowledge on the solutions to address these challenges has been developed.

Once the need to embed sustainability aspects into built environment practices was realised (to some extent) in the 1990s (see Chapter 2), it is remarkable how quickly new terminology, design and construction methods were adopted. In hindsight, this rapidly changing mindset marked an exceptionally productive period of creativity, where the solutions and techniques for sustainable development proliferated and cross-pollinated each other to achieve rapid progress. As Section 4.2 shows, there have been significant efforts and advancements made in defining and driving forward responses to carbon and net zero goals since that time. The imperative to minimise carbon emissions gained pace and became the primary focus of attention for many frameworks – sometimes at the expense of other sustainability aspects.

In recent years, there is a sense that another threshold has been reached. Those who still question the fact of global climate change are increasingly marginalised and there is no serious debate about the role of human action as the cause. At the same time, tangible evidence of changes already happening to weather and world systems are becoming evident to people across the globe. In the built environment, all participants are now aware of the need for action and there is a plethora of different action pathways available. This very abundance of opportunity presents a challenge to practitioners; which of the options to choose, which is most appropriate, which priorities are the most relevant?

Similarly, the rapidity at which technology and knowledge is evolving is exciting and inspirational, with new materials, technologies for transport, energy generation and control systems, for example, evolving at breakneck speed and all offering immense potential to benefit environmental, social and economic drivers. The process of development has not undergone a similar change of pace, with the planning, inception, design and construction timescales relatively little changed. This creates further challenges for practitioners who may need to rapidly respond to new opportunities within a relatively inflexible linear design and implementation process; this option might be the best choice now, but will it still be the best choice at the end of the design phase in a year's time?

In a very real sense, the enormity of the challenge can pose obstacles to action in that practitioners can feel paralysed in choosing suitable and sufficient goals to work toward and a lack of confidence in the decisions that they are making. Within a multi-disciplinary design and construction team, there may be differences of opinion or at least, a different perception of risk that can lead to delay or avoidance in making more sustainable choices. Across the world, there exists a plethora of different sustainability frameworks that can be selected by teams to guide decisions, but their proliferation and the variety of choice can actually form a barrier to making purposeful sustainable choices that respond to project need and place.

4.3.1 Decision making under uncertainty

Decision making in uncertain circumstances, where the effects of the decision are not clear, or where external factors may impact the outcomes in unpredictable ways has long been a subject of study. Different mechanisms to assist in such decisions differ by sector with some of the most mature in statistical and economic sectors.

Decision theory in statistics provides mechanisms whereby the different options are set out as probabilistic outcomes, with each outcome assigned a utility value by the decision maker. Based on this theory, the best decision is therefore the outcome that maximises the expected value. This approach can work well if the value of each outcome can be directly compared – for example, if the possible outcomes can all be valued by their likely financial benefit (monetary profit or loss). This type of decision is already applied to the assessment of financial business cases in the built environment, such as the optimal inclusion of renewable energy technologies in a building design. The familiar cost/benefit analysis (CBA) often used in business cases for development projects is a common use of this technique.

As might be expected, investors have led the way in developing mechanisms to determine optimal decision making to aid their capital investment timing. Methods such as Real Options Analysis (or Real Option Valuation) allow investors to make decisions under uncertain circumstances, such as to allow for volatility in energy prices. There are some drawbacks to this methodology due to the need to quantify the uncertainty to determine probabilities. This is not always possible as an explicit understanding of assumptions, values and contributing factors would be required. In application to the built environment, there are key assumptions that can materially affect the outcome of the appraisal – for example, the life of the asset being considered will be a key contributor to the outcome, since the uncertainty of the future value will increase alongside its assumed lifespan. The UK Government's Green Book recommended a framework for developing business cases to support decision making (HM Treasury, 2022). Since the economic value of an asset will depend on its ability to retain its value to the community (environmental, social and cultural value), it follows that assessing just the financial value in isolation will not be adequate (d'Amato and Kauko, 2012). More complex and perhaps competing values that cannot be easily compared make this process unwieldy and much less useful.

The conclusion must be that while these types of purely economically based decision-making approaches have significant benefits in the financial and investment sphere, they do not deal well with the need for the built environment to act as a social and environmental good as well as an economic good (Lorenz *et al.*, 2017).

Investment professionals have considered a range of ways to account for sustainability in appraisals, attempting to align environmental and social considerations with economic returns. It is acknowledged that these methodologies often fall short, meaning that current asset valuation practices seldom adequately reward investment in sustainable buildings and infrastructure (Lützkendorf and Lorenz, 2005). Framing sustainable inclusion as 'value' is a vital aspect; adding 'value' is seemingly not perceived as adding 'cost' in contrast to the perceptions often associated with building in sustainability (Karunasena *et al.*, 2016). Integrating value planning (the ratio of perceived benefit to perceived cost) with sustainability concepts aims to demonstrate investment benefits alongside those gained in environmental and social areas. This implies the need for sustainability criteria to be explicitly included in the value requirements of a project before the value management process (value planning, value engineering and value analysis) commences.

When considering the multiple factors that apply in sustainable decision making (social, environmental and economic), it is evident that this process of embedding factors in the same framework is fraught with difficulty. There are now examples of real options analysis used to assess investments to address climate challenges to long-term infrastructure capital investments and similar projects (Stroombergen and Lawrence, 2022) by adopting a more flexible approach, but their focus on financial outcomes still provides significant limitations. Perhaps more appropriate to the sustainable development challenge is the concept of 'multi-criteria decision making/analysis', which considers different qualitative and quantitative criteria in the decision-making process (Taherdoost and Madanchian, 2023) according to a weighting on the relative value of each criterion. There have been various different methodologies trialled by practitioners (da Mota Passos Neto *et al.*, 2023) that aim to provide a structured approach to evaluating and comparing alternative choices or outcomes to help select the best approach. The array of methodologies that have been adopted or trialled is somewhat overwhelming, demonstrating enormous complexity, huge diversity and variety without a single methodology finding pre-eminence.

Factoring wider sustainability criteria into decision making is a problem that has faced policy makers as well as investors. Many such methodologies were investigated and reviewed by the Intergovernmental Panel on Climate Change as part of their work (New *et al.*, 2022), particularly with respect to climate adaptation potential scenarios. The working group looking at this aspect reviewed the methodologies available and concluded that a more flexible and less precise approach, known as Robust Decision Making, best fit their needs (IPCC, 2023a). Robust Decision Making is a framework for identifying strategies that assesses any weaknesses, gaps or consequences in a proposal, thereby informing decision making (rather than directly comparing particular choices or decisions directly, or assigning any 'value' to the different options available). The report of the working group concludes that use of this methodology, using pathway analysis to determine 'no regrets' options among trade-offs, is a useful starting point, while acknowledging its limitations.

As comforting as it may be to ground any decisions on fixed 'facts' or values that can be determined based on evidence and certainty, it is evident that this approach is insufficient to encompass the breadth and complexity of sustainability considerations. The IPCC concluded that to support decision making across this wide context, a set of available methods, tools and processes would be required (IPCC, 2014). Accessibility is an often overlooked factor in such methodologies; even the use of investment and economic terminology that is inevitable in some of the value-based methodologies renders the methodology difficult to apply and may significantly hinder understanding of the outcomes it delivers. If embedding an understanding of shared values is vital to delivering a successful decision-making methodology, it follows that participation of multiple disciplines will be required to successfully apply a methodology to built environment projects.

Developing common purposes across a multi-disciplinary team would be one of the key benefits of utilising such a methodology (Halla and Merino-Saum, 2022) and, therefore, utilising a model that hinders understanding between different disciplines would be highly counterproductive. Inevitably, therefore, the decision-making methodology will combine qualitative and quantitative criteria and considerations, since some disciplines will have quantitative data available and some will not. Combining these aspects together is a considerable challenge and most multi-criteria decision making/analysis frameworks develop some sort of weighting of the different criteria in order to embed 'value' and guide prioritisation.

Many of the popular frameworks in existence (see Section 4.4) such as BREEAM and LEED utilise weightings of criteria to indicate relative importance and, by implication, value. The means by which such weightings are allocated vary from method to method, with efforts made to ensure a democratic participation to gain a variety of views and perspectives. Inevitably the nature of the (construction-focused) framework means that views are more likely to be canvassed from those who are already familiar with construction challenges and may already be practitioners or participants, or even 'expert groups'. Potentially, this may result in an element of sampling bias, whereby those with different experiences or views may not be heard; the 'one BREEAM solution suits all' approach may not reflect the 'value' that a development wishes to prioritise and achieve.

A recent systematic review of sustainability frameworks developed to guide sustainable construction projects found that in most cases, the criteria used entwined environmental aspects such as biodiversity with social sustainability aspects (da Mota Passos Neto *et al.*, 2023). These social aspects ranged from consideration of local community consultation and incorporation of their views, community quality of life and long-term social cohesion, and facilities and cultural heritage, for example. The synthesis suggests that the aim of such frameworks should be to foster a built environment that enhances social wellbeing, promotes social inclusion and contributes to present and future generations' quality of life. The emphasis placed on these social aspects in so many of the sustainability frameworks used by practitioners implies that they are perceived to have a high value and should therefore be highly prioritised; however, these aspects may not be given the highest priority in any weighting exercise. The importance of listening to and incorporating a variety of voices including decision makers and community voices rather than only 'expert' practitioners in developing such frameworks is therefore highlighted (da Mota Passos Neto *et al.*, 2023; Halla and Merio-Saum, 2022). The inclusion of varying views and interests will help to fulfil the objectives of the decision-making methodologies; assigning meaningful value and determining the optimal 'no regrets' option that best meet the present and future need in the face of complexity.

4.3.2 Purposeful sustainable development

The concept of defining a 'purpose' to an activity is by no means new but has been gaining adoption and application to a wider range of activities. The dictionary definition of purpose is 'the reason for which something is done or created, or for which something exists' and considering this definition, it is easily assumed that all activity has an intrinsic purpose. Nonetheless, in a shared project it is entirely possible that different individuals or groups have different purposes in mind and may not necessarily be aware of the purpose envisaged by others.

The UK standard on Purpose-Driven Organisations (BSI, 2022) summarises the core values of such organisations as those that intentionally protect and enhance wellbeing over the longer term, for everyone. What this means in practice will naturally vary depending on the organisation and on the nature of their business; however, the practice of defining and determining this meaning is an important framework for action, embedding purpose into the activity, whatever it may be.

The benefit in defining a framework that embeds the purpose of the endeavour, its values and the criteria that will be used to guide decision making during development has great power. Not only does such a framework enable the defined purposes as a shared endeavour of the delivery team, it also enables clear understanding to be established from the earliest stages and fully informs those who may join the project during the development phase.

A robust framework will help to guide decision making and will provide confidence in choosing a robust pathway despite the inevitable complexities and trade-offs that exist. Acting with purpose does not imply that a framework will always result in the 'right' decision, since in complex and changing circumstances the best way forward can never be absolutely defined. The process of assembling a framework will acknowledge and allow for this limitation but does not allow uncertainty to paralyse positive action.

4.4. Framework planning to guide sustainable development

The usefulness of a framework process to guide sustainable built environment projects is evident, but practical application of this concept requires practitioners to overcome uncertainties in widening out the existing environmental standards and tools to take account of new challenges including social and wellbeing interactions.

The original purpose of environmental frameworks was to provide a simple and transparent method to guide, categorise and focus the efforts of built environment professionals. The earliest frameworks set out to define what was meant by good and best practice and thereby drive improvement. The transparency and ease of understanding that these early frameworks set out to achieve was necessary, because knowledge of environmental measures was not widespread at that time. Similarly, simplicity of approach with a 'pass/fail' system of implementing best practice helped those unfamiliar with the concepts to easily incorporate environmental measures into their designs. This methodology means that nuanced approaches and emerging measures could not be easily included in the early frameworks. Indeed, BREEAM Assessment methodologies published in the early 1990s included appendix lists of 'emerging issues' that the authors proposed to include in future updates, flagging to users of the framework that there were more issues to consider and helping to raise awareness of wider considerations.

The early environmental frameworks achieved their key purpose incredibly successfully. Their simplicity and accessibility meant that their use spread remarkably quickly across different countries, being easily adaptable to local regulation and other construction sectors. Over time, awareness of wider environmental and social considerations meant that the frameworks were modified, sometimes to be more specific to a particular geography or sector. Reporting in 2007, one group of researchers (Ness *et al.*, 2007) found it beneficial to set out a methodology to categorise many tools then available for sustainability assessment. The three categorisation areas were as follows.

Indicators and indices, further broken down into those that integrate nature-society parameters and those that do not. Using the authors' terminology, an index in this case was defined as an aggregate of separate indicators, their aggregation allowing tracking of long-term sustainability trends over time. This category would include indicators such as ecological footprint or indices such as BREEAM, LEED and other sustainability benchmarking tools, for example.

Product related assessment tools that focus on the material and/or energy flows of a product or service. This category would include LCA and WLCA, for example.

Integrated assessment, being a collection of tools usually focused on policy of project implementation. This category would include cost benefit analysis and multi-criteria analysis, for example.

The authors (Ness *et al.*, 2007) concluded that sustainability assessment tools must be able to integrate nature and society, must be applicable at different scales or spatial levels and must address both long- and short-term perspectives. By the time that this article was written in 2007, around 15 years after the first building assessment methods were launched, it was already evident that sustainability assessment tools were struggling to combine transparency (accessibility) with the ability to be specific to particular challenges. The increasing number of parameters that practitioners were attempting to combine for consideration simultaneously was important but added even more complexity.

A more recent investigation of the integration of sustainability assessment into building projects (Goubran and Cucuzzella, 2019) found more than 600 varying assessment methods available to design teams. Writing in 2022, an investigation specifically into urban sustainability indicator frameworks found 67 such frameworks published between 2010 and 2019 alone (Halla and Merio-Saum, 2022). The authors also found that, while the frameworks were very helpful in organising different sustainability initiatives into a coherent structure, this did not mean that they are comprehensive, and, by their nature, these frameworks are generic and not specific to place. Very few of the frameworks reflected the relationship between the location of the project and their surroundings either locally or globally. This plethora of available methods would suggest that built environment practitioners are still struggling to find a suitable framework that addresses the challenges to their satisfaction. Despite numerous iterations, the fundamental challenges remain.

Despite the obvious difficulties in applying sustainable development assessment frameworks to the built environment, outside of this sector, their usefulness was being recognised and adopted at a global scale.

The UN Sustainable Development Goals or SDGs (Chapter 1), consist of 169 indicators (targets) across 17 different categories (goals) (United Nations Department of Economic and Social Affairs, 2015). This remarkable synthesis of planetary challenges acts as an index (using the definition above) of indicators across a whole suite of social, economic and environmental considerations. Despite the widespread acceptance of the SDGs, there is still limited integration of the SDGs into sustainable building design practices (Goubran and Cucuzzella, 2019). There is a considerable body of literature identifying synergies between the leading sustainability assessment methodologies, and each of the owners of the leading methodologies has made effort to demonstrate the ways in which their assessment methodology delivers aspects of the SDGs. This is increasingly helpful as more private companies align their own sustainability reporting practices to the SDGs. Implementation of the SDGs directly within built environment projects is still a barrier, partly due to the very wide focus of the indicators and the sheer number of them, which can make the SDGs seem unwieldy as an implementation tool.

Nevertheless, the SDGs demonstrate that a strong vision and purpose for a desired outcome can be developed collaboratively. Indeed, the SDG framework accomplishes the three main aspects identified as important in a sustainability assessment tool; to integrate nature and society, to be applicable at different scales or spatial levels and to address both long- and short-term perspectives (Ness *et al.*, 2007). The SDGs also fulfil an important role in indicator frameworks (Halla and Merino-Saum, 2022) as having a strong visual identity to communicate focus and purpose to the target audience.

There remains a gap between the existing assessment methodologies that have many benefits, not least that they are available and designed to apply specifically to built environment design and construction within a specific geography, and the high-level ambitions and scope of the SDGs. It is possible that, in future years, a methodology that successfully synthesises these elements into a single applicable framework will be available. In the present, built environment professionals can still follow an approach that strives to bridge these gaps and achieve a purposeful sustainable development. An example of such an industry-led framework is the One Planet Living framework (Bioregional, n.d.). It claims that in comparison to other sustainability frameworks, the One Planet Living framework puts people first – grounded in the reality of how people live their lives rather than box ticking without local context.

One Planet Living® framework

Bioregional created the One Planet Living sustainability framework – comprising ten simple principles and detailed goals and guidance based on learning from the BedZED eco-village (see Chapter 5 for more details). The framework was developed together with WWF.

The framework comprises ten simple principles that cover all aspects of social, environmental and economic sustainability. The framework also provides detailed goals and guidance documents that include advice and ideas on the goals, actions, targets and indicators that users could set for their own 'One Planet Action Plan'.

The ten One Planet Living principles are as indicated in Figure 4.7.

Figure 4.7 One Planet Living Framework (reproduced with kind permission of Bioregional (n.d.))

- Health and happiness – Encouraging active, social, meaningful lives to promote good health and wellbeing.
- Equity and local economy – Creating safe, equitable places to live and work that support local prosperity and international fair trade.
- Culture and community – Nurturing local identity and heritage, empowering communities and promoting a culture of sustainable living.
- Land and nature – Protecting and restoring land for the benefit of people and wildlife.
- Sustainable water – Using water efficiently, protecting local water resources and reducing flooding and drought.

- Local and sustainable food – Promoting sustainable, humane farming and healthy diets high in local, seasonal organic food and vegetable protein.
- Travel and transport – Reducing the need to travel, encouraging walking, cycling and low carbon transport.
- Materials and products – Using materials from sustainable sources and promoting products that help people reduce consumption.
- Zero waste – Reducing consumption, reusing and recycling to achieve zero waste and zero pollution.
- Zero carbon energy – Making buildings and manufacturing energy-efficient and supplying all energy with renewables.

The One Planet Living framework is designed to support range of different audiences including developers, local authorities, businesses and schools as well as individuals, to achieve their sustainability ambitions by helping to create a 'One Planet Action Plan' that acts as a route map towards a more sustainable future for their organisations.

In the United States and Canada, an initiative to embrace an integrated design and delivery process (IDP) in the built environment has been gaining significant ground. In 2015, a report by the Commission for Environmental Cooperation (Malin *et al.*, 2015) outlined five main steps for success in implementing an IDP approach aligned to environmental values that integrates with existing construction and design practices. This aligns well with research that assesses the existing sustainability indicator frameworks and aims to overcome the known challenges, by proposing a similar five-step process (Halla and Merino-Saum, 2022). The convergence of proposed approaches despite the different language and intended audience provide a useful broad process for framing purposeful sustainability in the built environment along a systematic series of steps, as shown in Figure 4.8.

Figure 4.8 Framing purposeful sustainable development (inspired by Halla and Merino-Saum, 2022)

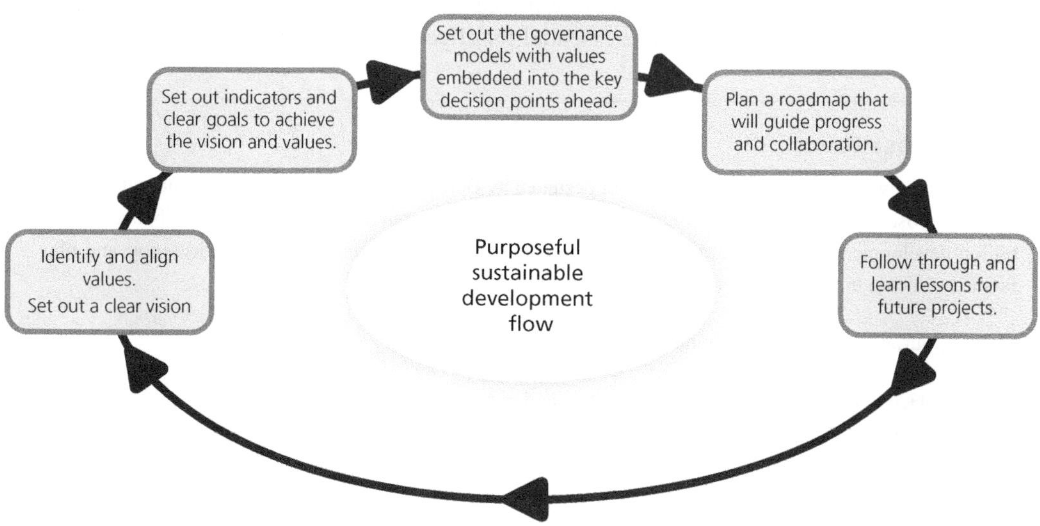

4.4.1 Identify key topics and align values

Aligned values to agree on a clear path forward at the earliest stages of a project avoid ambiguity and misaligned understanding, which can waste time and resources. Early alignment avoids later changes and makes it easier to have a meaningful impact on project direction. While the term 'values' is still not commonly used by clients and project teams, the questions set out by an IDP enable investigation of the core purposes that the project is to achieve

- What are you trying to accomplish by building this project?
- After it is built and occupied, what will define its success? What will have been achieved?
- How will the project continue to evolve as a contributing member of its community and place (as well as the immediate occupants or users)?

This implementation methodology aligns well with the driving concepts of purpose-driven organisations (BSI, 2022), whose core values set out their clear objectives, factoring in short and longer timeframes and widen the scope of the beneficiaries beyond the immediate stakeholders.

Setting out a purpose and a set of values for a built environment project will have different considerations than an organisational scale purpose, but applying the principles are equally beneficial (as well as equally challenging). The PAS 808:2022 standard (BSI, 2022) recommends gathering valid and reliable information as a starting point for value-based discussions; gathering this information will require drawing on the expertise of the wider professional team as well as other stakeholders. In the context of the built environment, where buildings and constructions are grounded in a particular geographic place, information on the constraints and opportunities associated with this place will also be vital.

Starting from this point overcomes some of the constraints of existing sustainable assessment methodologies, which have been criticised for being too generic and less sensitive to the needs of the specific location to which they are applied. Gathering information from a wide range of sources provides the diversity of approach that is recommended for purpose-driven organisations and helps to broaden the learning and understanding that is an essential part of this stage.

The IDP states that this stage of value definition creates a foundation that makes future decision making easier and builds the trust that is essential for effective collaboration.

Brent Cross town

Brent Cross town is a new urban development designed by Barnet Council and Related Argent with a strong initial purpose to maximise social connectivity, environmental sustainability, health and wellbeing. The £8 billion development will include a new station, homes, workspaces, retail, schools, parks, sport and play facilities (shown in Figure 4.9).

The core purposes were translated into four key goals.

(*a*) The place in London for participation in sport and play.
(*b*) A town where all can flourish.
(*c*) A net zero carbon town.
(*d*) Strengthening connections.

Figure 4.9 Brent Cross town (photo credit to © Visualisation One, courtesy of BXS Limited Partnership)

To embed the goals into the development, project consultants Buro Happold and University of Manchester assembled an indicator framework called the 'Flourishing Index'. The index considers both individual and community wellbeing and goes beyond typical measures of economic wellbeing (such as employment or living standards). An initial baseline gathered data before the development began and enabled the team to track progress utilising ongoing feedback.

This baseline identified a number of important areas of focus that the project could seek to address to improve value. These included social aspects such as job and neighbourhood satisfaction, loneliness and participation, and environmental aspects such as air pollution and low rates of active travel. There were also some key strengths identified that the project could seek to strengthen and build upon, such as a strong existing sense of belonging.

The project team has consciously built in learning to this long-term project to improve outcomes over time. The index outcomes will be reviewed at strategic points and will guide actions and decisions. The team is committed to publishing all insights to improve transparency and wider learning.

The index is fully embedded into the project decision making and is used to determine actions such as early delivery of new parks, rail station, cycle lanes and community centres help to front-load social value in the project. These social value indicators are embedded into a wider project indicator framework that includes environmental criteria including whole-life carbon emissions, climate mitigation and ecological richness.

4.4.2 Set indicators and clear goals

This stage of the process enables each of the participants in the project to set out more specific indicators and targets that will help them to focus on achieving the high-level values.

Indicators and goals can be reflective of the client's original brief and can also be taken from the value-creation process, from local design codes and requirements, from environmental impact assessments or indeed from any of the available sustainability assessment frameworks. The decision on which indicators and goals are most appropriate for the project should not be determined by each individual discipline involved in the project but determined as an integrated team to get the best outcomes. The IDP notes that mutual definition of goals and indicators encourages project potential to be realised and to reduce the risk of misunderstanding (Malin *et al.*, 2015). Deciding which indicators best match the project values by discussion will ensure a variety of voices are heard and will ensure a range of expertise and knowledge is included.

Adopting a value-driven approach to setting indicators and goals often widens the scope of such discussions to include post-occupancy outcomes, something that many built environment professionals are already considering but without necessarily including occupants, the existing neighbouring communities or other post-construction stakeholders in the goal-setting process (Karunasena *et al.*, 2016). They highlight the benefits of learning, discussing, explaining and clarifying sustainable and value issues among key stakeholders, resulting in a set of aspects that can be added into the project brief as 'sustainable value criteria', which can then be explored creatively and evaluated by the project team.

Depending on the type of built environment project and its location, there may be an 'off the shelf' existing framework (or set of frameworks) that could be directly applicable. Other sections in this chapter show that these generic frameworks, while very useful, may not cover all of the goals and indicators that will be important to achieving the values articulated for the specific project. Most of the existing framework models orient indicators and goals around a series of 'themes' such as resource use, energy consumption, water use, adaptation to climate change and so forth (Halla and Merino-Saum, 2022). Some apply this theme-based approach to social and wellbeing in communities in addition, or as a separate set of goals (such as the RIBA social value framework (Samuel, 2020)), broadly aligned to the social baseline outlined in the Doughnut Economic model (Raworth, 2013).

The themes often consider attributes covered in the SDGs. This approach is now beginning to gain ground in application to sustainable design and construction, such as piloting approaches to using the SDGs to identify relevant focus points for the project as part of the value and goal setting process (Goubran and Cucuzzella, 2019). The World Green Building Council has been through a similar process to identify the SDGs that they feel are most relevant to their work, identifying 11 (out of 17) goals for particular focus (World Green Building Council, 2022) and underlining the relevance of SDGs to built environment projects. The UN Environment Programme Finance Initiative (a UN-convened network of banks, insurers and investors committed to accelerating sustainable development) recognised that a holistic approach was needed to understand these sustainability topics, in a way that was easily understandable to their target audience of finance professionals. They proposed an Impact Radar (UNEP, 2022), shown in Figure 4.10 that compiles and depicts impact areas and topics across the three pillars of sustainable development.

Figure 4.10 UN Environment Programme Finance Initiative Impact Radar (©UNEP Finance Initiative (UNEP, 2022))

Radar maps such as these that depict a range of indicators and themes can be incredibly helpful in promoting engagement across disciplines. Transparency and accessibility are often overlooked components of indicator frameworks but when engaging multi-disciplinary teams, visual models ensure shared understanding of particular priorities. This can help teams scan their proposed indicator set for potential gaps, redundancies and comprehensiveness, ensuring that an indicator set is balanced in line with the purposes and values of the project (Halla and Merino-Saum, 2022). The technique of displaying performance against different indicators on a single chart has been used by various built environment frameworks very successfully.

Arup SPeAR

A need for a tailored framework to inform decision making on sustainability led consultancy Arup to design their SPeAR (Sustainable Project Appraisal Routine) tool in the early 2000s. Designed to monitor and evaluate project performance against a baseline and an agreed set of targets, it was envisaged as a way of comparing design options and opportunities to select the most sustainable way forward. Importantly, it contains environmental, economic and social parameters, but these can be modified to reflect project context by assessing the relevance or materiality of each indicator at the outset. An example of the view of the tool is given in Figure 4.11.

Materiality as defined by Arup can include potential areas of risk, emerging legislation, client policies and goals or societal trends, as well as acknowledged best practice. Assessing materiality as a team at the outset of a project reinforces a shared understanding of common goals.

Tracking progress throughout the project is undertaken against the suite of indicators and can be qualitative or quantitative, with a sliding scale of achievement to appraise performance having been set by the collaborative team early in the process.

Arup emphasise that the SPeAR process is iterative, allowing changes to be accommodated during the project – for example, in response to new information becoming available or in response to feedback from stakeholders.

Figure 4.11 Arup SPeAR, an example of analysis by Sarah Grice (licensed under CC BY-SA 2.0)

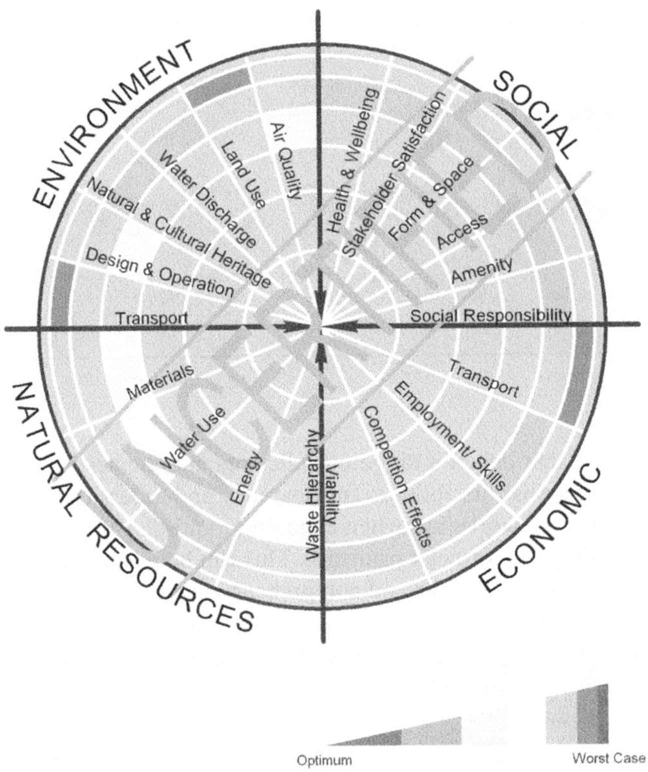

An important aspect of SPeAR is the accessible and simple presentation of outcomes. Performance against each indicator is shown using a simple 'traffic light' colour grading (red for 'sub-standard', green for 'exemplary' and shown in a different area of a radar chart). This easy-to-understand visualisation enables comparative assessments of options to be viewed immediately side by side and helps performance to be understood by non-technical stakeholders. It also allows the inevitable trade-offs between economic, social and environmental drivers to be depicted and understood.

Many of the sustainability framework assessments outlined in this chapter utilise a form of weighting of indicators to impose a value-based significance to particularly important issues. These weightings change depending on geographic area (usually by country) but this does not allow smaller scale significance of values to be included – example, in the UK, despite the small scale of the country, the south is highly water restricted while the north is generally not. A single weighted value attached to water consumption that applies UK-wide may not accurately represent the value to the specific project. By considering the high-level value of each SDG indicator and the relevance to the whole project over its life cycle, it is possible to utilise existing sustainability assessment frameworks but to embed them within a wider set of indicators and goals agreed by the whole project team. This allows bespoke tailoring of goals and indicators to more precisely match the project and importantly, the place for which the project aims to develop. Evidently some goals will be set with the responsibility of particular roles against leading and supporting roles within the project team, but sharing a common set of values leads to a supportive environment and greater understanding of a common purpose.

This mechanism of 'common but differentiated responsibilities' is the same mechanism recommended by the foundational sustainable development document 'Our Common Future' (Brundtland, 1987), which has been proven as an effective and unifying mechanism for purposeful action.

4.4.3 Governance (business models and contract structures) to embed values

While the design and construction team for a built environment project may be fully integrated into a value-driven and purposeful process, other important aspects of the successful project can be overlooked. As discussed earlier, financial, risk and business appraisals often use tools that struggle to accommodate non-monetary values, causing potential difficulties in decision making during the project delivery process.

The IDP (Malin *et al.*, 2015) acknowledges that the choice of contract structure can have a major influence on how possible full integration of the team can be in value and goal setting. Some delivery models and governance structures are easier to embed into a purpose-driven model than others, but in all cases, there are potential mechanisms for sharing risks and rewards that could be considered. This is especially an important issue for a large complex masterplan development where it may take many years to deliver involving master-developers, and various plot developers as well as other stakeholders.

Even where the contract and governance structure does not readily lend itself to this value-driven methodology and shared risk/reward is not feasible, governance structures can be set up to ensure that key gateways include value derived metrics – example, 'non-specialists' from a wider integrated team can be included in decision-making gateways (particularly financial gateways). Setting out a template to present recommendation documents and gateway metrics so that they reflect the values originally proposed for the project will, at the least, enable business decisions to be made with a more complete understanding of the 'value-benefits' rather than simply 'cost-benefits'. This reiteration of jointly developed values as key gateways and transitions is one of the more successful strategies in the case of a finance-driven contract arrangement.

4.4.4 Plan a roadmap to guide progress

An integrated team working to a set of common values, goals and indicators will determine responsibility for specific tasks and smaller groups that will discharge each responsibility. Building in regular review points where progress against the indicator framework can be reviewed and discussed will help the team to find those synergies that are so common in sustainable built environment themes. For this to happen successfully, transparency between different sub-groups and flexibility to work together to discharge responsibilities will be essential.

The fact that innovation in sustainable design and construction is moving so quickly means that regular review of goals and even high-level values can ensure that opportunities are not overlooked and that the collective experience of the organisations involved in the project can be gathered. A 'roadmap' keeps in mind the value-based outcomes that the project team are striving to deliver and ensures that all the reviews are approached consistently with these goals in mind.

Collaborative methods of working such as co-location and virtual online collaboration are now more common and enable team clusters to work effectively together on components of the roadmap. Setting out from the start of the project the expectation of joint and collaborative working to deliver not only the completed project but also the wider goals (which may cover a timescale longer than the construction stage) establishes clear expectations for all participants. A roadmap for achieving the values associated with the project rather than solely hand-over of the building project, helps to harness the team's knowledge to the benefit of future occupants and users. It also encourages the team itself to think ahead and be future-aware, supporting innovation and continual learning; essential in such a fast-changing environment.

4.4.5 Follow through and learn lessons

Equally important is recording the information that has been tracked – without reporting mechanisms, future decision-makers will be unable to learn valuable lessons and strive to improve performance. Decision-making processes often include formal reporting mechanisms but these are less regularly applied to tracking progress and learning lessons. The outcomes from such reports provide vital information to inform future decision-making comparing multiple criteria. Transparency of approach and documentation improves the credibility of this information and helps integrated teams to understand the value of particular goals and choices (Taherdoost and Madanchian, 2023).

Sustaining collaboration and a continued strong focus on values and goals is difficult and requires great effort throughout the whole project process. A well planned and strong roadmap with regular review points minimises the risk of losing momentum, but having an individual or organisation whose responsibility it is to maintain this focus, report progress and capture important learnings will maximise benefits. In effect, the feedback loops created by this process enable a self-learning system to be created that can better respond to challenges, changes and new information. This type of adaptive management has been incorporated into sustainability decision making for some time, originally defined by the IPCC (Intergovernmental Panel on Climate Change) in 2014 for decision making considering the risks of extreme events resulting from climate change (IPCC, 2014). Acknowledging the need for regular monitoring, reflection and feedback with a clear audit trail to inform the process is a sensible and pragmatic way to approach sustainability in design and construction practices, just as it is with scientific analysis; it enables rapid readjustments to be made if necessary.

Acknowledging and embedding adaptability into the roadmap encourages consideration of the longer term, beyond the design and construction process, indeed, beyond the assumed built environment life cycle. Despite many buildings in UK towns and cities having been in constant use for over a century, building life cycles are still assumed to be measured in decades. Furthermore, a study of city centre buildings found that 60% had changed their purpose over their life cycle and the frequency of changes had been accelerating in recent years (Manewa *et al.*, 2016).

Since many of the existing sustainability frameworks and methods are designed for specific building types (offices, homes etc.), one lesson that can already be learned is that a widened viewpoint, both sectorally and geographically, would benefit the project.

4.5. Social value frameworks

The above tools are mechanisms for fact- and evidence-led metrics that can be used to guide designers and built environment practitioners. While environmental and, particularly, carbon measures are becoming better understood, with a considerable effort underway to ease implementation, wider social aspects are not well integrated into the existing assessment methodologies and frameworks. A gap in guidance and indicators continues to persist although there are indications that teams are now starting to embed these wider considerations into their work, utilising tools from other sectors in some cases.

4.5.1 Doughnut economics

As discussed in Chapter 1, it is possible to envisage a conceptual model of planetary environmental boundaries, whereby excessive impacts to any of these boundaries represents a risk of major disruption to earth systems. The conceptual framework (Rockström *et al.*, 2009) defines nine environmental boundaries

(*a*) climate change
(*b*) ocean acidification
(*c*) chemical pollution
(*d*) nitrogen and phosphorus loading
(*e*) freshwater withdrawals
(*f*) land conversion
(*g*) biodiversity loss
(*h*) air pollution
(*i*) ozone layer depletion.

The planetary boundaries model provides an upper limit to the harm that can be inflicted upon any of these nine systems without risking excessive damage and can act as an 'early warning' that a quantified planetary boundary is approaching, allowing course correction to stay below the boundary threshold.

In 2012, the planetary boundaries framework that defines the safe operating space for humanity was brought together with a social boundary framework model (Raworth, 2013). Adding a series of thresholds across 11 social boundaries, below which humans would experience deprivation, provides a lower limit. Together, the upper 'environmental' limits and the lower 'social' limits define

Figure 4.12 The 'Quantified Doughnut' Framework showing the social foundation and ecological ceiling (reproduced with kind permission of Fanning and Raworth (2025) under CC BY-SA 4.0)

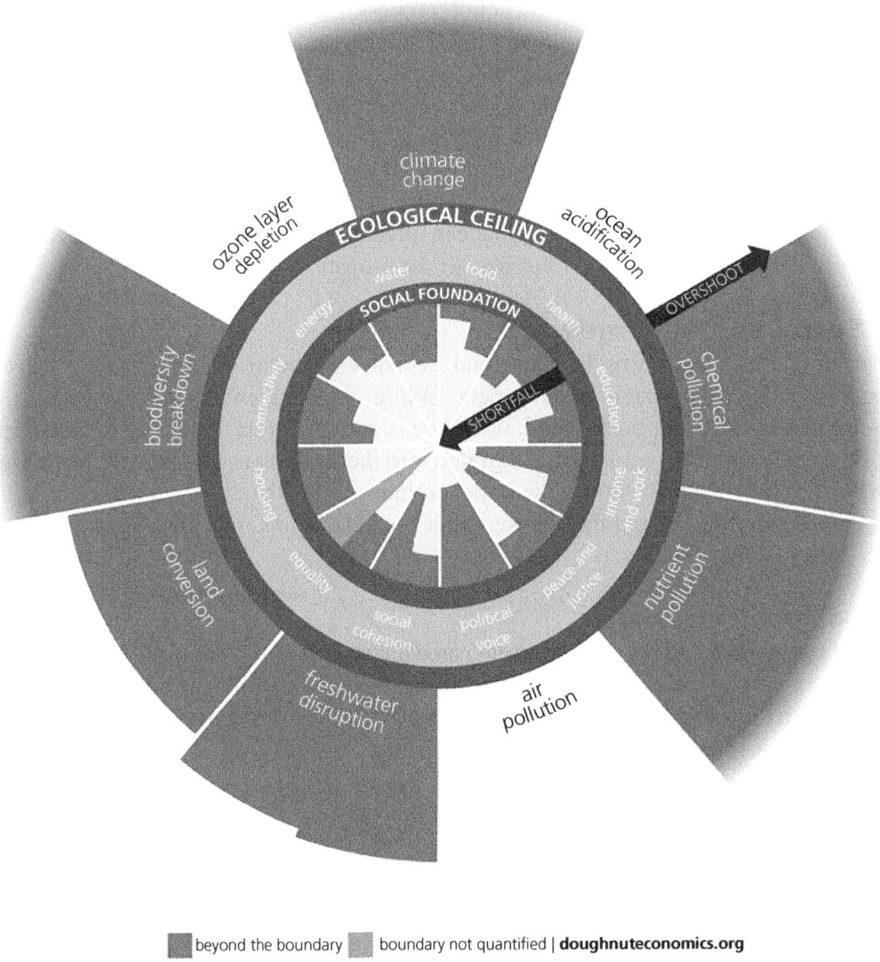

a 'safe and just space for humanity', shown in Figure 4.12 by the central, circular portion of the image that gives the 'doughnut' model its name.

This model acknowledges that some environmental impacts will be experienced if humans are to meet their social needs but determines that this is possible to achieve while remaining under the crucial environmental planetary boundaries.

The Doughnut model sets out the key priorities for the social foundation that forms the inner circle. Created before the SDGs, they were developed at a similar time and drew on the emerging consensus on these matters. Therefore, there is good agreement between the SDGs and the proposed foundation priorities, focusing on a fundamental drive for all people to be well (food secure, adequate income, health care etc.), productive (educated, resilient to shocks etc.) and empowered (gender equality, social equity etc.).

Since the doughnut framework was proposed, further work has been undertaken to further quantify and strengthen the boundaries and indicators captured in the framework, helped by improved data monitoring and evidence. The 12 principles of the social foundation have coalesced as

(*a*) food
(*b*) health
(*c*) education
(*d*) income and work
(*e*) peace and justice
(*f*) political voice
(*g*) social cohesion
(*h*) equality
(*i*) housing
(*j*) connectivity
(*k*) energy
(*l*) water.

The initial analysis that underpinned the Doughnut suggested that humanity was not living within planetary boundaries at a global scale. Further investigation and refinements undertaken since that time have refined and expanded the data available on both social and environmental/ecological aspects. The latest analysis and update (at time of writing) concludes that while there has been a small improvement in social shortfall since 2000, the ecological overshoot has worsened by a considerably greater amount (Fanning and Raworth, 2025). This is a global overview, however, and individual countries or regions' performances will vary, with the group of higher income countries having a much greater ecological overshoot. While naturally not directed at the built environment sector, this concept of combining social and environmental requirements in a single framework, which can be monitored and supported by robust data and can be applied to a wide range of scales and purposes, is a powerful tool. It is also very much assisted by clear visualisation methods that help non-experts to overcome the complexity of understanding the key messages of the evaluation.

4.5.2 Wellbeing economics (reframing prosperity)

The Doughnut framework described above has a social foundation that rests on the fundamental drive for all people to have a basic ability to be 'well' – that is, to have sufficient food, an adequate income to meet their needs, to have their health needs met and so on. A broader definition of 'wellbeing' goes beyond this fundamental level of having basic human needs met; the dictionary definition is 'the state of being comfortable, healthy or happy'. Wellbeing therefore encompasses quality of life indicators and implies a broader understanding of what constitutes an individual, community or a society to achieve a positive state. This definition is certainly not in conflict with the Doughnut model, indeed, the 12 principles of the social foundation form a 'wellbeing' foundation that go beyond basic human need. They broaden the standard wellbeing definition to cover other less physical needs that provide humans with a sense of meaning and purpose, enabling the realisation of individual and community potential.

The World Health Organisation believes that 'wellbeing societies provide the foundations for all members of current and future generations to thrive on a healthy planet, no matter where they live' (WHO, 2022). Their emphasis on the need for wellbeing to be integrated into all decision making is based on the experience of the Covid pandemic of 2020, which 'exposed fractures in society and highlighted. . . inequities within and between social groups and nations'. They argue that changes in investment and social structures to support people to achieve better health and wellbeing will enable societies and nations that are more resilient to shocks and future crises.

The World Health Organisation believes that application of wellbeing as a policy concept can be helpful in developing resilient communities and nations (WHO, 2021). They recommend several mechanisms that although framed as policy objectives are equally applicable to built environment practitioners. These include identifying relevant regional, national or local goals and objectives, and adopting processes that ensure open collaboration and accountability.

The Organisation for Economic Co-operation and Development (OECD)'s Economy of Wellbeing (Llena-Nozal *et al.*, 2019) offered an economic foundation for WHO's conclusions, providing an understanding of the linkages between wellbeing and economic growth. The concept provides an understanding of how this type of framing can result in a 'virtuous circle' where individual wellbeing and long-term economic growth can be mutually reinforcing. It may seem self-evident that improving the health of individuals and communities will result in greater economic benefit (by reducing the pressure on the healthcare system and enabling people to remain economically active); however, these social benefits are often excluded from economic discussions.

Considering just mental health alone, a recent study found the economic and social costs of mental ill health in England in 2022 was estimated at £300 billion, comprising economic costs (such as lost work days and productivity), human costs resulting from lost quality of life, and health and care costs to the national healthcare system (Cardoso and McHayle, 2024). The biggest share of the economic impact is not borne by the healthcare system but by businesses, each bearing the cost of staff turnover and presenteeism individually but aggregating to a large economic impact.

In the context of the built environment, it has been rare for the economic cost of social aspects to be considered in assessing business cases, or indeed in briefing for a proposed development. This is changing somewhat, with more recognition of the benefits to business of a workplace that supports mental and physical wellbeing of staff, as more buildings are adopting WELL Buildings and WELL Communities schemes (see Chapter 5 for more discussion of these frameworks). Because these social aspects are unquantified in terms of economic benefit, there is still a tendency to view them as cost and not value, as previously discussed. This makes them vulnerable to underestimation in any value appraisal or value engineering process. In the UK, with an aim of quantifying the health impact due to a built environment project, Health Impact Assessment (HIA) has been introduced as a parallel exercise to an EIA for a development project planning application (Public Health England, 2020).

The OECD are clear that a wellbeing approach does not mean abandonment of GDP or economic measures of performance (Llena-Nozal *et al.*, 2019). Rather, focusing on wellbeing provides a broader picture that will improve strategic decisions and outcomes in the longer term.

4.5.3 Social value frameworks

The term 'social value' is often used interchangeably with the term 'wellbeing' and, in embedding these issues into mainstream economic decision making, this can be useful. A term such as 'social value' explicitly recognises the important 'value' of these social and wellbeing aspects, that might otherwise be viewed as lower priority or unmeasurable.

Some sectors have proven adept at integrating social aspects into core business; contractors in the UK now regularly commit to the employment of apprentices, use of local staff and supply chain contractors and to supporting the professional development of staff. Such commitments are generally presented as 'social value' interventions; however, by supporting learning, skills and retention of employment in the locality of a development, there will be an economic benefit to the local community, albeit seldom quantified.

The lack of quantifiable measures for social value (or indeed, wellbeing) on the scale of a development programme is a real challenge. There is a tendency to omit measures that cannot be valued and captured in numerical terms. This is likely why existing social value measures in the built environment context tend to be focused on *number* – how many apprentices employed per £m capital spend, how many homes are within *x* metres of greenspace and similar measures.

One of the major challenges in quantifying social value in the built environment is that values change both spatially and over time. Samuel and Watson (2023) highlight a key point often overlooked in the consideration of social value – for whom is the social value being judged, since decisions made regarding the present built environment will impact those in future generations as well as the current users? They comment that there is reasonable common agreement in what constitutes social value relative to the built environment, but little agreement on how it can be demonstrated and monitored in a way that appears meaningful given the prevailing focus on quantifiable benefits.

RIBA has produced a Social Value Toolkit for Architecture (Samuel, 2020), which responds to this challenge by seeking to 'define and capture the social value of architecture and good design' with reference to housing (but applicable to the wider built environment).

The authors propose the social value of architecture to lie in the following

- Fostering positive emotions – such as connecting with nature or providing opportunity for a more active lifestyle.
- Connecting people and the environment.
- Providing freedom and flexibility to pursue different lifestyle choices (autonomy).
- Supporting community participation in design and creation of their place.

It is clear from these main principles that there is a strong spatial and wider community aspect to social value as applied to the built environment. Spatial aspects of social value are particularly pertinent wherever a built environment intervention is imposed on an existing settlement. In practice, a new built environment without any kind of impact on existing communities or place is virtually impossible; it follows, therefore, that considering social value must include consideration of the place in which it is to be delivered. The importance of factoring spatial considerations, as well as economic, into social value calls into question the recent UK tendency to monetise social value using social return on investment (SROI) (Samuel and Watson, 2023).

The RIBA toolkit authors therefore recommend gathering data to baseline social value from those directly impacted by the built environment intervention (such as occupants of a building to be refurbished, for example) but also those in the community that will be affected by the change. At an appropriate time after completion of the intervention, the same questions are asked and the change can be measured and reported as social value data. This data can then be used as a proxy for monetised value, utilising financial values such as those contained in the UK Social Value Bank. This is a proven and robust technique that provides a methodology for capturing SROI and can be used to predict the likely social value created by an intervention and allow evaluation afterward.

Engaging with 'value' by converting social benefits to financial benefits is pragmatic and avoids these more intangible (but very important) aspects being sidelined in favour of easy-to-measure metrics. The SROI Network who publish *A Guide to Social Return on Investment* (SROI, 2012) used as the basis for many of these methodologies note that 'SROI is about value, rather than money. Money is simply a common unit and as such is a useful and widely accepted way of conveying value'. This is a crucial point and in striving to embed social (wellbeing) value into frameworks it is vital to bear in mind that not everything valuable can be conveniently measured. Nevertheless, applying social value within assessment frameworks is still likely to result in a meaningful engagement with these issues, however imperfect.

A range of tools and frameworks are increasingly available to assist organisations and project operators to integrate social value factors into decision making and valuation. The UK Social Value Bank was originally set up to assist primarily social housing organisations to articulate, measure and utilise social value in their investment decisions but their methodology is equally appropriate to other sectors. This framework focuses on individual wellbeing metrics but does include environmental outcomes, enabling a value calculation based on wellbeing, health impacts and, where appropriate, an Exchequer value representing savings that accrue to the state (wider societal financial benefit).

In the UK, this was driven by public policy such as the Public Services (Social Value) Act 2012, which encouraged (although stopped short of requiring) publicly funded contracts to embed these issues in their procurement process. In 2018, the UK Government determined that government departments would be required to evaluate social value and report upon the social impact delivered as part of their work.

This requirement was embedded into the work of the UK Government and other public bodies in several ways that have considerable impact upon the built environment, from procurement of delivery contracts (using a centrally developed Social Value Model) through to embedding social value into the processes required to assess business cases as part of decision making.

In the UK, HM Treasury (the UK Government department responsible for public spending management and economic policy) publishes guidance on financial appraisal of all policies, programmes and projects. Known as the 'Green Book', the guidance is used by all publicly funded bodies and public service deliverers and thereby sets a standard methodology for benefit valuation. The Green Book (HM Treasury, 2022) uses a methodology known as the 'Five Case Model', which sets out to achieve a broad understanding of the objectives and constraints of any proposal in five ways

(*a*) Strategic case – what is the case for change and what outcomes are expected compared with the existing situation?
(*b*) Economic case – what is the net value to society (the social value) of the intervention compared to the existing situation?
(*c*) Commercial case – what is the commercial benefit and risk and who owns them?
(*d*) Financial case – what is the impact on both capital and revenue budgets?
(*e*) Management case – how will the proposal be managed in a realistic and robust way?

The 'economic case' is the area directly relating to social value cost and benefits and the Green Book recommends using Social Cost Benefit Analysis (Social CBA) where valuation can be done in monetary terms. If quantification is not possible for the different options being considered, then Social Cost-Effectiveness Analysis (Social CEA) is recommended.

Social CBA, as defined by the Green Book and its associated guidance, still focuses on economically tangible costs and benefits albeit on a wider and more complex basket of measures. While it quantifies the wider societal costs and benefits of an intervention, it does not factor in less tangible wellbeing metrics or perceptions, unlike SROI methodologies. Commentators have noted that the Social CBA methodology treats societal financial gain as an end, rather than a means to achieving societal well-being (Stiglitz *et al.*, 2009; Vardakoulias, 2013). Objective measures of wellbeing such as health, education, personal activities and environmental conditions can be considered, but other measures of life satisfaction such as equality of choice, political voice and insecurity are not tangibly measured and therefore obscured. Social CEA is a comparative methodology whereby the benefits offered by two or more potential avenues are compared; however, the accompanying Treasury guidance (HM Treasury, 2022) comments that 'the lack of benefits in monetary metrics limits its value compared to cost-benefit analysis'.

It is evident that to be considered of comparable importance to other decision-making metrics and valuations, social value must be at least partially quantified and quantifiable. The emerging social value frameworks brought forward by various sectors (such as the UK Social Value Bank referenced above) allow this to happen by drawing on evidence-based metrics gathered through application and robust monitoring and evaluation processes. Not all of these available frameworks are directly applicable to the built environment but they share a common basis and enable shared language and understanding to be built between disciplines and sectors. By building this consensus, consolidated into Government guidance and implementation mechanisms such as the HM Treasury Green Book and public sector procurement processes (Blears *et al.*, 2019), including national measurement frameworks, there is no doubt that social benefit is more likely to be delivered.

Nevertheless, such broadly applicable social value frameworks do not account for the specific challenges of built environment interventions; particularly the spatial aspect which is a crucial component of this sector. This lack of consideration for spatial impacts and opportunities calls into question the relevance of these methodologies (Samuel and Watson, 2023). If the construction and built environment sector is to fully realise the potential benefits of a social value/wellbeing aligned approach, a framework that recognises the crucial aspect of place is required.

In 2021, the UKGBC published a framework specifically designed to assist built environment professionals to consider and embed social value in their projects (UKGBC, 2021a). The framework draws on the work to date on embedding social value and proposes a methodology that works within built environment design and management structures. The framework sets out a high-level methodology but, importantly, it includes many of the components considered to be of greatest value in previous commentary, such as a focus on people (personal wellbeing) in terms of stakeholder engagement across the wider community, thereby grounding the outcomes in the place where they will be delivered. There is also a focus on measurement of impact and additionality (benefits delivered over and above those that would have happened anyway, without the built environment intervention).

Following the publication of the framework, UKGBC concluded that empowering built environment teams to fully embed these concepts required a more delivery-focused approach. A guide to delivering social value (including valuable case study examples) and a targeted 'delivery checklist' were therefore published (UKGBC, n.d.). These approaches can be used alongside public sector social valuation methods, or financial metric-based methods such as those discussed above, where necessary.

The UKGBC approach acknowledges many of the challenges that exist in embedding social value (wellbeing) into built environment projects. The great variety of opportunities that exist in any spatial intervention, the long timescales for realising benefits and the complex delivery structures are all acknowledged. The nature of these challenges has been recognised by other organisations similarly seeking to embed a social value and wellbeing approach in construction projects (Squarebird, 2024). These organisations all conclude that the use of case studies is hugely useful, clearly showing that despite the variety of project types, embedding social value is both possible and demonstrably beneficial. A growing number and range of case studies appropriate to the built environment is becoming available and wider application of these real-life studies will help to empower other teams to embed social value as a standard project consideration. It follows, therefore, that teams applying social value and wellbeing frameworks into built environment projects should seek to capture their work, the benefits and lessons learned, and make the knowledge more widely available.

4.6. Tools for corporate social responsibility reporting

Within the built environment sector, tools for social value, wellbeing and associated measurement and reporting are only now slowly starting to be integrated into standard practices. Other sectors, such as corporate and organisational governance, have been utilising these reporting practices for some time and they are now reaching a level of maturity where there are significant lessons that can be learned for the built environment sector.

Across major economies, environmental, social and governance (ESG) reporting is fast evolving, starting with establishing a transparent data reporting and benchmarking framework. Transparent measurement and disclosure of sustainability performance is now considered to be a fundamental part of effective business management, and essential for preserving trust in business as a force for good. Yet, the complexity surrounding sustainability disclosure has made it difficult to develop a comprehensive solution for corporate reporting.

Since 2020, there has been pivotal change towards corporate social responsibility reporting, integrating not only financial performance but also sustainability-related factors. Businesses are now starting to prioritise the health and safety of their employees and customers above the immediate financial success. ESG indices (or green indices) provide a comprehensive framework for evaluating and identifying companies based on the sustainability of their business practices.

4.6.1 Greenhouse gas emission protocol

GHG emissions reporting is an essential part of the ESG of a company. The GHG Protocol Corporate Accounting and Reporting Standard (GHG Protocol Team *et al.*, 2011) provides requirements and guidance for companies and other organisations preparing a corporate-level GHG emissions inventory. GHG emissions reporting boundary covers Scopes 1, 2 and 3. Figure 4.13 illustrates the GHG protocol scopes

- Scope 1 emissions are direct carbon emissions produced and caused by the direct operation of the business activities including the company's owned vehicles, company facilities such as boilers using the fuel for the facilities.
- Scope 2 emissions are indirect carbon emissions produced by the indirect activities of the business such as the purchased electricity, steam and heating/cooling for own use.
- Scope 3 emissions are indirect carbon emissions as a result of all the value chains where the company may have an influence, including upstream activities such as leased assets, purchased goods and services, waste generated in operation, business travel, employee commuting, and downstream activities such as use of sold products, leased assets, processing of sold goods.

Figure 4.13 Overview of GHG protocol scopes (reproduced with kind permission of UK Green Building Council (UKGBC, 2024))

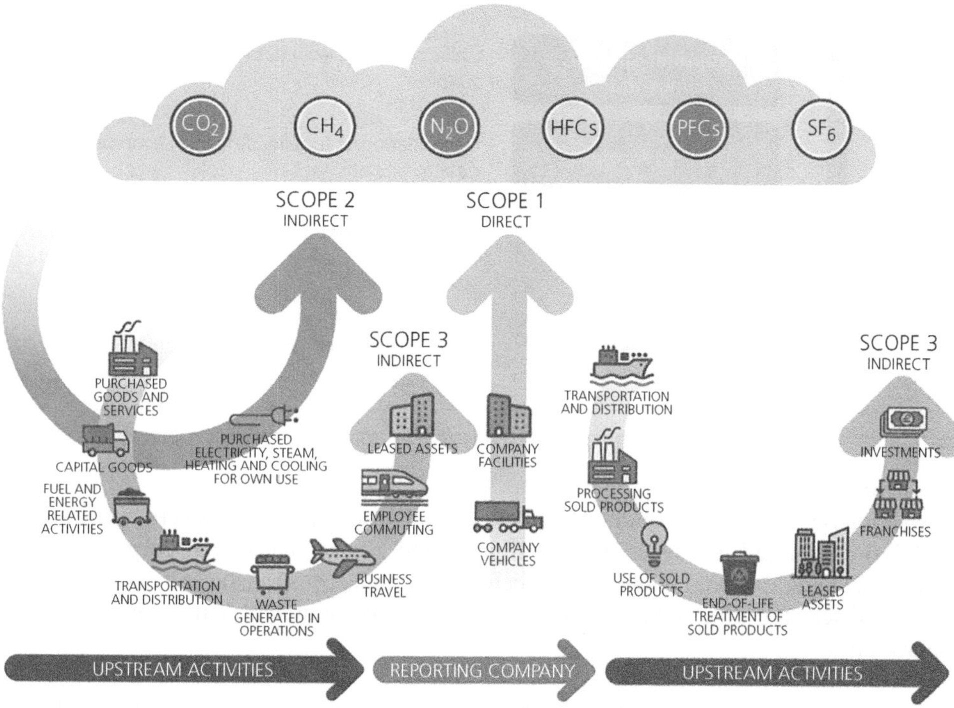

Scopes 1 and 2 have been the focus for operational carbon reduction for many companies in the past. Whereas, Scope 3 emissions are receiving increased focus, with most companies starting to develop inventory to improve data accuracy and extending the reporting boundary.

Scope 3 carbon emission calculation has mostly been based on spend method, the least accurate method, whereas project-based calculation using real product level data directly from suppliers is the most accurate way of assessing Scope 3 in relation to the purchased goods, as shown in Figure 4.14.

4.6.2 ESG frameworks and tools

Businesses globally are already using a mixture of frameworks and standards to provide stakeholders with robust, effective information to drive better decision making and capital allocation. Common ESG standards are described as follows.

4.6.2.1 Climate Disclosure Project (CDP)

CDP is a not-for-profit charity, established in 2000. CDP runs the global disclosure system for investors, companies, cities, states and regions to manage their environmental impacts. CDP evaluates and assesses evidence of ESG performance and is more specific due to their thematic focus on the environmental matters. CDP hold comprehensive global collection of environmental data from companies, cities, states and regions, based on which CDP publishes annually an A list of ratings for the companies for their leadership in environmental transparency and performance on climate change, deforestation and water security.

Figure 4.14 Calculation methods for Scope 3 assessing purchased goods from value chain

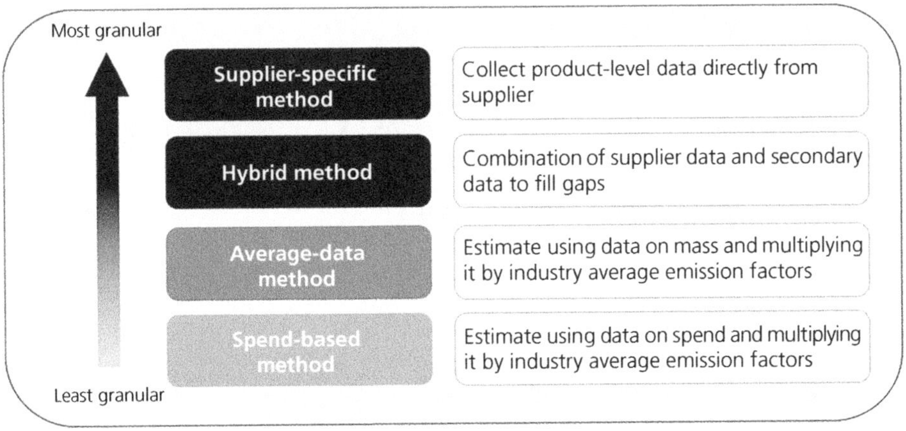

4.6.2.2 Climate Disclosure Standards Board (CDSB)

CDSB is an international consortium of business and environmental non-governmental organisations (NGOs) committed to advancing and aligning the global mainstream corporate reporting model to equate natural capital with financial capital. It is industry agnostic and designed to facilitate effective disclosure of a company's natural capital, environmental and climate-related risks and opportunities.

CDSB provides the companies with a framework for reporting environmental information with the same rigour as financial information. This enables companies to use the CDSB framework to incorporate climate change and natural capital-related information in mainstream financial reports. In turn, this helps them to provide investors with decision-useful environmental information by way of the mainstream corporate report, enhancing the efficient allocation of capital. Regulators also benefit from compliance-ready materials.

4.6.2.3 Global Reporting Initiative (GRI)

The Global Reporting Initiative (GRI) is an international not-for-profit organisation, with a network-based structure. GRI produces free sustainability reporting standards, to enable all companies and organisations to report their economic and ESG performance covering a wide range of topics (GRI, 2022a, 2022b).

GRI also offers services, tools and training to guide reporters through the reporting process, including their materiality assessment.

4.6.2.4 Sustainability Accounting Standards Board (SASB)

SASB Standards are designed to identify and standardise disclosure for the sustainability issues most relevant to investor decision making in each of 77 industries, including consumer goods, financials, minerals processing, infrastructure industries and so on. Real estate is contained as a sub-industry of the infrastructure.

SASB Standards help companies disclose relevant sustainability information to their investors. SASB Standards identify the sustainability-related risks and opportunities most likely to affect an entity's cash flows, access to finance and cost of capital over the short-, medium- or long-term and the disclosure topics and metrics that are most likely to be useful to investors. It can be used in conjunction with principles-based frameworks, such as the Taskforce on Climate-related Financial Disclosure (TCFD).

As of August 2022, the International Sustainability Standards Board (ISSB) of the International Financial Reporting Standards (IFRS) Foundation assumed responsibility for the SASB Standards. The ISSB has committed to maintain, enhance and evolve the SASB Standards and encourages preparers and investors to continue to use the SASB Standards.

4.6.2.5 Taskforce on Climate-related Financial Disclosure (TCFD) and Taskforce for Nature-related Financial Disclosure (TNFD)

Taxonomies are definitions of sustainable finance that aim to be comprehensive classification systems. Sustainable finance taxonomies are playing a pivotal role in mobilising private capital towards environmentally sustainable investments. Under EU Sustainable Finance Taxonomies regulation in 2020, taxonomy regulation covers six environmental objectives, but detailed technical screening criteria have so far only been issued for the first two: climate change mitigation and adaptation. The associated classification of climate-related hazards comprises four hazard groups, related to water, temperature, winds and mass-movements. Whereas, TCFD taxonomy divided climate-related risks into two major categories: risks related to the physical impacts of climate change (Climate Change Physical Risk) and risks related to the transition to a lower-carbon economy (Climate Change Transition Risk) (TCFD, 2017) – for example, the risks and financial impact for a real estate company could include the following

- Policy and legal risks – building regulations and EPC minimum energy efficiency requirement for existing private rented assets.
- Physical acute or chronic risks – overheating.
- Financial impact on revenues – rental income.
- Financial impact on assets, liabilities, capital – property value, carbon tax, retrofits.

TCFD focuses specifically on how climate-related issues affect financial performance. TCFD was introduced in 2016, chaired by Mark Carney, at that time the Governor of the Bank of England, as a voluntary reporting exercise. It is now becoming a policy for all listed companies to assess and report TCFD. Around the globe, the TCFD recommendations are being incorporated into corporate reporting requirements. Figure 4.15 illustrates TCFD risks, opportunities and financial impact.

TCFD recommends companies should provide assessment based on climate-related relevant metric categories as they are important proxies for measuring climate-related risks and opportunities, forming the basis for estimating climate-related financial impact. The cross-industry metric categories include (TCFD, 2021) the following

Figure 4.15 TCFD risks, opportunities and financial impact (created from information in TCFD Final Report (TCFD, 2017))

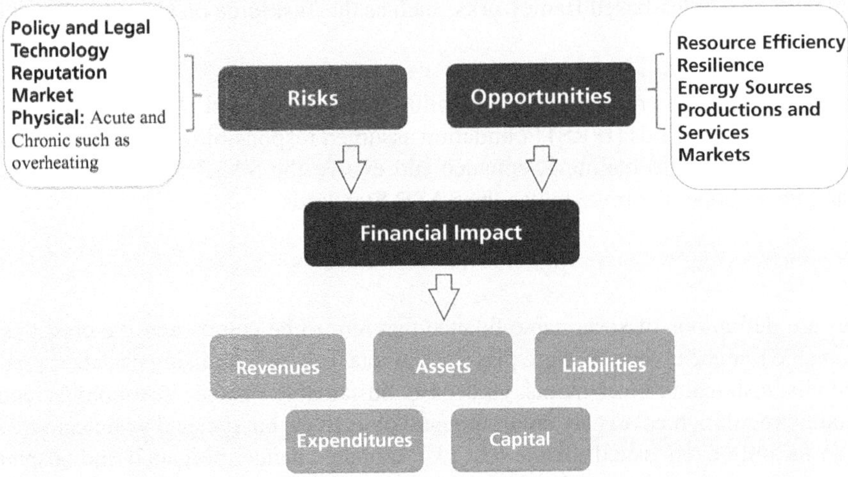

- GHG emissions – Absolute Scopes 1, 2 and 3 emission intensity, using MT of carbon dioxide equivalent.
- Transition risks – Amount and extent of assets or business activities vulnerable to transition risks,
- Physical risks – Amount and extent of assets or business activities vulnerable to physical risks
- Climate-related opportunities – Proportion of revenue, assets or other business activities aligned with climate-related opportunities.
- Capital development – Amount of capital expenditure, financing or investment deployed toward climate-related risks and opportunities.
- Internal carbon prices – Price on each ton of GHG emissions used internally by an organisation, using price in reporting currency per MT of carbon dioxide equivalent .
- Remuneration – Proportion of executive management remuneration linked to climate considerations. This policy is an important indicator to show organisational goals and objectives related to governance, oversight and managing climate-related issues.

In July 2023, the Financial Stability Board (FSB) announced that the work of the TCFD has been completed, with the ISSB Standards taking on the responsibility of monitoring the progress of TCFD work. Having fulfilled its remit, TCFD disbanded in October 2023 (IFRS, 2024b).

As another market-led global initiative aiming to provide financial institutions and corporates with a complete picture of their environmental risks and opportunities, TNFD was formally introduced in 2023 focusing on nature-based solutions affecting financial performance. Although TNFD is at an early stage of developing guidance and specific targets, it has endorsement and early adopters in major financial institutions, corporates and governments.

TNFD has developed a set of disclosure recommendations and guidance that encourage and enable business and finance to assess, report and act on their nature-related dependencies, impacts, risks and opportunities. The aim is to support a shift in global financial flows away from nature-negative outcomes and toward nature-positive outcomes, aligned with the Global Biodiversity Framework

(UN Environment Programme Finance Initiative, 2024). The Science Based Targets Network (SBTN, n.d.) provides guidance for science-based targets for nature, including a five-step approach to setting science-based targets for nature, and insight into how SBTN and TNFD guidance fit together.

4.6.2.6 Morgan Stanley Capital International (MSCI)

MSCI is a leading provider of critical decision support tools and services for the global investment community for climate related investment decisions. Originating in the USA, MSCI has developed a suite of tools including MSCI ESG Ratings (MSCI, n.d.), which aim to measure a company's management of financially relevant ESG risks and opportunities. MSCI use a rules-based methodology to identify industry leaders and laggards according to their exposure to ESG risks and how well they manage those risks relative to peers. MSCI ESG ratings range from leader (AAA, AA), average (A, BBB, BB) to laggard (B, CCC). They also rate equity and fixed income securities, loans, mutual funds, exchange-traded funds (ETFs) and countries using the tools.

MSCI ESG Ratings can provide an index for every portfolio, with tools for every investment decision so that the data and insights can address investment considerations like ESG, climate, and risk and return to help investors to build better portfolios.

4.6.2.7 Global Real Estate Sustainability Benchmark (GRESB)

GRESB as an industry-led membership organisation, is an independent not-for-profit organisation providing validated ESG performance data and peer benchmarks for investors and managers to improve business intelligence, industry engagement and decision making. GRESB seeks to score and benchmark global real estate and infrastructure. GRESB provides various tools (GRESB, n.d.) to synthesise and benchmark data and provide actionable insights to members, partners and investors. GRESB has been an active member of the Carbon Risk Real Estate Monitor (CRREM) project and has served as a resource to its development.

GRESB engagement tools and reporting solutions aid over 150 institutional and financial investors and as well as 3000 funds, companies and asset operators to make informed decisions based on ESG data.

Financial institutions that create the investment tools often use GRESB to provide the underlying ESG data of listed real estate and infrastructure companies. It has been reported that all the funds on the Open Ended Diversified Core Equity (ODCE) index had actively adopted GRESB Real Estate Assessment reporting (GRESB, n.d.). ODCE index tracks open-end funds pursuing a core investment strategy, generally characterised by low risk, low levered, stable properties geographically diversified across US markets.

4.6.2.8 CRREM

CRREM provides the real estate industry with transparent, science-based decarbonisation pathways aligned with the Paris Climate Goals of limiting global temperature rise to 2°C, with ambition towards 1.5°C. It was originally funded by the European Commission's Horizon 2020 Programme. CRREM has now been seen the expansion of the CRREM pathways to include major real estate markets outside the European Union (EU), as well as the residential real estate sector (CRREM, n.d.). As Fernanda Amemiya, Sustainability Reporting Manager, Landsec put it: 'CRREM is a helpful tool that can support companies in assessing the required reduction in carbon emissions at asset level' (CRREM, 2020).

CRREM is a useful tool for a global investor to understand where the portfolio sits against the pathway that aligns with the Paris Agreement. The CRREM tool aligns closely with the performance data GRESB collects on real estate assets and is adopted by Partnership for Carbon

Accounting Financials (PCAF)'s latest project, which addresses the need to mobilise the financial industry to accelerate the transition of European buildings towards net zero by 2050 (PCAF, n.d.).

4.6.2.9 International Integrated Reporting Council (IIRC), International Financial Reporting Standards (IFRS) Foundation and International Sustainability Standards Board (ISSB)

The five framework- and standard-setting institutions of international significance, the Carbon Disclosure Project (CPD), Climate Disclosure Standards Board (CDSB), Global Reporting Initiative (GRI), US Integrated Reporting Council (IIRC) and the Sustainability Accounting Standards Board (SASB), have co-published a shared vision that includes both financial accounting and sustainability disclosure, connected by way of integrated reporting of the elements necessary for more comprehensive corporate reporting. Taking GRI, SASB, CDP and CDSB together, these organisations guide the overwhelming majority of sustainability and integrated reporting.

Given global concerns about the diverse range of sustainability standards, standard setters are joining forces to strive for consistency in corporate reporting. The IFRS Foundation Trustees are establishing the ISSB, which will propose financial and non-financial reporting standards related to climate change.

The IIRC is committed to working together with key actors, including International Organization of Securities Commissions (IOSCO) and the IFRS, the European Commission and the World Economic Forum's International Business Council with an aim to an increased clarity and to drive a holistic system for reporting across the value chain with the integrated reporting framework that connects sustainability disclosure to reporting on financial and other capitals.

The IFRS Foundation has taken over the monitoring of the progress on companies' climate-related disclosures from the TCFD. The IFRS Foundation has published the Progress on Corporate Climate-related Disclosures – 2024 Report in November 2024 (IFRS, 2024b). IFRS recently launched IFRS S2 recommendations for climate-related disclosure. While there are differences between IFRS S2 and the TCFD recommendations, IFRS S2 are consistent with the four core recommendations and 11 recommended disclosures published by TCFD, with some differences in wording, and additional details required by IFRS S2 – for example, one of the differences is IFRS S2 sets out a Scope 3 measurement framework to provide guidance for preparing Scope 3 GHG emissions disclosures (IFRS, 2024a).

Focus area

It is increasingly evident that non-financial information such as environmental and climate-related disclosures are becoming equally important requirement for companies and government alongside the financial disclosure. Demand for a consistent standard to cover a wide range of corporate reporting standards is high and the search for common global definitions and metrics has gathered pace. IIRC, IFRS Foundation, European Commission and World Economic Forum's International Business Council have joined forces aiming to establish one global set of definitions, reporting standards and metrics.

While complete convergence will take time, real estate companies and built environment professionals are required to understand the relevant guidelines, tools and standards. In particular, CRREM, GRESB are becoming a popular suite of choices, which are aligned with TCFD. A synopsis of these tools is provided in Figure 4.16 to help ease of reference.

Figure 4.16 A synopsis of ESG reporting tools

	Tool	Description	
Tools and guidance	Climate Disclosure Standards Board (CDSB)	*CDSB provides the compantes with a framework for reporting environmental information with the same rigour as financial information*	
	Global Reporting Initiative (GRI)	*GRI produces free Sustainability Reporting Guidelines*	
	Climate Disclosure Project (CDP)	*CDP is a not-for-profit charity, runs the global disclosure system for investors, companies, cities, states and regions to manage their environmental impacts*	Wider Sustainability
	Sustainability Accounting Standards Board (SASB), now under International Sustainability Standards Board (ISSB)	*SASB is designed to identify and standardise disclosure for sustainability issues, risks and opportunities that are most relevant to investor decision-making covering wide range of sectors*	
	International Integrated Reporting Council (IIRC)	*IIRC aims to include both financial accounting and sustainability disclosure, connected by way of integrated reporting of the elements, having joined forces with CDP, GRI, SASB, and CDSB*	
	Carbon Risk Real Estate Monitor (CRREM)	*CRREM provides the real estate industry and investors a useful tool with science-based decarbonization pathways aligned with the Paris Climate Goals. Useful to understand where the portfolio sits*	For Real Estates
	Taskforce Climate-related Financial Disclosure (TCFD)	*TCFD focuses specifically on how climate-related issues affect financial performance. Now TCFD disbanded with the monitoring role taken by ISSB Standards under IFRS Foundation*	
	Taskforce Nature-related Financial Disclosure (TNFD)	*TNFD focuses on nature-based solutions affect financial performance*	Climate only
Benchmarking and ratings	Morgan Stanley Capital International (MSCI)	*MSCI is a leading provider of support tools and services for the global investment community for climate-related investment decisions including tools for MSCI ESG Ratings.*	
	Global Real Estate Sustainability Benchmark (GRESB)	*GRESB is an independent not-for-profit organization providing validated ESG performance data and peer benchmarks for investors and managers to improve business intelligence, industry engagement and decision-making*	For Real Estates

4.6.3 Science-based targets (decarbonisation pathways)

Science-based targets provide companies with a clearly defined path to reduce emissions in line with the Paris Agreement goals. Targets are considered 'science-based' if they are in line with what the latest climate science deems necessary to meet the goals of the Paris Agreement (United Nations, 2015) – limiting global warming to 1.5°C above pre-industrial levels. Science-Based Targets initiative (SBTi) recommends that GHG emissions must halve by 2030 and drop to net zero by 2050.

More than 6000 businesses around the world are already working with the Science-Based Targets initiative (SBTi) (SBTi, n.d.a). Companies from all sectors and of all company sizes are encouraged to join the SBTi movement and commit to setting a science-based target. Businesses who sign the SBTi commitment letter are immediately recognised as 'Committed' on their website, as well as by the CDP, UN Global Compact, We Mean Business and WWF. The SBTi does not currently assess targets for cities, local governments, public sector institutions, educational institutions or non-profit organisations.

SBTi recommends five steps of setting targets and taking actions: commit, develop, submit, communicate and disclose, as illustrated in Figure 4.17. For a typical company, after the assessment following the carbon reduction pathways for a sector that the company is operated within, the target could look like this: the company commits to reduce Scope 1 and Scope 2 GHG emissions 25% by 2030 from a 2020 base year, and to measure and reduce its Scope 3 emissions.

Science-based target setting makes business sense. More and more built environment companies have established their own SBTi. At the project level, the UK Net Zero Carbon Buildings Standard aligns its building targets with SBTi.

As a major source of GHG emissions, the private sector is taking action by starting the science-based targets commitment and cutting emissions at scale.

Figure 4.17 Five steps of setting a science-based target (created from information from SBTi (SBTI, n.d.b))

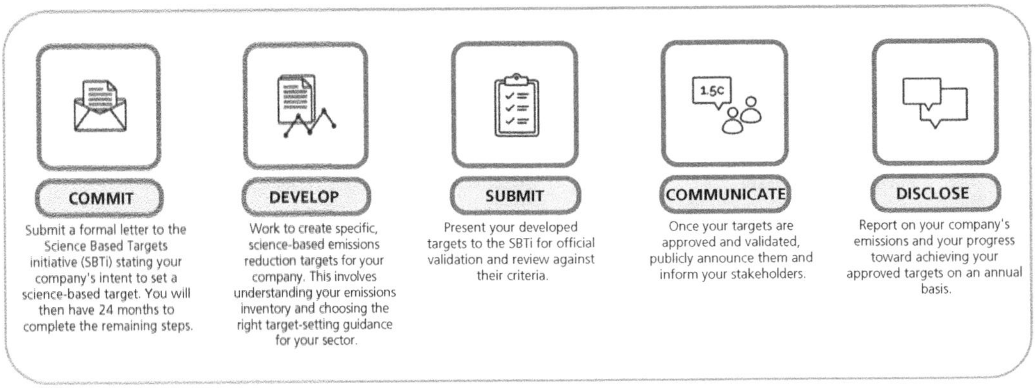

Science-based targets

The SBTi is a corporate climate action organisation having offices in Germany, Mexico, the UK and USA. SBTi has developed standards, guidelines and a target dashboard for organisations (SBTi, n.d.b).

SBTi publicly discloses temperature alignment based on the ambition of a company's Scope 1 and 2 targets. Scope 3 targets are also evaluated during the target validation process. Companies' Scope 3 ambition are evaluated to ensure it meets the temperature alignment or supplier engagement specifications outlined in the SBTi criteria. SBTi are carrying out a comprehensive review of the Scope 3 target setting methods and criteria to ensure they are fully aligned with the Net-Zero Standard.

The following lists three examples on the SBTi Target Dashboard to reveal the SBTi targets committed by the three global companies (data sourced from SBTi, n.d.c).

The following is noted.

- Organisations whose target status is 'committed' have made a public commitment to set a science-based target aligned with the SBTi's target-setting criteria within 24 months.
- Organisations adopting the Net-Zero Standard are required to set both near-term and long-term science-based targets.
- Long-term science-based targets are only set by organisations who are aligned to the SBTi's Net-Zero Standard. These organisations must produce close to zero emissions no later than 2050.
- Business Ambition for 1.5°C campaign – an urgent call to action from a global coalition of UN agencies, business and industry leaders, in partnership with the Race to Zero.

Microsoft Corporation, USA: Software and services. Data published 2019.

- Commits to Near Term target of reducing 1.5°C by 2030.
- Business ambition for 1.5°C Campaign member.
- Commits to reduce Scope 3 GHG emissions intensity per unit of revenue 30% by 2030 from a FY2017 base year and to avoid growth in absolute Scope 3 emissions.

Apple Inc. USA: Technology hardware and equipment. Data published/updated in 2021.

- Commits to Near Term target of reducing 1.5°C by 2030.
- Business ambition for 1.5°C Campaign member.
- Commits to reduce absolute combined Scope 1, 2 and 3 GHG emissions 62% by FY2030 from a FY2019 base year. Apple also commits to continue annually sourcing 100% renewable electricity through FY2030. The target boundary includes biogenic emissions and removals from bioenergy feedstocks.

Landsec: UK / Europe. Real estate business. Data published/updated in 2023.

- Commits to Near Term and Long-Term targets of reducing 1.5°C by 2030 and 2040, respectively.
- Commits to Net Zero by 2040.
- Business ambition for 1.5°C Campaign member.
- Overall Net-Zero Target: Landsec commits to reach net-zero GHG emissions across the value chain by 2040 from a 2020 base year. Near-Term Targets Landsec commits to reduce absolute scope 1, 2 and 3 GHG emissions 47% by 2030 from a 2020 base year. Long-Term Targets Landsec commits to reduce absolute Scope 1, 2 and 3 GHG emissions 90% by 2040 from a 2020 base year.

4.6.4 Reporting responsibility of Scope 3

It has been recognised there is overlap between Scope 3 reporting and embodied carbon assessments. The UKGBC has published guidance on embodied carbon and its inclusion within Scope 3 reporting, aiming to address the need for a cohesive approach, enabling organisations to accurately monitor, report and manage their carbon footprint across the life cycle of built assets.

Traditionally, Scope 3 reporting is based on spending to convert to carbon equivalent using spend-based factors. Using embodied carbon assessments directly within Scope 3 reporting would represent an easier and more detailed solution when compared with the traditional approach, since the calculation of embodied carbon is done centrally at the project level, from which the results can be shared with all relevant parties, rather than each organisation involved in the value chain calculating the associated emissions for their Scope 3 report. This would not only reduce the multiplication of work currently seen within emissions reporting, but also providing more accurate assessment when done at a project level using product-level emissions data (such as EPDs or LCAs) within embodied carbon assessments.

A spend-based method is still used in most of Scope 3 assessment. However, with embodied carbon assessments becoming widespread across industry, a hybrid approach can be used alongside the typical spend-based method.

The guidance has provided clarity and method of reconciliation between reporting of Scope 3 for an organisation (which takes place annually) and project-level upfront embodied carbon, which are typically undertaken at practical completion (as-built embodied carbon assessments). The aim is to align yearly Scope 3 reporting with project-level embodied carbon as realistic and practical as possible. Figure 4.18 illustrates the categories of total organisational emissions.

Figure 4.18 Spend-based GHG emissions against asset/project based embodied carbon (reproduced with kind permission of UK Green Building Council (UKGBC, 2024))

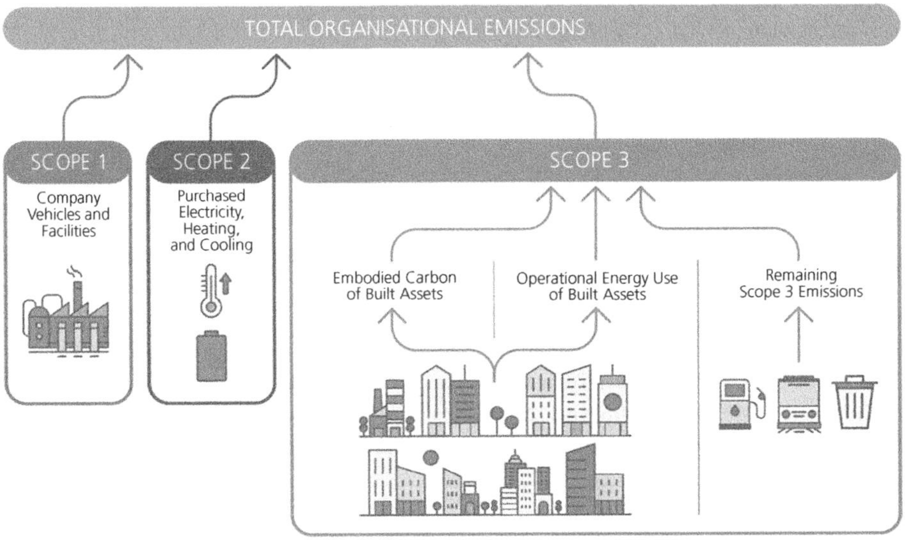

Throughout the value chain, different organisations play different roles, but all play a part in influencing embodied carbon of materials in a construction project – for instance, Figure 4.19 indicates the suggested 'scope' for reporting upfront embodied carbon for a typical asset life cycle. 'Designed embodied carbon' is not within the Scope 3 emissions of architects and engineers. However, UKGBC is developing a methodology for project reporting technique for design firms in the future.

4.7. Moving towards purposeful action

The sustainable built environment is at a pivotal moment. As the urgency of climate change intensifies, the sector has responded with a proliferation of frameworks, standards and tools aimed at achieving net zero carbon and broader sustainability goals. From the UKGBC's Net Zero Carbon Buildings Framework to the integration of social value and wellbeing metrics, the landscape is rich with opportunity as well as complexity.

This chapter has shown the scale and scope of the existing technical guidance and standards that already exist. While these are essential, they are not sufficient on their own. The real challenge for built environment professionals lies in navigating uncertainty, reconciling competing priorities, and embedding sustainability into every stage of decision making. Purposeful sustainability requires the industry to move beyond compliance, it requires improved clarity of intent, shared values and a collaborative approach that reflects the unique context of each project.

To move forward effectively, practitioners must begin by asking: *What are we trying to achieve?* Only by identifying the core values and desired outcomes of a project can teams select the most appropriate frameworks and tools to support those goals. Frameworks, standards and methodologies should serve the purpose, not define it. The emphasis should be on identifying shared purpose first, then selecting the tools that will be most relevant and capable of delivering the required impact.

Figure 4.19 GHG Protocol Scope 3 reporting boundary for stakeholders across an asset's lifetime (reproduced with kind permission of UK Green Building Council (UKGBC, 2024))

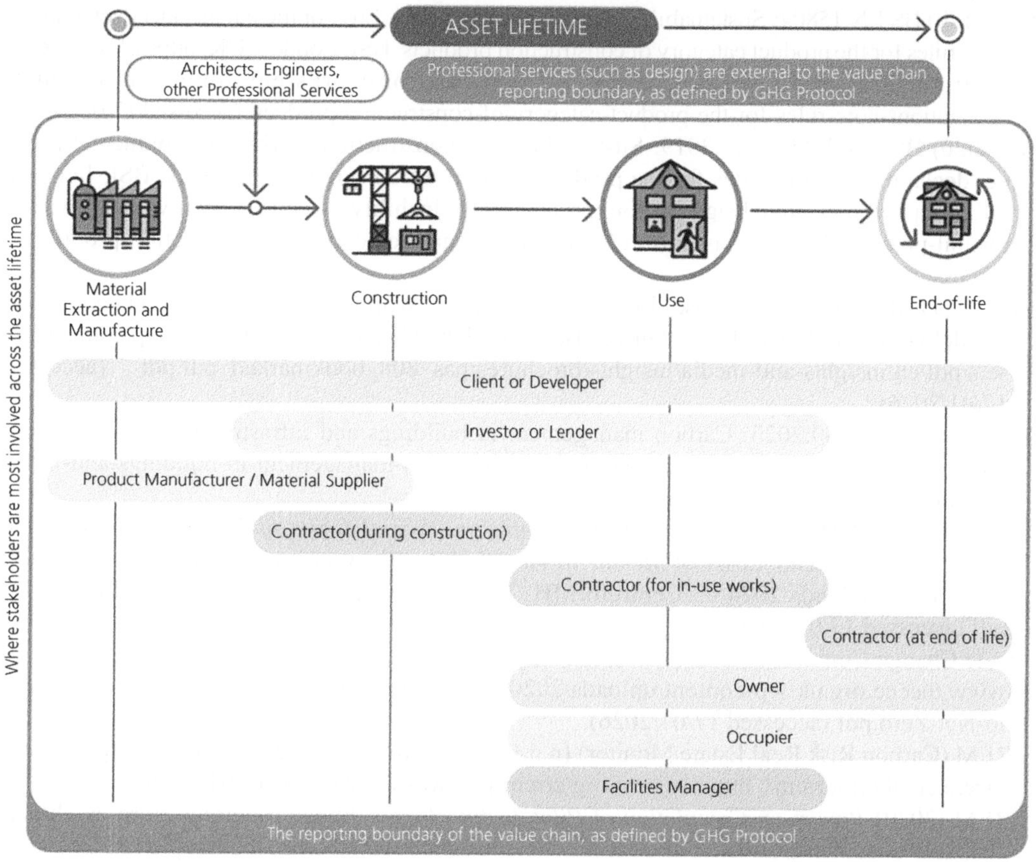

Having established a firm foundation, the next question is how to begin practical realisation. The following chapter focuses on tools for design implementation that can help translate these values and frameworks into real-world outcomes. Such tools help to move built environment projects from ambition to action, building on purposeful planning, governance and learning.

REFERENCES

Adil L *et al.* (2025) *Climate Risk Index 2025, Who Suffers Most From Extreme Weather Events?* Germanwatch e.V. https://www.germanwatch.org/sites/default/files/2025-02/Climate%20Risk%20Index%202025.pdf (accessed 17/01/2026).

Bioregional (n.d.) *One Planet Living® – A Sustainability Framework by Bioregional*. https://www.bioregional.com/one-planet-living (accessed 17/01/2026).

Blears H, White C and Battle G (2019) *National TOMS Framework 2019 for Social Value Measurement, National Social Value Taskforce*. https://www.local.gov.uk/sites/default/files/documents/National%20TOMs%202019%20Guidance%201.0.pdf (accessed 17/01/2026).

Brundtland GH (1987) *Our Common Future*. Report of the World Commission on Environment and Development. Geneva, UN-Dokument A/42/427. https://www.are.admin.ch/en/1987-brundtland-report (accessed 19/01/2026).

BSI (2012a) BS EN 15978: Sustainability of construction works – Assessment of environmental performance of buildings. BSI, London, UK. https://standardsdevelopment.bsigroup.com/projects/2023-01815 (accessed 17/01/2026).

BSI (2012b) BS EN 15804: Sustainability of construction works. Environmental product declarations. Core rules for the product category of construction products. BSI, London, UK. https://knowledge.bsigroup.com/products/sustainability-of-construction-works-environmental-product-declarations-core-rules-for-the-product-category-of-construction-products (accessed 17/01/2026).

BSI (2020) BS EN 15804+A2:2019: Sustainability of construction works. Environmental product declarations. Core rules for the product category of construction products. BSI, London, UK. https://knowledge.bsigroup.com/products/sustainability-of-construction-works-environmental-product-declarations-core-rules-for-the-product-category-of-construction-products-1 (accessed 17/01/2026).

BSI (2022) PAS 808:2022: Purpose-driven organizations – Worldviews, principles and behaviours for delivering sustainability – Guide. BSI, London, UK. https://www.bsigroup.com/siteassets/pdf/en/insights-and-media/insights/brochures/pas_808_bookmarked_pdf.pdf (accessed 17/01/2026).

BSI (2023) PAS 2080:2023: Carbon management in buildings and infrastructure. BSI, London, UK. https://knowledge.bsigroup.com/products/carbon-management-in-buildings-and-infrastructure (accessed 17/01/2026).

Cardoso F and McHayle Z (2024) *The Economic and Social Costs of Mental Ill Health, the Economic and Social Costs of Mental Ill Health*. https://www.centreformentalhealth.org.uk/wp-content/uploads/2024/03/CentreforMH_TheEconomicSocialCostsofMentalIllHealth-1.pdf (accessed 17/01/2026).

Climate Change Committee (2020) *The Sixth Carbon Budget: The UK's Path to Net Zero*. https://www.theccc.org.uk/wp-content/uploads/2020/12/The-Sixth-Carbon-Budget-The-UKs-path-to-Net-Zero.pdf (accessed 17/01/2026).

CRREM (Carbon Risk Real Estate Monitor) (n.d.) *Carbon Risk Real Estate Monitor*. https://www.crrem.eu/about-crrem/; https://crrem.org/crrem-pathways/ (accessed 17/01/2026).

CRREM (2020) *Report on Completion of Pilot Testing Phase*. https://ec.europa.eu/research/participants/documents/downloadPublic?documentIds=080166e5cf6b5fb2&appId=PPGMS#:~:text=flexibility%20allows%20EVORA%20to%20project,assets%20are%20at%20future%20risk (accessed 17/01/2026).

Da Mota Passos Neto G, Alencar LH and Valdes-Vasquez R (2023) Multiple-criteria methods for assessing social sustainability in the built environment: a systematic review. *Sustainability* **15(23)**: 16231, 10.3390/su152316231.

D'Amato M and Kauko T (2012) Sustainability and risk premium estimation in property valuation and assessment of worth. *Building Research* and *Information* **40(2)**: 174–185, 10.1080/09613218.2012.655069.

Fanning AL and Raworth K (2025) Doughnut of social and planetary boundaries monitors a world out of balance. *Nature* **646**: 47–56, 10.1038/s41586-025-09385-1.

Forster PM *et al.* (2025) Indicators of Global Climate Change 2024: annual update of key indicators of the state of the climate system and human influence. *Earth System Science Data* **17(6)**: 2641–2680, 10.5194/essd-17-2641-2025.

GHG Protocol Team *et al.* (2011) *Corporate Value Chain (Scope 3) Accounting and Reporting Standard*. https://ghgprotocol.org/sites/default/files/standards/Corporate-Value-Chain-Accounting-Reporing-Standard_041613_2.pdf (accessed 17/01/2026).

Goubran S and Cucuzzella C (2019) Integrating the sustainable development goals in building projects. *Journal of Sustainability Research* [preprint], 10.20900/jsr20190010.

GRESB (n.d.) GRESB Real Estate Assessment. https://www.gresb.com/nl-en/real-estate-assessment/; https://www.gresb.com/nl-en/the-link-between-gresb-participation-and-financial-returns/ (accessed 17/01/2026).

GRI (Global Reporting Initiative) (2022a) *GRI Standards for Reporting Publicly on a Range of Economic, Environmental and Social Impacts*. https://www.globalreporting.org/how-to-use-the-gri-standards/gri-standards-english-language/ (accessed 28/09/2025).

GRI (2022b) *The Global Reporting Initiative – An Overview*. https://www.globalreporting.org/about-gri/#:~:text=GRI%20(Global%20Reporting%20Initiative)%20is,language%20to%20communicate%20those%20impacts (accessed 12/06/2024).

Halla P and Merino-Saum A (2022) Conceptual frameworks in indicator-based assessments of urban sustainability—an analysis based on 67 initiatives. *Sustainable Development* **30(5)**: 1056–1071, 10.1002/sd.2300.

HM Treasury (2022) *The Green Book: Appraisal and Evaluation in Central Government* (2022 edn; Updated 2024). https://www.gov.uk/government/publications/the-green-book-appraisal-and-evaluation-in-central-government/the-green-book-2020 (accessed 17/01/2026).

IFRS (International Financial Reporting Standards) (2024a) *Comparison – IFRS S2 Climate-related Disclosures with the TCFD recommendations*. https://www.ifrs.org/content/dam/ifrs/supporting-implementation/ifrs-s2/ifrs-s2-comparison-tcfd.pdf (accessed 17/01/2026).

IFRS (2024b) *Progress on Corporate Climate-related Disclosures—2024 Report*. https://www.ifrs.org/content/dam/ifrs/supporting-implementation/issb-standards/progress-climate-related-disclosures-2024.pdf; https://www.ifrs.org/sustainability/tcfd/ (accessed 17/01/2026).

IPCC (Intergovernmental Panel on Climate Change) (2014) Foundations for decision making. In: Part A: global and sectoral aspects. In *AR5 Climate Change 2014: Impacts, Adaptation and Vulnerability. Working Group II Contribution to the Fifth Assessment Report of the Intergovernmental Panel on Climate Change*. Cambridge University Press, Cambridge, UK, pp. 195–228. https://www.ipcc.ch/report/ar5/wg2/ (accessed 17/01/2026).

IPCC (2023a) 17 – Decision-making options for managing risk. In *Climate Change 2022 – Impacts, Adaptation and Vulnerability. Working Group II Contribution to the Sixth Assessment Report of the Intergovernmental Panel on Climate Change*, Cambridge University Press, Cambridge, UK, pp. 2539–2654, 10.1017/9781009325844.026.

IPCC (2023b) Summary for policymakers. In *Climate Change 2021: The Physical Science Basis. Contribution of Working Group I to the Sixth Assessment Report of the Intergovernmental Panel on Climate Change*. Cambridge University Press, Cambridge, UK, pp. 3–32, 10.1017/9781009157896.001.

ISO (International Organization for Standardization) (2006) ISO 14025:2006: Environmental labels and declarations. ISO, Geneva, Switzerland. https://www.iso.org/standard/38131.html (accessed 17/01/2026).

Karunasena G, Rathnayake RMNU and Senarathne D (2016) Integrating sustainability concepts and value planning for sustainable construction. *Built Environment Project and Asset Management* **6(2)**: 125–138, 10.1108/bepam-09-2014-0047.

LETI (London Energy Transformation Initiative) (2020) LETI Embodied Carbon Primer. https://www.leti.uk/_files/ugd/252d09_8ceffcbcafdb43cf8a19ab9af5073b92.pdf (accessed 17/01/2026).

Llena-Nozal A, Martin N and Murtin F (2019) The economy of well-being. *OECD Statistics Working Papers* [preprint], 10.1787/498e9bc7-en.

Lorenz D, Dent P and Kauko T (eds) (2017) Value in a changing built environment. In *Wiley eBooks*, 10.1002/9781119073666.

Lützkendorf T and Lorenz D (2005) Sustainable property investment: valuing sustainable buildings through property performance assessment. *Building Research* and *Information* **33(3)**: 212–234, 10.1080/09613210500070359.

Malin N *et al.* (2015) *Improving Green Building Construction in North America: Guide to Integrated Design and Delivery* In Commission for Environmental Cooperation (Kirk D, David J and Fortson J (eds)). http://www.cec.org/files/documents/publications/11661-improving-green-building-construction-in-north-america-guide-integrated-design-en.pdf (accessed 17/01/2026).

Manewa A *et al.* (2016) Adaptable buildings for sustainable built environment. *Built Environment Project and Asset Management* **6(2)**: 139–158, 10.1108/bepam-10-2014-0053.

Mann M (2023) *Our Fragile Moment: How Lessons from the Earth's Past Can Help Us Survive the Climate Crisis*. Scribe, London, UK.

MSCI (Morgan Stanley Capital International) (n.d.) https://www.msci.com/data-and-analytics/sustainability-solutions/esg-ratings (accessed 07/2025).

Ness B *et al.* (2007) Categorising tools for sustainability assessment. *Ecological Economics* **60(3)**: 498–508, 10.1016/j.ecolecon.2006.07.023.

New M *et al.* (2022) Decision-making options for managing risk. In *Climate Change 2022: Impacts, Adaptation and Vulnerability. Contribution of Working Group II to the Sixth Assessment Report of the Intergovernmental Panel on Climate Change.* Cambridge University Press, Cambridge, UK, 10.1017/9781009325844.026.

Partnership for Carbon Accounting Financials (PCAF) (n.d.) https://carbonaccountingfinancials.com/en/partners#accredited-partners (accessed 07/2025).

Public Health England (2020) *Health Impact Assessment in Spatial Planning*. https://assets.publishing.service.gov.uk/government/uploads/system/uploads/attachment_data/file/929230/HIA_in_Planning_Guide_Sept2020.pdf (accessed 17/01/2026).

Public Services (Social Value) Act 2012 (2012). https://www.legislation.gov.uk/ukpga/2012/3/enacted (accessed 17/01/2026).

Raworth K (2013) Defining a safe and just space for humanity. In *Island Press/Center for Resource Economics eBooks*, pp. 28–38, 10.5822/978-1-61091-458-1_3.

RIBA (Royal Institute of British Architects) (2021) *RIBA 2030 Climate Challenge Version 2 (2021).* https://www.architecture.com/about/policy/climate-action/2030-climate-challenge?srsltid=AfmBOooYQElkbM2YV0iANV5IbICdiJlR8pL0liP-Pnx-YU7KpI8A8M-O (accessed 20/06/2025).

RICS (Royal Institution of Chartered Surveyors) (2024) *RICS Professional Standard: Whole Life Carbon Assessment for the Built Environment*, 2nd edn, Version 3. https://www.rics.org/profession-standards/rics-standards-and-guidance/sector-standards/construction-standards/whole-life-carbon-assessment (accessed 17/01/2026).

Rockström J *et al.* (2009) A safe operating space for humanity. *Nature* **461(7263)**: 472–475, 10.1038/461472a.

Samuel F (ed.) (2020) *RIBA Social Value Toolkit for Architecture*. RIBA and University of Reading, UK. https://riba-prd-assets.azureedge.net/-/media/GatherContent/Social-Value-Toolkit-for-Architecture/Additional-Documents/RIBAUoR-Social-Value-Toolkit-2020pdf.pdf?rev=8ce05f7e8f7649d09116a13dd3fb9f60 (accessed 17/01/2026).

Samuel F and Watson K (2023) Social value of the built environment. *Buildings and Cities* **4(1)**: 938–947, 10.5334/bc.399.

SBTi (Science Based Targets Initiative) (n.d.a) *Climate Action Milestone: 6,000 + Companies Adopt Science-Based Targets*. https://sciencebasedtargets.org/blog/climate-action-milestone-6-000-companies-adopt-science-based-targets (accessed 17/09/2024).

SBTi (n.d.b) *Science Based Targets Initiative – How to Set Science Based Targets*. https://sciencebasedtargets.org/how-to-set-science-based-targets (accessed 06/10/2025).

SBTi (n.d.c) *Target Dashboard*. https://sciencebasedtargets.org/companies-taking-action (accessed 06/10/2025).

SBTN (Science-Based Targets Network (n.d.) *Biodiversity within SBTs for Nature*. https://science-basedtargetsnetwork.org/companies/take-action/set-targets/biodiversity-targets/ (accessed 08/2025).

Social Return on Investment (SROI) Network (2012) *A Guide to Social Return on Investment*, https://socialvalueuk.org/standards-and-guidance/ (accessed 17/01/2026).

Squarebird (2024) *The Official Quality of Life Framework*. Quality of Life Foundation. https://www.qolf.org/framework/ (accessed 17/01/2026).

Stiglitz J, Sen A and Fitoussi J (2009) *Report of the Commission on the Measurement of Economic Performance and Social Progress (CMEPSP)*.

Stroombergen A and Lawrence J (2022) A novel illustration of real options analysis to address the problem of probabilities under deep uncertainty and changing climate risk. *Climate Risk Management* **38**: 100458, 10.1016/j.crm.2022.100458.

Taherdoost H and Madanchian M (2023) Multi-criteria decision making (MCDM) methods and concepts. *Encyclopedia* **3(1)**: 77–87, 10.3390/encyclopedia3010006.

TCFD (Taskforce Climate-related Financial Disclosure) (2017) *Recommendations of the Task Force on Climate-Related Financial Disclosures*. https://assets.bbhub.io/company/sites/60/2021/10/FINAL-2017-TCFD-Report.pdf (accessed 17/01/2026).

TCFD (2021) *Task Force on Climate-related Financial Disclosures – Implementing the Recommendations of the Task Force on Climate-related Financial Disclosures*. https://assets.bbhub.io/company/sites/60/2021/07/2021-TCFD-Implementing_Guidance.pdf (accessed 17/01/2026).

The London Plan 2021 (2021) https://www.london.gov.uk/programmes-strategies/planning/london-plan/new-london-plan/london-plan-2021 (accessed 17/01/2026).

UKGBC (UK Green Building Council) (2019) *Net Zero Carbon Buildings: A Framework Definition*. Initial definition document. https://ukgbc.org/wp-content/uploads/2019/04/Net-Zero-Carbon-Buildings-A-framework-definition-print-version.pdf (accessed 17/01/2026).

UKGBC (2021a) *Framework for Defining Social Value*. https://ukgbc.org/wp-content/uploads/2021/02/Framework-for-Defining-Social-Value.pdf (accessed 17/01/2026).

UKGBC (2021b) *Net Zero Whole Life Carbon Roadmap*. https://ukgbc.org/our-work/topics/whole-life-carbon-roadmap/ (accessed 17/01/2026).

UKGBC (2022) *A Guide for Delivering Social Value on Built Environment Projects*. https://ukgbc.org/wp-content/uploads/2022/03/UKGBC_Social-Value_2022_Digital_All.pdf (accessed 17/01/2026).

UKGBC (2024) *Embodied Carbon: Scope 3 Measurement and Reporting*. https://ukgbc.org/wp-content/uploads/2024/03/Embodied-Carbon-Scope-3-Report.pdf (accessed 17/01/2026).

UKGBC (n.d.) *Delivery Checklists*, pp. 10–11. https://ukgbc.org/wp-content/uploads/2022/03/UKGBC_Social-Value_Delivery-Checklists_March-2022.pdf (accessed 17/01/2026).

UKNZCBS (2025) UK Net Zero Carbon Buildings Standard Pilot Version rev 2. https://www.nzcbuildings.co.uk/_files/ugd/6ea7ba_1ef36b6835de46668f2ad8b589ff1b93.pdf (accessed 09/05/2025).

UNEP (UN Environment Programme) (2022) UN Environment Programme finance initiative. *Impact Radar 2022*. https://www.unepfi.org/publications/unep-fi-impact-radar-2022/#:~:text=The%20Impact%20Radar%2C%20a%20compilation,help%20operationalise%20this%20holistic%20approach (accessed 17/01/2026).

UN Environment Programme Finance Initiative (2024) *The Global Biodiversity Framework – How Has Financial Policy and Regulation Evolved to Support Ambition on Nature Action?* https://www.unepfi.org/themes/ecosystems/the-global-biodiversity-framework-how-has-financial-policy-and-regulation-evolved-to-support-ambition-on-nature-action/ (accessed 17/01/2026).

United Nations (2015) *Paris Agreement*. https://unfccc.int/files/essential_background/convention/application/pdf/english_paris_agreement.pdf (accessed 17/01/2026).

United Nations Department of Economic and Social Affairs (2015) *Transforming Our World: The 2030 Agenda for Sustainable Development | Department of Economic and Social Affairs*. https://sdgs.un.org/2030agenda (accessed 17/01/2026).

Vardakoulias O (2013) *Social CBA and SROI*. New Economics Foundation. https://www.nefconsulting.com/wp-content/uploads/2014/10/Briefing-on-SROI-and-CBA.pdf (accessed 17/01/2026).

World Green Building Council (2022) *Sustainable Built Environments & the UN's Sustainable Development Goals – World Green Building Council*. https://worldgbc.org/sustainable-development-goals/ (accessed 17/01/2026).

WHO (World Health Organisation) (2021) *Towards Developing WHO's Agenda on Well-Being*. https://www.who.int/publications/i/item/9789240039384 (accessed 17/01/2026).

WHO (2022) *Geneva Charter for Well-Being*. https://cdn.who.int/media/docs/default-source/health-promotion/geneva-charter-4-march-2022.pdf?sfvrsn=f55dec7_21&download=true (accessed 17/01/2026).

Navigating an Evolving Sustainable Built Environment: Delivering Net Zero Carbon and Climate Resilience

emerald PUBLISHING ice Publishing

Mei Ren and Rachel Waggett

ISBN 978-1-83662-853-8
https://doi.org/10.1108/978-1-83662-850-720261008

Chapter 5
Sustainable processes and design tools

5.1. Introduction

Earlier chapters have discussed framework planning in the context of policies and published guidance and the importance of developing a purposeful project with sustainability at its core. Chapter 4 explored how frameworks, governance structures and decision-making tools can be used to establish the values that will underpin a project. Given the current media reporting on environment, social and governance (ESG) for organisations, be it financing, trading or public relations, this may prompt questions about whether ESG may supersede sustainability in the building industry. It is important to clarify that ESG frameworks offer organisations a benchmark for public reporting or marketing purposes, while sustainability principles are fundamental in the implementation of built environment initiatives. Having established *why* integrated, values-led frameworks are essential, this chapter focuses on *how* designers, planners and project teams can deliver on these principles, exploring useful processes and tools for design implementation in the built environment.

5.2. Sustainability key design issues for development projects

5.2.1 Passive design

Resource efficiency is often a key design issue, but how far can designers push for minimisation of energy, water and waste? In most cases, a growing economy with an increasing population often involves building more, which leads to consuming more resources. How can a well thought through balance be reached?

Passive design measures, commonly referred to as a 'fabric first' approach, aim to make the building envelope work harder, achieving better performance by cutting down the heating or cooling requirement, shielding from the outdoor environment, while still making a building breathable when required. An energy hierarchy explains this stepped approach to resource efficiency design: be lean, be clean and be green.

This energy hierarchy, illustrated in Figure 5.1, is a useful tool to guide the design process to reduce energy wastage. *Be lean* is mainly about optimising the design of a building so that it uses less energy, which includes careful massing, fabric performance, envelope and shading design, natural light and ventilation. *Be clean* means all the services systems are designed with maximum efficiency by using heat recovery where possible, higher efficiency lighting, good automatic controls, demand control of ventilation, efficient heating and cooling generation and distribution systems to ensure efficient operation and reduced operational costs. *Be green* is the final step of applying renewable energy and seamless integration with the building.

The energy hierarchy should be applied in any design project to identify opportunities for carbon emissions reductions. The key question is how far the 'lean' approach can be taken. A forward-thinking developer would require demonstration of the operational cost savings (opex) and carbon reductions achieved for different options, set against any additional capital expenditure (capex) required.

Figure 5.1 Energy hierarchy

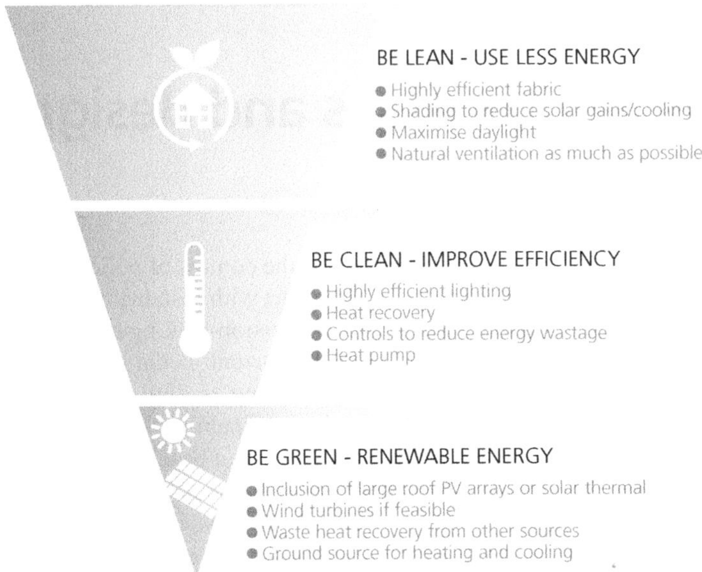

BedZED, London

An early example of adopting excellent passive design principles is the BedZED development. BedZED in south London is deemed the UK's first large-scale, inspirational low-carbon, environmentally friendly housing community development.

Figure 5.2 BedZED eco-village (Bioregional, 2016) (photo credit to One Planet Living in Sutton licensed under CC BY 2.0)

Passive design measures such as good effective layout design, passive solar heating design, high levels of insulation and airtightness resulted in major energy savings with an overall 27% reduction in electricity consumption and 36% reduction in gas consumption between 2012 and 2015 (Bioregional, 2016), including electricity savings provided by solar photovoltaic arrays integrated into the buildings.

- **High insulation**. The accommodation blocks are carefully oriented, arranged in east–west rows with double glazing covering all of the south-facing facades, while smaller windows facing north, east and west are triple glazed. BedZED's levels of insulation were much higher than those required by mandatory building standards at the time when it was built and completed in 2002. The external walls have a 300 mm layer of rockwool insulation between an outer brick. It was designed to have high levels of airtightness to reduce heat loss.
- **Solar design**. Most of BedZED's homes have conservatory-like 'sunspaces' behind big, south-facing glazed facades. These sunspaces are found on all three storeys and are 1.2 m deep. The air within them is warmed by the sun; once a comfortable temperature is reached in the sunspace, the double-glazed doors and windows that separate it from the rest of the home can be opened to allow this warmed air to flow through the rooms.
- **Passive ventilation**. The key feature is a ventilation system using roof wind cowls to bring in fresh, external air into the homes and workspaces and expel stale air effectively without wasting heat in the process. Each of these cowls has a vane and can rotate, so that it always faces into the wind. At the front, a small opening collects fresh, exterior air that moves down into the building and emerges into the homes and workplaces through. At the back of the cowl is a larger hole through which the stale, outgoing air (collected through vents in the homes' ceilings) exits. This effectively enables heat recovery without the need for any electrically powered fans. In hot weather, windows can be opened to increase ventilation; many of the homes have windows opening on different sides, allowing for cross ventilation.
- **Low carbon heat network**. While much of BedZED's heat demand is met by the warmth from its occupants and by solar gain, its homes and workplaces still require some additional heat – mostly to provide household hot water. BedZED has its own small-scale district heating system that circulates hot water to provide heat to each household through an underground network of insulated pipes. Every home has its own large hot water tank, placed in a vented cupboard, warmed by the heat network. A wood-burning combined heat and power (CHP) plant was installed to power the heat network, although later on, due to maintenance issues, the CHP plant was decommissioned and replaced by gas fired boilers. Due to passive solar heating principles, the heat demand is low enabling significant running cost savings.

The Passive House (German: Passivhaus) standard was originated and founded in 1988 by Bo Adamson of Lund University, in Sweden, and Wolfgang Feist of the Institut für Wohnen und Umwelt (Institute for Housing and the Environment), in Darmstadt, Germany. Passivhaus adopts a whole-building approach with clear, measured targets, focused on high-quality construction, certified through an exacting quality assurance process. It recently received major interest from many real estates and developers and institutions. By 2016, there were approximately 60 000 such certified structures of all types worldwide (Passivhaus Trust, n.d.).

The principle of the standard is to make buildings as 'passive' or as energy efficient as possible so that it results in ultra-low energy buildings that require less energy for space heating or cooling. Adoption of such a standard and a full certification needs a complete buy-in from every design team member and client to ensure quality of the build; it often requires a dedicated skilled designer to advise and ensure a building is 'sealed' to meet air tightness standards. It often results in higher construction cost due to the level of attention needed to meet ultra-high standards for air tightness and insulation, shown in Table 5.1 alongside the other key Passivhaus criteria. Passivhaus is not a net zero building; however, it is a significant step advancement in meeting carbon targets.

Table 5.1 Passive house key design parameters, created using data from Passive House Institute criteria for buildings (Passive House Institute, 2023)

	Passivhaus Criteria
Airtightness n50	≤ 0.6 ACH @50Pa
Space Heating Demand (SHD)	≤ 15 $kWh/m^2.a$
Peak heating load (alternative criteria)	≤ 10 W/m^2
Primary Energy Renewable (PER)	≤ 60 $kWh/m^2.a$
Space cooling demand	≤ 15 $kWh/m^2.a$
Surface temperature	≥17 °C
Overheating	Max 10% ≥ 25°C

The House at Cornell Tech in New York

The Passivhaus certified House at Cornell Tech in New York is a good example described here (Figure 5.3).

Figure 5.3 (a) The House at Cornell Tech in New York (House at Cornell Tech, 2023) (photo credit to Tdorante10 licensed under CC BY-SA 4.0; (b) a cluster of Cornell Tech buildings (photo credit to to Rhododendrites licensed under CC BY-SA 4.0)

(a) (b)

Developed by a joint partnership between Hudson Companies, Related Companies and Cornell University, The House at Cornell Tech (House at Cornell Tech, 2023) is a student and faculty residential tower with 272 000ft^2 over 26 floors on Roosevelt Island in New York City. Completed in 2017, it is the tallest building in the world built to Passive House standards and is Leadership in Energy and Environmental Design (LEED) Platinum certified.

Cornell Tech demanded a precedent-setting building with a focus on sustainability, energy performance and occupant comfort. The 26-storey residential building on the northern edge of campus consists of approximately 350 rental apartments for students and faculty, and is part of the campus' urban island neighbourhood of new buildings totalling more than 2 million ft^2 of space. Key design features include the following (Pearson, 2021).

- A high-performance facade integrated with highly efficient indoor environmental systems and floor layouts that take full advantage of the building's unique orientation is a key factor in the design to reduce energy use and carbon emissions.
- The building envelope is a crucial aspect of passive design – The House is made airtight in order to prevent air infiltration and to therefore reduce energy consumption for heating and cooling. Prefabricated metal panels help to ensure a higher quality envelope along with additional insulation and continuous water/vapour barriers. All windows are triple glazed, and all facade joints and connections are sealed to minimise heat transfer. The efficient concrete structure is detailed to eliminate thermal bridging at the facade, roof, foundation and exterior elements such as the entry canopy. The engineer on this project, Buro Happold, worked closely with the owner, the architect, Handel Architects, and the New York City Department of Buildings to design heating, ventilation and air conditioning (HVAC), lighting and structural systems that support the Passive House standard.
- The building is ventilated centrally with high efficiency heat recovery systems. Additional heating and cooling is provided by the use of Variable Refrigerant Flow (VRF) system.
- The building was a certified Passive House, having been designed with Passive House principles. The building consumes 73% less than that of the average house energy use in New York (444kWh/m^2 per year median for residential buildings of more than 20 000m^2 in New York City, based on 2013 benchmark data), as shown in Figure 5.4. A whole-building pressure test measured airtightness of just 0.13 air changes per hour at 50 pascals – against the Passivhaus maximum 0.60.

Figure 5.4 Energy Use Intensity (EUI) for the House at Cornell Tech (Pearson, 2021)

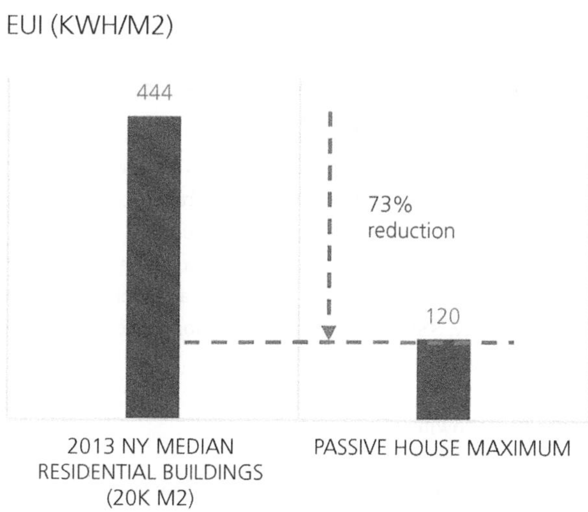

5.2.2 Sustainability planning tools

In the past two decades, various efforts have been made to develop sustainability tools aiming to measure the efficacy of sustainability solutions. Apart from Building Research Establishment Environmental Assessment Method (BREEAM) for buildings and Civil Engineering Environmental Quality Assessment and Awards Scheme (CEEQUAL) for infrastructure (now BREEAM Infrastructure), urban planners and masterplanners developed bespoke sustainability indicator framework system tools to guide their design decisions.

Traditionally, designers have developed design elements of a master plan in light of their personal experiences. However, there is a great need for evidence-based, scientific, scenario-analysis techniques that are based on accurate and meaningful data, considering the life span of the urban development. It is also crucial to produce a model that includes life cycle costing as well as carbon as an aid to the client's decision-making process.

In light of predicted climate change, it is imperative that we plan for and create communities that can robustly meet this challenge. New master plans and developments must be extremely resilient and designed to cope with, and adapt to, future uncertainty. It is vital to ensure that masterplanning is based on accurate and meaningful information.

This means that all of the elements that constitute a master plan, such as urban landscape, open space, utility infrastructures and transport corridors, will need to be 'remodelled' to deal with increasing temperatures, changing rainfall patterns and rising sea levels. Evidence of such 'climate-proofing' can not only enhance an organisation's reputation with its stakeholders, but can also protect investments, as well as reducing health risks and insurance costs.

Two examples are provided here with varying degrees of focus: Sustainable Systems Integrated Model (SSIM) developed by EDAW/AECOM and Arup's Sustainable Project Appraisal Routine (SPeAR).

SSIM – AECOM Sustainable Systems Integrated Model

SSIM is a geographical information system (GIS)-based land use spatial planning tool that quantifies and compares a range of sustainability and climate change parameters for a master plan and evaluates the sustainability merits of alternative urban form solutions. The SSIM framework and models make it possible to create and test multiple scenarios across various performance levels and cost-benefit outcomes (Srivastava and Thompson, 2018). SSIM is a fully scalable climate change/land planning tool that addresses key drivers affecting planning, such as climate change and carbon emissions, and includes information on energy planning and urban heat islands. Crucially, SSIM also defines key performance indicators (KPIs) that will ensure any subsequent design and construction adheres to the master plan.

In summary, SSIM performs the following.

- It identifies the most sustainable master plan among several alternative options.
- It creates a set of sustainability initiatives that can be tested against desired outcomes and design standards at a master plan, infrastructure, buildings and public realm scale.

- It enables effective decision making through the generation of quantitative and qualitative outputs that detail key performance parameters such as reductions in greenhouse gas (GHG) emissions and energy and water consumption.
- It provides initial, recurring and life-cycle cost data and cost-benefit analysis for alternative sustainable planning and design strategies.
- It allocates costs and savings to various cost 'centers' (government, developer, end-user, etc.)

The model used to demonstrate sustainable outcomes for projects is based on a range of KPIs (Sofian *et al.*, 2015). A typical SSIM will incorporate the following themes, each with its own set of KPIs

- building energy
- public realm energy
- potable water and storm water
- waste water
- green building
- ecological services/carbon sequestration
- sustainable transportation
- social and cultural
- urban heat island effect
- total carbon footprint.

It is worth noting that key drivers for urban masterplanning vary from one country/project to another. The SSIM typically follows four steps, as shown in Figure 5.5. After the sub-optimisation of each theme developing 'good', 'better practice' and 'best practice' designs during the second step, SSIM will then combine the individual core themes (i.e. energy, water, transportation, etc.) into a set of comprehensive, all-systems master programme alternatives at the fourth step, with the outputs related to environmental benefits, capital costs and life-cycle costs.

Figure 5.5 SSIM processes to support achievement of planned outcomes

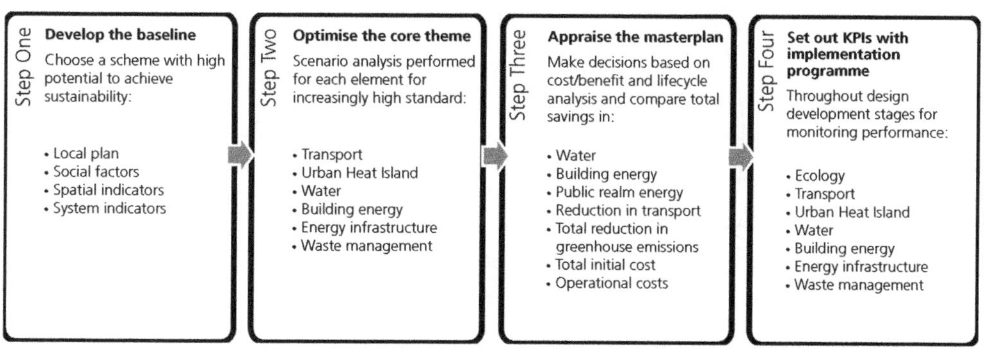

KPIs are set out to facilitate monitoring of sustainability performance and targets. These KPIs include key design standards for buildings and associated facilities in order to design an infrastructure suitable for future demand, taking into account climate risks.

A case study of applying SSIM is for the 890-acre (360-hectare) Marina Bay and Greater Southern Waterfront (AECOM, 2012), one of Singapore's key development areas commissioned by the Urban Redevelopment Authority to seamlessly extend the downtown district and further support the city-state's continuing growth as a major business and financial hub in Asia. SSIM was applied to incorporate expertise from landscape, masterplanning, environment, transportation and building engineering to deliver a sustainable development framework for the planning and design of Marina Bay as well as the adjacent Greater Southern Waterfront. SSIM was used to assess and quantify the possibilities of sustainability performance. Based on the model, design solutions were provided relating to specific aspects of urban design and urban microclimate, landscape engineering, water management and sustainable transportation, together with planned outcomes for the total water savings, total building energy savings, total public realm energy savings, total reduction in transport, total reduction in GHG emissions, total initial costs and total on-going monthly costs.

SPeAR – Arup Sustainable Project Appraisal Routine

As described in Chapter 4, SPeAR (Sustainable Project Appraisal Routine) was developed by Arup to help clients' sustainability goals (Arup, 2017). The SPeAR tool appraises projects based on key themes such as transport, biodiversity, culture, employment and skills. It provides a holistic basis for appraising infrastructure projects based on 21 indicators within the social, environmental and economic pillars of sustainability. Figure 5.6 shows an example of SPeAR tool. It is focused on reducing environmental impact and is primarily suited to developed country contexts.

Figure 5.6 Arup SPeAR, an example of analysis (image created by Sarah Grice licensed under CC BY-SA 2.0)

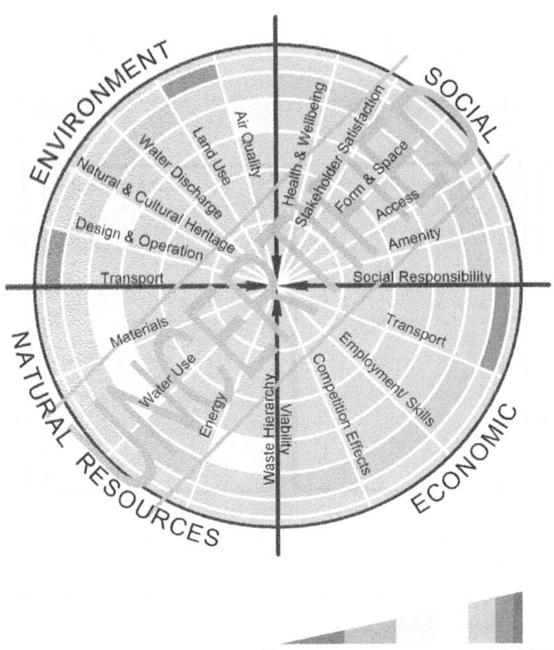

Using this bespoke sustainability tool, SPeAR can visually display the estimated outcomes of a project, using a traffic-light system to highlight the anticipated performance in key areas such as transport, biodiversity, culture, employment and skills. The company uses SPeAR to maximise sustainability benefits and reduce risks and cost for clients.

The tool is suitable for buildings, infrastructure and master plan development. The software can be used for sustainability appraisal throughout a development project's whole life cycle, from initial design to operation – for example, SPeAR was used in developing a surface water management project in Grangetown, Cardiff, which has resulted in significant water, energy and carbon savings as well as enhanced use of green space, improved social interaction and engagement with local artists.

Zargarian *et al.* (2016) believe that SPeAR has many advantages, including adaptability, transparency of the assessment methodology, identification of improvement opportunities, optimisation of the indicators and measures, and the possibility to trace all assessment steps easily and clearly.

Although it covers a wide range of sustainability issues, it is very broad and descriptive, which also leaves a wide gap for the subjective judgment of the sustainability practitioner. Many sustainability issues can be quantitatively measured and can have numerical ranges for the accepted levels. More designers and clients are now adopting a bespoke sustainability framework with topics and targets using the recognised design matrices, which are discussed later.

5.3. Energy planning applications

Energy planning is a relative term compared with urban planning. In essence, energy planning aims to develop long-term energy policies to balance energy supply and demand in a strategic way. It has various scales for managing energy systems at local, regional, national, or global levels considering factors such as the environment, economy, society and technology to meet energy needs.

The energy market in the UK is fragmented with various stakeholders such as government agencies, local authorities, utility companies, electrical grid companies and energy customers. Market drivers and factors influencing growth include the increasing pressure of decarbonisation, geopolitical drivers and government policies to facilitate the transition to net zero carbon emissions by 2050.

Traditional centralised energy systems have been inefficient and unadaptable. Energy planning is undergoing a revolution in the supply, storage and use of energy considering multi-vector energy systems with the aim of transitioning to a future of flexible, clean energy systems, designed around consumers. Multi-vector energy systems refer to integrated energy infrastructures that combine various energy vectors like electricity, heat (and cooling), gas and hydrogen to enhance flexibility, efficiency and resilience in the energy network.

Energy is a significant cost for many companies, who are now compelled to invest in decarbonisation of their assets and to navigate highly competitive supply chains to be able to develop the skills to deal with their decarbonisation journey. There have been major investment and technology advancements in recent years with supportive policy environments, and a shift in thinking, especially in heat network and smart, integrated local energy systems that use the latest advances in electric vehicles, renewable generation, batteries and digital data technology.

To achieve the full potential of integrated energy planning (Collins *et al.*, 2023), the UK Government is now beginning to require local authorities (municipalities) to produce Local Area Energy Plans (LAEPs). Led by local government and developed collaboratively with defined stakeholders, LAEPs are fully costed, spatial plans that identify the changes needed to the local energy system and built environment, detailing 'what, where and when and by whom', in order to meet the 2030 target of a 68% reduction in emissions, the 2035 target of a 78% reduction in emissions, and net zero by 2050.

LAEPs are recognised as the leading method for translating national net zero targets into local energy system action with plans that are collaborative, data driven and cost effective.

LAEPs provide the level of detail for an area that is equivalent to an outline design or master plan; additional detailed design work is required for identified projects to progress to implementation. LAEPs define a long-term vision for an area but should be updated approximately every 3–5 years (or when significant technological, policy or local changes occur) to ensure the long-term vision remains relevant.

5.3.1 Low carbon heat networks

Heating buildings accounted for approximately 33% of the UK's carbon dioxide emissions in 2018. In the UK, most UK homes and commercial buildings are heated using natural gas boilers, so the UK will not meet its climate change targets without a revolution in heating.

The European Union (EU)'s heating and cooling needs in 2020 accounted for 50% of its total gross final energy consumption (EEA, 2023). This significant share has persisted for over a decade, despite extensive EU and national efforts to lower heating and cooling needs (EEA, 2021, 2022). While heating buildings accounts for the largest share of energy used for heating and cooling in the EU, with space and water heating accounting for about 60%, industrial heat demand accounts for about a third of heating and cooling needs. The remaining energy is consumed in agriculture and for cooling (EC, 2022).

For the EU and its member states, decarbonising heating and cooling represents a major challenge on the way to meeting climate targets and ensuring that fundamental energy needs, such as for residential heating, can be met more securely than they are today. However, national policymakers face very different challenges and opportunities in decarbonising heating and cooling, because the availability of sustainable energy resources and the demand for heating and cooling from buildings and industry vary significantly at the country and regional levels. There is an urgent need to move away from fossil fuels for heating systems to accelerate adoption of low carbon heat sources and renewables (Nijs *et al.*, 2021).

In the UK, heat networks have the potential to decarbonise the supply of heat in built up areas and are a key part of the Government's net zero policy. Heat networks supply heat and hot water from a central source to consumers, through a network of pipes. The use of a modern heat network provides a far more resource-efficient system due to load sharing and effective incorporation of renewable energy sources. The Climate Change Committee estimated that around 18% of UK heat could come from heat networks by 2050 to support cost-effective delivery of carbon targets (up from around 2% currently).

There are two types of heat networks.

- Communal heat network which supplies heat and hot water to a number of premises and customers within a single building. This is the most common form of heat network in the UK.
- District heat network which supplies heat to more than one building. District heat networks can cover a large area and supply many buildings.

Scandinavian and Baltic EU countries – where average heating needs are higher because of climatic factors – had already reached high shares, of more than 50%, of renewable sources in final energy consumed for heating and cooling in 2020, by utilising biomass extensively. The widespread use of modern district heating systems in these countries can facilitate the integration of low-temperature heat from geothermal and solar thermal sources and from recovered waste heat. In central and eastern Europe, many urban district heating systems are old and face existential challenges that question their future viability, ranging from inefficient heat production to network losses and a strong fossil-fuel dependence (EPG, 2022). One of the main challenges for these systems is overcoming the problems resulting from poor maintenance and abandonment.

In the UK, large-scale robust networks are beginning to appear. Until recently, heat was unregulated energy. Since the Energy Act 2023 was introduced, Ofgem is the regulator of heat networks across England, Scotland and Wales (Great Britain). The Energy Act 2023 is the initial component of a regulatory framework that seeks to provide heat network customers comparable protections to gas and electricity consumers. In preparation for regulation, it encourages all heat suppliers and heat network operators to follow existing good practice as set out by the Heat Trust. The Energy Security Bill, updated in March 2023 introduces a regulatory framework for heat networks and powers to enable heat network zoning in England. These measures will work together to reduce the cost of living by regulating prices and ensuring this sector can contribute to delivering net zero in the most cost-effective way (DESNZ, 2023).

With the decarbonisation of the electrical grid and move away from fossil fuel technologies, there is a growing movement towards lower temperature networks. This has the dual benefit of reducing heat loss by way of the distribution pipework and increasing efficiency of heat delivery by way of heat pump technologies. Historically, district heating would be powered by steam or high temperature hot water (c. 120°C). These types of heat network are known as 1st generation and 2nd generation, respectively. This system of categorisation is summarised below.

In the UK, 3rd and 4th generation networks represent current best practice, with 3rd generation often employed to allow traditional 80/60°C flow/return temperatures to be delivered to buildings, thus minimising impact on the building systems. Ambient loop systems are now described as 5th generation, with large scale examples currently only operational in Scandinavian countries.

- 1st generation – steam network.
- 2nd generation district heating – high pressure and temperature (>100°C) hot water system.
- 3rd generation district heating – mid-pressure with flow temperatures 70-95°C, temperature differential of 20K.
- 4th generation district heating – low pressure with flow temperature 70°C, temperature differential of 30K.
- 5th generation district heating – 'ambient loop' system with very low flow temperature (20°C flow) and temperature differential (5K).

1st generation and 2nd generation heat networks are commonly found in energy intensive campuses, such as universities and hospitals or pharmaceutical campuses, which require high initial investment and maintenance programmes. Whereas 3rd generation heat networks represent more modern system specification installed since the 1990s. A good example of such is the London Olympic Legacy heat network.

Olympic Park Energy

The construction of Queen Elizabeth Olympic Park (Olympic Park) for the London 2012 Olympic and Paralympic Games included a commitment to provide a combined cooling, heat and power (CCHP) generation facility and heating and cooling distribution networks on the site. This was part of the Sustainable Development Strategy for the London 2012 Games and, in particular, was designed to deliver substantial carbon savings in line with the London Plan and progressively tighter government legislation. The network was designed to serve the London 2012 Games and the legacy development of the park as well as Stratford City (ADE, 2023).

Olympic Park Energy Centre, shown in Figure 5.7, is one of the largest CCHP generating facilities built in the UK with an estimated investment of £113 million, including two energy centres and community energy network. It is located at the west-end of the Olympic Park and the utilities are partly housed in the historic Kings Yard Industrial Buildings. The project was developed by the Olympic Delivery Authority (ODA) and was opened in October 2010. A second energy centre has since been built in Stratford City. It will also serve the heating, cooling and electricity requirements of the Games venues and other facilities. Both the energy centres were constructed to serve the new metropolitan development developed in East London after the 2012 Games, with about five new neighbourhoods always intended for development in the future, after the Games (Power Technology, 2012).

Figure 5.7 Olympic Park Energy Centre (photo credit to *Oast House Archive* licensed under CC BY-SA 2.0)

The Community Energy Network was procured by ODA with a 40-year Concession Agreement under a finance, design, install and operate contract awarded to a subsidiary of ENGIE S.A, East London Energy Limited (ENGIE).

The large-scale district heating and cooling system consists of two interconnected energy centres, supplying energy within and around the park with over 20 km of pipes serving domestic residents and commercial buildings, including the London Stadium and Westfield Shopping Centre. The network is designed to be expandable and utilises CHP engines, chillers and a biomass boiler for energy production. The output from the energy centres is 46.5 MW of heating and 16 MW of cooling.

The network plays a key role in the Olympic legacy, contributing to energy efficiency and carbon reduction in the area with combined aims of achieving competitive energy prices for consumers while making it easier and more cost effective. It includes the following features.

Low carbon emissions. The use of biomass and CHP engines contributes to reducing carbon emissions in the area.

Reliable energy supply. The network provides a reliable and efficient energy supply to various buildings in the park.

Economic and social benefits. The network contributes to the regeneration of the area and creates jobs and opportunities.

Decarbonisation efforts. The network is actively involved in decarbonisation efforts through the use of low-carbon technologies and the expansion of the network.

East London Energy has installed a heat pump in the Stratford City Energy Centre as part of the first phase of the works required to decarbonise Queen Elizabeth Olympic Park's District Heating Network.

LLDC (the London Legacy Development Corporation) is exploring other opportunities to decarbonise the heat source, including a potential Water Source Heat Pump (WSHP) on the Stanstead Mill Stream and heat pumps at the Old Ford Water Recycling Plant. LLDC is also speaking to development partners about hosting heat pumps within future developments.

5.3.1.1 4th generation (4G) heat network

A 4th generation heating network with a flow temperature of 70°C and return of 40°C has an advantage such that it creates an opportunity for very low carbon technologies to be deployed at scale to share operational benefits and significant cost savings – for example, heat pumps.

The installation of such networks has implications on the building side 'secondary' heating system to make sure the heat is exchanged and transferred to the building side efficiently providing sufficient heating to users.

5.3.1.2 5th generation (5G) heat network

A 5th generation district heating network can exhibit significantly lower distribution losses than other types of network. The very low operating temperatures (typically 20–30°C) mean that heat losses from pipework to surroundings are low throughout the year. Additionally, very low carbon heat supply technologies such as heat pumps can be more readily deployed and operate more efficiently than a 4G network.

An example of a 5th generation network is known as an 'ambient loop', illustrated in Figure 5.8. A set of staged heat pumps situated locally to connected buildings draw heat from the network 'loop' (e.g. at 20°C) to generate the requisite temperatures for the secondary system (typically, this will be at 60°C to avoid legionella issues). The relatively low lift from 20°C to 60°C allows for efficient heat pump operation, reducing costs and carbon associated with heating. If underfloor heating were deployed in buildings, an even lower operating temperature (c.40°C) could be used, further improving heat pump efficiency (an additional heat pump would be required for domestic hot water generation at a higher temperature).

The temperature differential on an ambient loop is typically around 5°C. Given such the temperature regimes, low-cost plastic pipework could be used instead of steel pipework. An ambient loop system would have the benefit of enabling lower grade heat sources to be used, such as solar thermal or waste heat from electrical transformers.

Figure 5.8 Ambient loop system diagram

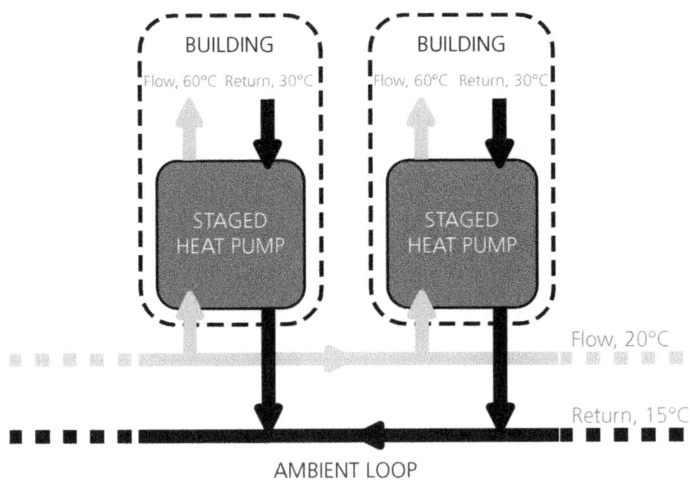

Energy sources

District heat networks can use a variety of heat sources

- waste heat recovery – any waste heat from data centres, underground tube, transformer and sewage plant, and waste heat from any industrial processes
- gas fired CHP

- fuel cell CHP
- biofuel CHP
- biomass CHP
- ground source heat pump (GSHP) – open loop or closed loop
- water source heat pumps from lakes or river
- deep geothermal
- solar thermal system
- biomass boiler
- natural gas boiler or electric boiler – often used as a secondary peak topping system.

These technologies can often work together – for example, waste heat sources such as that from industrial processes are generally of a lower temperature and are less suitable for district heating as a single energy source, where the low grade heat requires heat pumps to upgrade the heat to a suitable temperature for supply to district heating. One of the key benefits of using a heat network is the increased potential of harnessing low carbon energy sources.

These systems should be assessed for a particular application using guidance from CIBSE CP1 (CIBSE, 2020) on the basis of

- whole life costs
- contributions to carbon dioxide reductions
- future trends in energy prices and electricity decarbonisation
- level of technology risk
- local environmental impact
- space and access requirements
- sustainability credentials
- transport requirements
- security of supply.

Most of these energy technologies are established and documented in many guidance documents – for example, the Chartered Institution of Building Services Engineers (CIBSE) has published guidance on Ground Source Heat Pumps (CIBSE TM51: CIBSE, 2013). Most of the evaluation and guidance regarding these renewable technologies does not include embodied carbon primarily due to a lack of accurate data; however, efforts are now increasing to identify the evidence needed to assess embodied carbon impacts of energy technology over the life cycle (Finnegan *et al.*, 2018).

Another emerging technology with significant potential to contribute to city scale heat source is deep geothermal (BGS, 2024).

Deep geothermal

British Geological Survey (BGS) (BGS, 2024) published an assessment report of the current deep geothermal energy situation and future potential of deep geothermal energy in the UK. Deep geothermal energy already delivers environmental, economic and technical advantages in countries with similar geology to the UK, such as the Netherlands, Belgium and Germany. McCay *et al.* (2019) performed a life cycle assessment of the carbon intensity of deep geothermal heat systems and found that the carbon intensity of the heat produced is 9.7–14.0 kg(CO2e) MWh_{th}, which is

4.9–7.3% of the emissions from heat from natural gas. These values are compatible with Scotland's plans for long-term decarbonisation of heat in line with the UK emission reduction obligations.

Geothermal energy is energy produced and stored as heat in the subsurface. Most of the UK's onshore deep geothermal potential is found in deeply buried (deeper than 500 m) limestones and sandstones in sedimentary basins. Groundwater within these rocks may reach temperatures of more than 100°C but more data from deep wells and from operational geothermal projects is needed to estimate the economically usable portion of the energy. In some areas, hot granites provide a potential source for geothermal power. Deep geothermal has a potential of providing an ultra-low-carbon source for heating, cooling and power generation.

The current high costs of drilling restrict the use of geothermal energy to areas with certain geological settings. As technologies improve and new extraction methods are developed, more areas should become economically viable for geothermal exploitation.

There are currently two deep geothermal projects in development or operation in Cornwall for provision of heat and power, including the Eden Geothermal Project, and three other UK sites are currently using geothermal water from springs or shallower wells for heat.

Eden Geothermal Project, Cornwall

The Eden Geothermal Project aimed to decarbonise Eden Project's heat supply by replacing gas with renewable heat from a deep geothermal well. The heat network system includes a single well heat exchanger to extract heat from EG-1 and a new 3.8 km heat main to link the geothermal site with the heat loads on the Eden Project site supplying heat to its famous Biomes, greenhouses and offices.

The well EG-1 is the UK's longest geothermal well. The drilling finished in October 2021, and the completed well is 4871 m in total vertical depth and 5277 m in measured depth (Eden Geothermal, 2023).

The single well heat exchanger has been created by installing a coaxial circulation system in the well itself. Vacuum-insulated tubing was installed to a depth of 3850 m in the centre of the well. The tubing serves as the conduit for hot water being pumped to the surface.

At the surface, a plate heat exchanger extracts heat from the well water, and the newly cooled water travels back down the annulus (outer 'ring') of the well to be reheated at depth, in a continuous circulation.

The project has also extended to a research and development project led by University of Exter for a better understanding of the geothermal resource in Cornwall and the wider region, to unlock future development.

Eden Geothermal Limited was formed with partnership of Eden Project Limited, Bestec (UK) Limited, which is alliliated with Bestec GmbH based in Germany. Eden Geothermal was responsible for project design, planning and public procurement of the £20 million of contracts to complete the programme including specialist services, site works, drilling, completion and testing of the deep well, installation of a heat main to Eden, and heat production and supply to Eden. Figure 5.9 illustrates the Eden deep geothermal heat network.

Figure 5.9 Eden geothermal network (image reproduced with kind permission of ©Eden Geothermal Ltd 2021)

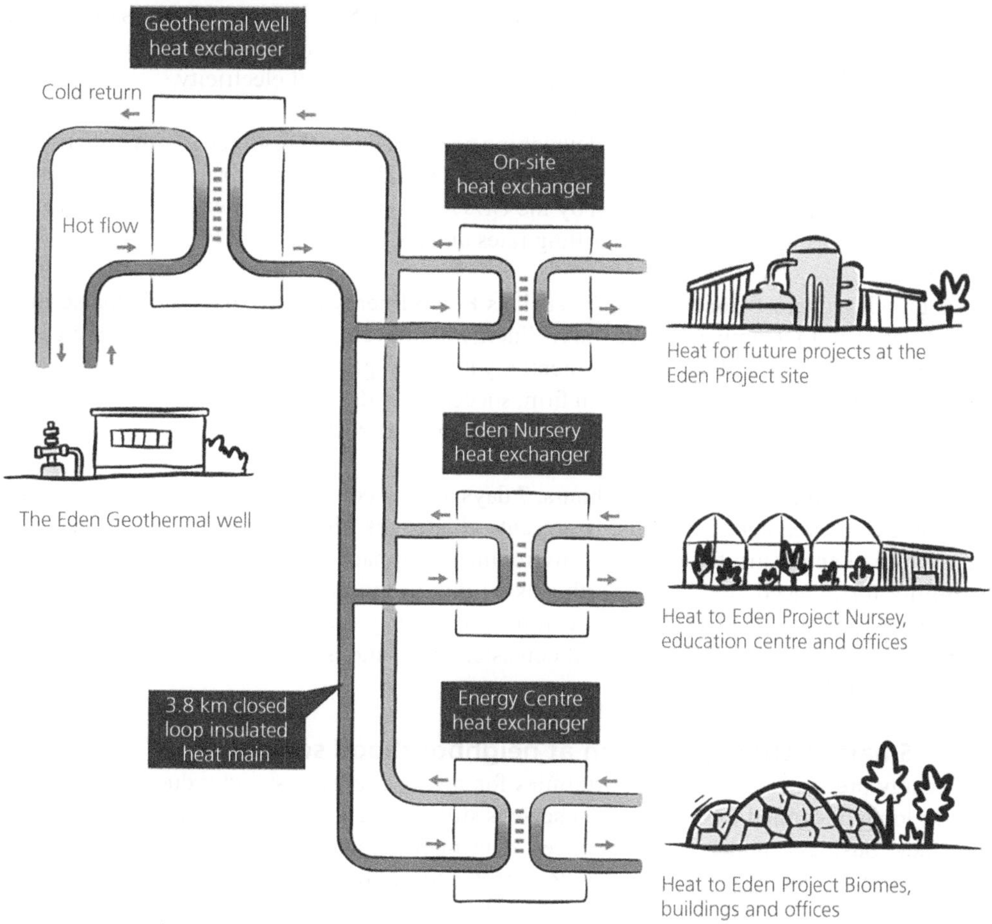

Well testing delivery partners include: Camborne School of Mines (part of the University of Exeter's College of Engineering, Mathematics and Physical Sciences) and the specialist contractors GeoScience Ltd (providing wellsite geology services).

Deep geothermal system is not only a good choice for supplying low carbon heat source for a heat network, but it also has potential for power generation. Geothermal power generation falls into two main categories: conventional (hydrothermal) and next-generation. Conventional resources rely on naturally occurring conditions and are relative rare, more concentrated in places including Iceland and New Zealand. Next-generation technologies, particularly Enhanced Geothermal Systems (EGS), use modern engineering to expand geothermal potential across diverse geological settings (Merino-Garcia *et al.*, 2024).

In the United States, a combination of new drilling techniques, government support and the pressing need to secure continuous clean power in an era of climate crisis means that geothermal energy is increasingly gaining interest from private sectors; it has been estimated that 10% of electricity could be generated by geothermal by 2050 aiming to deploy geothermal energy at scale (Milman, 2024). The technology is in the form of EGS (DoE, n.d.a). EGS technology involves drilling deep into the Earth, creating fractures in hot, dry rock, and then injecting water to create a geothermal reservoir. This allows for the extraction of heat and the generation of electricity.

The US has been actively exploring and developing deep geothermal resources, with initiatives like the US Department of Energy (DoE)'s FORGE project – The Frontier Observatory for Research in Geothermal Energy (FORGE), funded by the US DoE, is a major EGS demonstration project in Milford, Utah, focusing on improving drilling rates and rock stimulation (DoE, n.d.b).

There is growing interest from companies such as Fervo Energy (Fervo, n.d.) and Quaise Energy (Quaise Energy, n.d.). Companies like Fervo Energy are developing technology to make geothermal power more flexible and economic, potentially adding gigawatts of flexible geothermal capacity to carbon-free grids. Eavor, a Canadian firm, successfully drilled a more than 4800 m-deep hole in New Mexico to prove it could access heat in deep, granite rock (Eavor, 2023).

Although deep geothermal provides 24 hours, 7 days a week consistent and reliable energy source, not dependent on weather conditions, it presents challenges: the cost of the drilling and development can be significant; the potential environmental impact, especially seismicity, as there is a possibility for induced seismicity (earthquakes) due to the manipulation of underground rock (although this can be managed through careful monitoring and mitigation techniques) and the regulatory framework, requiring clear and consistent regulations for EGS development (Merino-Garcia *et al.*, 2024).

5.3.2 Smart energy application at neighbourhood scale

Smart energy presents significant opportunities for saving energy cost and reducing carbon emissions. Building integrated local systems at scale is still an emerging new technology, which would require new business models with advanced data management, with collaboration and investment to realise the potential. In 2018, UK Research and Innovation established a UKRI Industrial Strategy Challenge Fund programme and set up the Prospering from the Energy Revolution challenge programme (UKRI, 2022). UKRI invests around £104 million, matched by industry, to help businesses, researchers and local communities develop, test and prove smart local energy systems. Figure 5.10 illustrates such a smart energy vision.

The electricity grid is traditionally supply led, responding to demand. Modern power systems with higher complexity can be managed with a greater role for ramping up and down demand in response to system needs. As electrification accelerates, with higher numbers of heat pumps and electric vehicles, in order to meet growing demand, applying demand-side flexible solutions can be cost-effective to deploy. Demand-side flexibility means being able to shift the consumption of electricity at peak times, such as through 'smart charging' an electric vehicle, or time-shifting usage of other electricity use, offsetting new grid or grid upgrade and generation capacity needed across the system (Energy Transition Commission, n.d.).

The use of automated software solutions, which interacts with demand-side hardware, are increasing and will be critical to scaling demand-side flexibility. The creation of a digitalised database

Figure 5.10 Smart energy vision (image credit to 'Smart Grid' by portland general licensed under CC BY-ND 2.0)

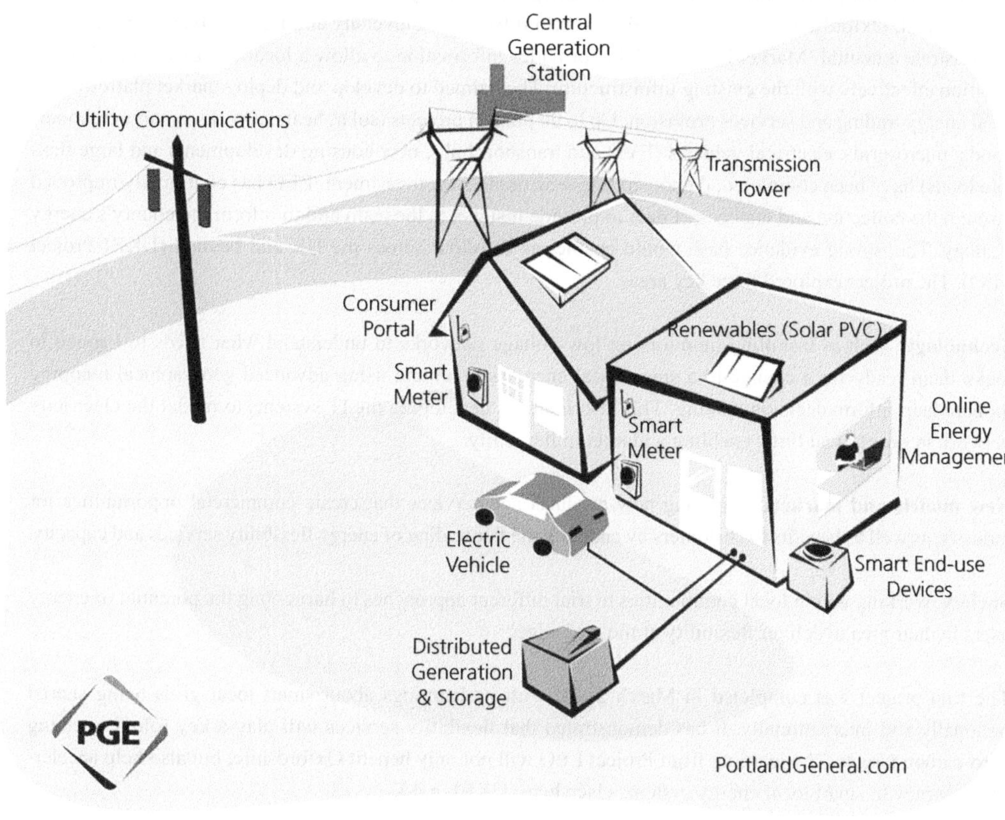

with the energy data capture at a granular level can be beneficial for the wider local distribution network, where there are opportunities to generate value for the distribution network (e.g. by providing flexibility). Data-driven smart grid solutions can offer customers both cost savings and carbon reduction.

Smart energy case study

Local Energy Oxfordshire (Project LEO, 2023) (UKRI, 2022) is one of the most wide ranging and holistic smart grid trials conducted in the UK. Funded by UKRI's Prospering from the Energy Revolution programme, the £15 million project ran trials in Oxfordshire to build a broad range of reliable evidence of the technological, market and social conditions needed to understand how new technologies and services can benefit local people, communities and the energy system, seeking to accelerate the UK's transition to a zero-carbon energy system. The project's focus is on understanding of flexibility at the 'grid-edge', the point, at which homes and businesses connect to the network, where people are using energy, aiming to find a solution for a greener, smarter and more flexible and fair electricity system.

The local power infrastructure is at capacity in Oxfordshire. Scottish and Southern Electric Power Distribution Plc, as lead of the project consortia with EDF Energy, Nuvve, Open Utility, University of Oxford, Oxford City Council, Oxfordshire County Council and Oxford Brookes University and Low Carbon Hub, set out to demonstrate a neutral 'Market Integrator' that provides information to allow a local energy market place to function effectively with the existing infrastructure. They aimed to develop and deploy market platforms for local energy trading and services provision. Up to 90 plug-in projects (solar, heat networks, smart neighbourhoods, micro-grids, electrical vehicles (EVs) and transport hubs, new housing developments and large flexible loads) have been completed, demonstrating new models for investment. LEO has continually improved through the collection and analysis of data to provide insights to the team and to inform the county's energy strategy. This strong evidence base would enable rapid rollout across the UK and beyond (UKRI Project LEO). The project explored three key areas.

Technology: such as installing monitors on low-voltage networks to understand what needs to happen to make them ready for a change to a smart local energy system and using advanced geographical mapping tools to help inform decision making. This also includes developing the IT systems to model the electricity network in near to real time, enabling a market in flexibility.

New models and markets: exploring new products and services that create commercial opportunities for industry, as well as benefits to customers by enabling the fair trading of energy flexibility services and capacity.

Society: working within local communities to trial different approaches to harnessing the potential of energy users in their area to deliver flexibility at the grid edge.

The trial project was completed in March 2023, with its learnings about smart local grids being shared nationally and internationally. It has demonstrated that flexibility services will play a key role in meeting zero-carbon targets. The learning from Project LEO will not only benefit Oxfordshire, but also help accelerate a journey to smart local energy systems elsewhere (UKRI, n.d.).

Battery technology will play an increasingly important role in balancing national electrical grids with large renewable installations, as it can enable all the services to be provided – frequency regulation, peak shaving, demand response, creating new business opportunities. Batteries are also a key tool in building smaller, localised versions of the power grid. In the US, grid scale battery storage is gaining momentum. In 2024, the US installed 12.3 gigawatts of energy storage (Irfan, 2025). New grid battery installations in 2025 are on track to almost double compared with the year before. Battery storage capacity now exceeds pumped hydro capacity, totalling more than 26 gigawatts. These localised energy storage installations with microgrids can power remote communities with reliable power and one day shift the entire power grid into a more decentralised system that can better withstand disruptions such as extreme weather. True grid-scale battery storage in combination with wind or solar farms is not just an enabler of clean energy, but a way to upgrade the power system for a new era.

5.4. Sustainability design matrices

Sustainability design matrices are tools used to assess and manage sustainability performance of designs or projects by considering environmental, social and economic impacts. They help to identify potential areas of improvement and guide decision making throughout the design process (Sánchez-Carracedo *et al.*, 2020). These matrices can range from simple checklists to more complex, multi-dimensional frameworks.

Such matrices can cover a wide range of themes and can be daunting for designers. While BREEAM and LEED are among the most established green building assessment methods in the built environment for a design project, they may be regarded by some as being less focused and less tailored to a particular project or less advanced for a subject area. There has been significant advancement in our understanding of emerging and specialised subjects, such as health and wellbeing, public realm and mobility, for which design matrices have been developed to provide design guidance for decision making. The following table (Table 5.2) summarises the matrices outlined in this section – readers should note that this list is not exhaustive but aims to provide an understanding of the breadth of different available matrices.

Table 5.2 Sustainability design matrices (continued on next page)

Sustainability design matrices	Subject area
Living Building Challenge	The Living Building Challenge defines the most advanced measure of sustainability in the built environment and promotes design to generate 'positive' impact, instead of reducing 'negative' environmental impact of a project. It covers seven 'petals': Place, Water, Energy, Health and Happiness, Materials, Equity and Beauty.
WELL	The WELL Building Standard is a vehicle for buildings and organisations to deliver more thoughtful and intentional spaces that enhance human health and wellbeing, aim to advance human health through design interventions and operational protocols and policies and foster a culture of health and wellbeing.
Active Travel	Mobility, design and assessment tool to drive active travel design provisions. Four rating and scoring tools are introduced: ActiveScore focuses 100% on active lifestyle ratings for built environment. ModeScore evaluates all aspects of mobility, including public transport, electric vehicles, and sustainable logistics, assessing a building's overall connectivity and accessibility, encompassing all modes of transportation. CyclingScore rates buildings according to an official set of standards that determine a building's micro-mobility friendliness. Urban Mobility Scorecard tool helps a city to benchmark and measure key sustainability and mobility goals and their progress.
Flourishing Index	The Flourishing Index focuses on measuring the psychological and social aspects of flourishing, encompassing domains like happiness, health, meaning and relationships, as well as financial stability. The Flourishing Index can be used to measure and quantitively demonstrate social impact of urban development.
Urban Greening Factor	Urban Greening Factor (UGF) as part of the Green Infrastructure Framework developed by Natural England, is a planning tool to improve the provision of green infrastructure particularly in urban areas. It is currently voluntary and can be used to increase urban greening and contribute to Biodiversity Net Gain (BNG). Urban Greening Factor is a tool to evaluate the quality and quantity of urban greening. It enables major developments to demonstrate how they have included urban greening as a fundamental element of site and building design.

(*Continued*)

Table 5.2 (Continued)

Sustainability design matrices	Subject area
SITES	SITES is a rating system that guides, evaluates and certifies a project's sustainability in the planning, design, construction and management of landscapes and other outdoor spaces. SITES supports landscape architects, planners and others in implementing nature-based solutions. SITES projects enhance biodiversity and mitigate climate change, while conserving resources, improving public health and protecting critical ecosystems.
BREEAM Infrastructure	BREEAM Infrastructure is the evidence-based sustainability assessment, rating and awards scheme for civil engineering, infrastructure, landscaping and public realm projects. It was formerly called CEEQUAL, launched in 2003, originally developed by a team led by the Institution of Civil Engineers (ICE) with financial support from the UK Government. CEEQUAL became a world-leading benchmark for high performing infrastructure. In 2015, BRE acquired CEEQUAL and it has become part of BREEAM suite of tools, bringing together all sustainability assessment methods for buildings, masterplanning and infrastructure.
Sponge City	The Sponge City Programme promotes permeable pavements, green infrastructure to address challenges of flood protection, access to clean water, and pollution of receiving water bodies as a result of China's rapid urban growth, exacerbated by climate change impacts. The Sponge City Programme includes an implementation framework that enables local government bodies to adopt Sponge City measures and set up design guidance and targets in relation to rainwater run off and pollution control based upon the regional conditions.
WiredScore, SmartScore	WiredScore is the global digital connectivity rating system for real estate. WiredScore works with landlords and developers to understand, benchmark, improve and promote the digital connectivity in their buildings and developments. Launched by WiredScore in 2021, SmartScore is the global smart building certification that evaluates, improves, benchmarks and promotes cutting-edge smart technology in the built world. SmartScore identifies best in class smart buildings that deliver an exceptional user experience, drive cost efficiency, meet high standards of sustainability and are fully future-proof.
UK Net Zero Carbon Building Standard	Net Zero Carbon Buildings Standard is the UK's cross- industry initiative to bring together Net Zero Carbon requirements for all major building types for the UK built environment, including the majority of common building typologies for both existing and new buildings (e.g. Homes, Offices, Education, Industry, Retail, Hotels, Healthcare etc.). The approach taken is evidence-based assessment to set the science-based targets for buildings standard. Adoption of the Standard is currently anticipated to be voluntary.

5.4.1 Living Building Challenge (zero carbon, zero energy, zero water)

The Living Building Challenge is an attempt to dramatically raise the bar from a paradigm of doing less harm to one in which we view our role as a steward and co-creator of a true Living Future. The Challenge defines the most advanced measure of sustainability in the built environment today and acts to rapidly diminish the gap between current limits and the end-game positive solutions sought. Figure 5.11 indicates the Living Building Challenge philosophy.

As the Challenge (Living Future, n.d.) states: 'The Living Building Challenge is a philosophy, certification, and advocacy tool for projects to move beyond merely being less bad and to become truly regenerative.'

The Challenge aims to transform how we think about every single act of design and construction as an opportunity to positively impact the greater community of life and the cultural fabric of our human communities.

The Living Building Challenge consists of seven performance categories, or 'Petals'. Each Petal is subdivided into Imperatives, for a total of 20 Imperatives in the Challenge (see Table 5.3). The Imperatives can be applied to every building project, of any scale and any location. It is versatile and applies to different project scopes, or Typologies: New Building, Existing Building, Interior, and Landscape or Infrastructure (this typology is for any project that does not include an enclosed structure as part of its primary programme. Projects may be parks, roads, bridges, plazas, sports facilities or trails).

Figure 5.11 Living Building Challenge philosophy (reproduced with kind permission of Living Building Challenge 4.0. Copyright International Living Future Institute)

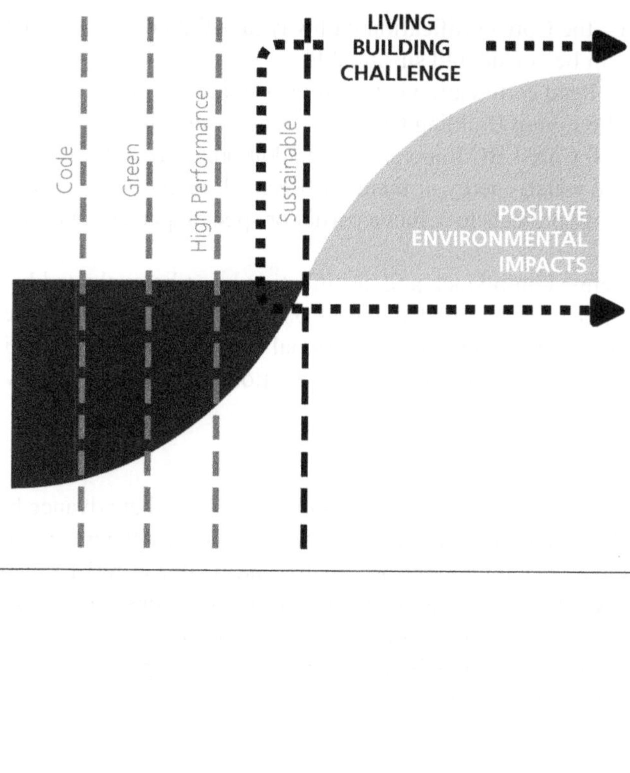

Table 5.3 Living Building Challenge petals and imperatives

7 Petals	20 Imperatives	
Petals	Imperatives	
Place	1 Ecology of Place 3 Habitat Exchange	2 Urban Agriculture 4 Human Scaled Living
Water	5 Responsible Water Use	6 Net Positive Water
Energy	7 Energy and Carbon Reduction	8 Net Positive Energy
Health and Happiness	9 Healthy Interior Environment. 11 Access to Nature	10 Healthy Interior Performance
Materials	12 Responsible Materials 14 Responsible Sourcing 16 Net Positive Waste	13 Red List 15 Living Economy Sourcing
Equity	17 Universal Access	18 Inclusion
Beauty	19 Beauty and Biophilia	20 Education and Inspiration

LBC 4.1 is the seventh version of the Living Building Challenge Standard. The Living Building Challenge is an ever-evolving programme shaped by the experiences and feedback from the industry. Living Building Challenge offers three main certification options.

- Living Certification – living certification is for projects striving for the highest level of health, equity, sustainability and regenerative design. A project achieves Living certification by attaining all Imperatives assigned to its relevant typology. All 20 Imperatives are required for New Buildings.
- Core Certification – the Core certification pathway lays the holistic foundation for building performance that can be considered sustainable.
- Petal Certification – Petal certification is for projects that want to dive deeply into one or more areas, or Petals of the Living Building Challenge. This certification requires, at minimum, the achievement of all 10 Core LBC Imperatives, in addition to the achievement of all Imperatives in one or more of the water, energy or materials Petals. It offers Petal and Zero Energy certifications, allowing projects to achieve partial or specific performance targets.

Living Building Challenge compliance and certification must be evaluated by a third party, must show proven holistic performance, and is based on actual, rather than modelled or anticipated performance. Therefore, projects must be operational for at least 12 consecutive months prior to audit to verify Imperative compliance. Chapter 7, Section 7.4 provides the Living Certified Santa Monica City Hall Building as a case study example.

5.4.2 WELL

WELL aspires to transform buildings and organisations in ways that advance health and wellbeing to help people thrive. It is the first standard focused exclusively on the impact buildings, and nearly everything in them, have on human comfort, health and wellness. WELL was launched in 2014 based on an evolving evidence-based understanding of the relationship between the physical environment and human health. The WELL Standard is a library of over 500 evidence-based, design, policy and operational strategies that, when implemented, can improve the health and wellbeing of people and communities (WELL, n.d.a).

WELL certification is administered by the International WELL Building Institute (IWBI). The WELL Building Standard version 2, WELL v2 (WELL, n.d.b), includes a set of strategies, backed by the latest scientific research, that aim to advance human health through design interventions and operational protocols and policies and foster a culture of health and wellbeing.

The standard is based on scientific and medical research that has found an inextricable link between the built environment and the health and wellness of its occupants. Considering that most humans now spend more than 90% of their time in some type of building, it makes sense that the built environment has a great impact on health and wellbeing. As the IWBI describes it, WELL is a 'performance-based system for measuring, certifying, and monitoring features of the built environment that impact human health and wellbeing, through air, water, nourishment, light, fitness, comfort, and mind' (WELL, n.d.b).

The Green Business Certification Inc. (GBCI) manages and administers third-party certification for the WELL Building Standard, including oversight and training of WELL assessors, who perform the third-party review of WELL certification on behalf of GBCI. Every WELL project is verified through on-site testing of building performance. This practice is fundamental to high-performing buildings. The process for on-site assessments and testing is called Performance Verification completed by an authorised WELL Performance Testing Agent, who usually spend 1–3 days in the building to validate the project's documentation and complete a series of performance tests, spot-checks and measurements.

WELL v2 certification can offer two options, determined primarily by ownership type: owner-occupied, or WELL Core, used when the project owner occupies a small portion of the project area and rents/leases most of the space to one or more tenants.

There are a number of different performance categories that the standard is assessed against, shown in Figure 5.12. These include air, water, nourishment, light, fitness, comfort and mind. Each category is assessed against a range of possible features. About 15% are precondition features, which must be achieved. The remaining features are optional. Depending upon the number of points obtained for each feature, projects achieve WELL Silver, Gold or Platinum rating.

Figure 5.12 Ten concepts in WELL v2 (created using information from WELL v2 (WELL, n.d.b)

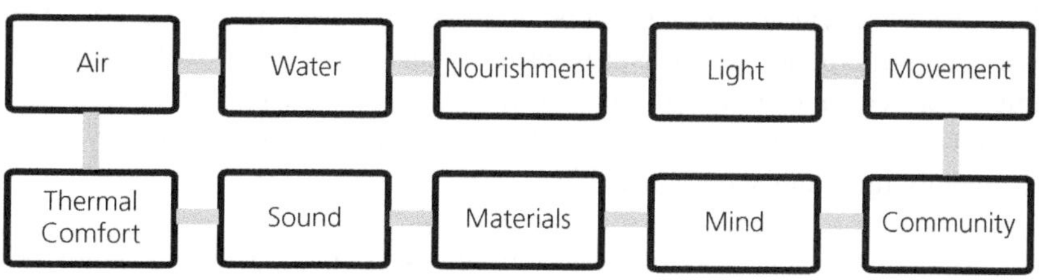

5.4.3 Active Travel

Transport accounts for 26% of carbon dioxide emissions in the UK and contributes 14% of PM2.5 - fine particulate matter (2.5 microns or smaller) that can penetrate deep into the lungs (Department of Transport, 2023). Active mobility includes many forms: cycling, running, walking, wheeling using mobility aids, along with emerging modes like e-bikes, e-scooters and scooters. It is good for the environment and good for people's health and wellbeing. According to research, active commuters take fewer sick days,1.3 days less per worker annually (Hendriksen *et al.*, 2010). In a 2019 Cyclescheme study, 82% reported feeling less stressed at work (Bikedock, 2023). This finding highlights a potential positive impact of cycling on employee wellbeing, suggesting that it can contribute to reducing stress levels in the workplace. Daily physical activity is widely recognised to lower the risk of depression by up to 30% (Public Health England, 2020) along with a range of health benefits, such as reducing the risk of disease, managing existing conditions, and developing and maintaining physical and mental function.

Travel patterns have changed significantly especially since Covid. Modes of transportation have also evolved. Work from home and hybrid work has become accepted as a viable business model post Covid. There have been fewer cars on the roads on certain working days. Ride sharing services such as Uber continue to evolve, and is another driver reducing the need for parking spaces in the city. Peak building occupancy rates may still be high, but average occupancy rates over any given work week are lower. There is now a large focus on EV cars or motorcycles and EV chargers across many countries, which will reduce direct pollution across our cities and neighbourhoods from the traditional fossil fuel vehicles. Four related rating systems are introduced here: ActiveScore, ModeScore, CyclingScore and Urban Mobility Scorecard tool.

5.4.3.1 ActiveScore

ActiveScore (Activescore, n.d.) focuses 100% on active lifestyle ratings for the built environment. It originated from a need within real estate, as landlords, developers and agents began to believe that health and wellbeing facilities mattered almost as much to occupiers as rent and internet connectivity.

Launched in 2017 by Active Travel Partners Ltd., ActiveScore Certification is the definitive accreditation for Active Travel provision in real estate. The scheme assesses and rates buildings using a transparent, scientific methodology developed in partnership with urban planners and experts from the Active Travel and cycling industry. An ActiveScore rating communicates a building's future-proofed credentials to prospective tenants, partners and occupiers, offering world-class facilities and highlighting sustainable design, tenant wellbeing as well as ESG standards.

ActiveScore assessment includes a building-specific Clean Air Calculator that highlights the average quantities of carbon dioxide and PM2.5 emitted per mode of local transport. This calculation provides a baseline from which the total Scope 3 emissions saved by Active Travel can be modelled. The assessment methodology measures three aspects of Active Travel: infrastructure, occupier engagement services and future proofing. Ratings are designed to cover from an entry-level Certified, Silver, Gold, Platinum, to a top class Platinum 100. The ActiveScore system is designed to encourage amenity upgrades where possible.

Although the certification is a stand-alone solution, the assessment criteria benchmark against the other major certifications that touch on the subject of active travel (WELL, BREEAM, BCO etc.). ActiveScore recommendations can help buildings earn credits for other green certifications such as WELL, BREEAM, LEED and GRESB. ActiveScore goes beyond other green building certifications' sustainable transport requirements and encourages clients to consider other forms of active travel beyond cycling (Tellvik, 2024). More than 250 projects worldwide are ActiveScore certified. A range of real estate companies have adopted the scheme including Lendlease, Brookfield Properties, GPE and Bruntwood.

5.4.3.2 ModeScore

ModeScore expands on ActiveScore by including all modes of transportation and promoting wider sustainability efforts (Modescore, n.d.). ModeScore is a certification that rates buildings on their surrounding environment, infrastructure and transport services. It is a more comprehensive approach than that of ActiveScore as ActiveScore focuses specifically on facilities and services supporting active travel including walking and cycling. ModeScore incorporates ActiveScore within its assessment criteria, meaning a building that is ModeScore certified will also have a strong ActiveScore rating.

ModeScore evaluates all aspects of mobility, including public transport, electric vehicles and sustainable logistics, assessing a building's overall connectivity and accessibility, encompassing all modes of transportation (Modescore, n.d.). Its goal is to assess and certify the sustainability of transport facilities and services, promoting a wider range of sustainable transportation options, helping developers and investors integrate multi-modal transport into their projects, improving ESG performance (Tellvik, 2024). The rating process is relatively simple and scalable (Modescore, n.d.).

5.4.3.3 CyclingScore

CyclingScore (Cyclingscore, n.d.) rates buildings according to an official set of standards that determine a building's micro-mobility friendliness. These standards were developed in conjunction with industry professionals, travel planners, tenants and executives in cycle/micro-mobility orientated businesses. CyclingScore Certification demonstrates how actively commuter-friendly a building is.

Landlords of commercial offices and residential property developers choose to be certified by CyclingScore because it propels their assets in the market and makes them highly sought after for their high-quality active commuter facilities to prospective clients, as well as improving their overall cycle/micro-mobility friendliness.

All buildings are eligible for CyclingScore certification. The certification follows a simple two-step process, with ratings from CyclingScore Certified, Silver, Gold to Platinum.

5.4.3.4 Urban mobility scorecard tool

The mobility matrices above are at the scale of a single building, which is helpful but limited. To help cities benchmark and measure key sustainability and mobility goals and their progress, the Global New Mobility Coalition (GNMC) – an initiative from the World Economic Forum – has developed the Urban Mobility Scorecard tool in collaboration with three test cities (World Economic Forum, 2023). Produced in deep collaboration with the public and private sectors, the mobility tool aims to help cities benchmark their journey towards sustainable mobility. Through assessments covering key issues in urban mobility, the mobility tool allows cities to understand their strengths and weaknesses, recognise good practice and identify areas for action. The mobility tool aims to provide resources and options to set and inspire actions on new policies in cities to advance the transition to sustainable, inclusive mobility.

This initiative covers stakeholders from across sectors – including shared mobility, electrification, active mobility and payments providers – as well as several mobility NGOs. The Urban Mobility Scorecard tool investigates urban mobility through three pillars.

- Governance. The tool explores how to create more nimble, adaptable and future-ready urban mobility ecosystems through institutional coordination (planning, financing and resource allocation), regulation (environmental, urban vehicle access and data management) and innovation (shared mobility, pilot projects and processes, and electrification).

- Resilience. This pillar investigates how cities allocate space for different uses (such as on-street parking) and how mobility fits into those planning efforts. This pillar comprises issues of space allocation (for cycling, micro-mobility, walking and public/shared modes), as well as kerbside management issues (parking and delivery logistics management).
- Connectivity. This pillar investigates issues including the integration of mobility (multimodal integration, mobility payments and ticketing and informal transport), as well as looking at safe and equitable access.

The Scorecard tool was tested by one of the cities, Curridabat in San José, Costa Rica. The city has ambitions to take its leadership on nature-based solutions even further by embedding biodiversity in its plans for walking and cycling schemes, while doubling pedestrian paths in the next 4 years. The Scorecard Tool has incorporated the recommendations from the private sector, public sector and NGOs. It has found that a number of wider structural challenges need to be addressed by cities and companies alike to achieve a rapid transition to more sustainable mobility. To see a transition to cities where mobility is sustainable, equitable and innovative, action is needed on several fronts. Fundamental issues remain to be addressed to see a step-change in action in urban mobility such as investment.

5.4.4 Flourishing Index

Urban planning is deeply implicated in both the planetary crisis of climate change and the personal crises of unhealthy lifestyles. The *Routledge Handbook of Planning for Health and Well-Being* (Barton *et al.*, 2015) authoritatively and comprehensively integrates health into planning, strengthening the hands of those who argue and plan for healthy environments including compelling case studies from four continents. Barton (2016)'s *City of Well-being* research into variations in active travel shows that different suburban forms and locations can support as much as 60% active travel or as little as 20%, challenging design professionals to think differently about masterplanning.

Going beyond physical health, *Creative Health: The Arts for Health and Wellbeing*, a UK cross-party report (APPG on Arts Health and Wellbeing, 2017), recorded that 85% of people in England agree that the quality of the built environment influences the way that they feel.

'On the promotion of human flourishing', by Professor VanderWeele, presented in the *Proceedings of the National Academy of Sciences* (VanderWeele, 2017), is a measurement approach to human flourishing, based around five central domains: happiness and life satisfaction, physical and mental health, meaning and purpose, character and virtue, and close social relationships. The programme aims to encourage the measurement and tracking of these various aspects of flourishing in workplace, medical, educational and governmental settings, and to better understand the determinants of flourishing, and policies to promote it.

The Flourishing Index focuses on measuring the psychological and social aspects of flourishing based on some empirical validation that are widely used in the wellbeing literature, encompassing domains like happiness, health, meaning and relationships, as well as financial stability. In its most simple form, flourishing is about human experiences of feeling good and life going well for both individuals and communities.

As discussed in Chapter 4, it is often difficult to measure the success and impact of such programmes to promote flourishing and social value. This is the main aim of the innovation programme adopted by Brent Cross Town Development, shown in Figure 5.13. Brent Cross Town Flourishing Index (Brent Cross Town, n.d.) measures flourishing at a community level as well as at an individual level. The Index sets out a comprehensive approach to mapping the present and potential strength of an area so that the social impact of regeneration can be quantitively demonstrated. In order to establish a robust quantitative baseline data assessment, a mix of research methods were used including conducting face to face and online surveys as well as movement data collected from using the latest sensor technology. The baseline work provides several opportunities for innovative 'before and after' analyses, going beyond Post Occupancy Evaluation. The innovative programme received research funding and was carried out by the University of Manchester in partnership with Buro Happold, commissioned by Related Argent and Barnet Council.

The Flourishing Index provides comprehensive, data-driven insights that pinpoint what is and is not working pre-regeneration. It also enables targeted interventions to create sustainable, healthy and equitable places, and measures to capture and demonstrate their value.

While the Flourishing Index acknowledges the importance of the environment and social contexts in which individuals live, it does not explicitly measure experiences. A positive and stimulating environment can contribute to overall wellbeing and potentially influence one's experience of flourishing. 5-sensory design is a design approach that consciously considers and incorporates all five senses (sight, sound, smell, taste and touch) to create a more holistic and impactful experience. It is often used in architecture, and seeks to enhance the overall user experience of a space – for example, a well-designed workplace that considers sensory inputs (like lighting, acoustics and smell) may foster a more positive and productive environment, which could indirectly impact an individual's sense of flourishing. As the workplace continues to evolve, a strong

Figure 5.13 Brent Cross Town (photo credit to © Visualisation One, courtesy of BXS Limited Partnership)

case is emerging for a sensory palette framework that supports a systems approach to building environmental design, enabling the integration of people's multi-modal sensory responses across different environmental settings. Clements-Croome *et al.* (2019) promoted the application of the established conceptual theories of 'Flourish' to move beyond comfort when designing the interiors and comfort controls towards a sensory impacts framework that considers a whole life costing (Clements-Croome *et al.*, 2019). Technological advances, in the form of wearables that monitor our physiological and stress responses offer the opportunity to capture empirical data, further enabling the investigation to see how a diverse range of environmental settings affect our physical, mental and social wellbeing.

5.4.5 Urban Greening Factor

Nature-based design essentially considers the ecosystems of development areas, with the cumulative and compounded benefits being delivered to biodiversity and human wellbeing. The IUCN (International Union for Conservation of Nature) Global Standard for Nature-based Solutions guides users to design and implement nature-based solutions in a way that allows nature to deliver its valuable ecosystem services including protection, management and restoration as well as infrastructure while offering human wellbeing and biodiversity benefits (IUCN, 2020a). In addition, the Global Standard sets clear benchmarks to measure the progress of nature-based solutions (IUCN, 2020b).

Significant progress has been made in quantifying the benefit and measuring design of built environments that harness green infrastructure. Green infrastructure is a network of multi-functional green and blue spaces and other natural features, urban and rural, which is capable of delivering a wide range of environmental, economic, health and wellbeing benefits for nature, climate, local and wider communities and prosperity (MHCLG, 2024).

Green infrastructure provides an opportunity to provide linkages into development-scale regulatory requirements and spatial landscape and nature network-scale planning. At a development scale, integrating the wider picture is vital, yet defining the criteria for good design and measuring success has been less well studied until recently.

Natural England, the UK Government's official adviser on the natural environment in England has developed principles and standards in the form of a Green Infrastructure Framework for England (Natural England, 2023). The Urban Greening Factor (UGF) for England, as one of a suite of five headline green infrastructure standards within the framework, is a planning tool to improve the provision of green infrastructure particularly in urban areas. It is currently voluntary and can be used to increase urban greening and contribute to recently implemented planning regulations to ensure development-scale biodiversity net gain (BNG). By enabling developers to demonstrate how they have included urban greening as a fundamental element of site and building design, the UGF is an effective tool to evaluate the quality and quantity of urban greening.

A concise set of UGF Target Scores and UGF Surface Cover Types for the model UGF for England was provided (Natural England, 2023). The scores for Surface Cover Types comprise a familiar set of green infrastructure elements used in development including vegetation and tree planting; green roofs and walls; sustainable drainage systems and water features; and paved surfaces. The UGF is designed to provide flexibility to enable an appropriate response to meeting local green infrastructure priorities, site specific constraints as well as to respond to different uses within a development.

The guidance with a schedule of UGF policies are currently used in UK planning practice, although it is voluntary in many authorities. The London Plan 2021 Policy G5 Urban Greening Factor guidance (Greater London Authority, 2023) requires applicants to refer to a borough's Green Infrastructure Strategy and its associated plans, which should be used to understand a site's context and to determine its current and potential role within the wider green infrastructure network. UGF factor is calculated as the weighted factor scores considering surface cover type, multiplied by the area of corresponding surface cover type to determine the scheme's UGF score.

Examples of how an urban greening scheme could demonstrate as being locally appropriate include the following.

- Providing green routes that promote active travel.
- Designing sustainable urban drainage systems (SuDS), which reduce surface water run-off into local water courses.
- Taking design cues from local habitat types or the Local Nature Recovery Strategy or Biodiversity Action Plan.
- Designing publicly accessible open space to reduce local deficiencies in access to open space.
- Providing other types of publicly accessible greening (such as small 'pocket' parks) where a reduction in deficiency to open space cannot be achieved.

UGF can contribute to delivery of Biodiversity Net Gain (BNG) on a development site. BNG is a way of creating and improving natural habitats and makes sure development has a measurably positive impact ('net gain') on biodiversity, compared with what was there before development.

BNG has been recently introduced into planning legislation and it is hoped that it will also pave the way for creating markets for green infrastructure and biodiversity investment. This facilitates the markets for adaptation outcomes across relevant legislation and policy programmes, including initiatives on carbon market integrity and environmental land management schemes, which provide incentives to farmers and other land managers for beneficial actions.

For developments with a low baseline level of biodiversity, an ecologically informed approach can create new areas of wildlife habitat to deliver biodiversity gains alongside other green infrastructure benefits. New habitats should be designed and managed to contribute to local biodiversity priorities and the Local Nature Recovery Strategy, if one has been produced, ensuring that the cumulative benefits across a wider area can be realised.

Where protected species or priority habitats or species are found on a development site or a proposed development may impact an already-identified Site of Importance for Nature Conservation, the impacts on biodiversity should be managed in addition to any relevant UGF targets. In these cases, a priority for any new urban greening should be to help avoid or mitigate the impacts of the proposed development on biodiversity as well as to provide locally relevant greening that complements the site's existing wildlife value and will achieve an overall gain in biodiversity.

5.4.6 Sustainable SITES

Sustainable SITES (SITES) (Sustainable SITES Initiative, n.d.) is a rating system that guides, evaluates and certifies a project's sustainability in the planning, design, construction and management of landscapes and other outdoor spaces. It was launched in 2009. The SITES certification is administered by Green Business Certification Inc. (GBCI). GBCI is the organisation responsible for providing third-party project certification to projects meeting the SITES v2 Rating System requirements, which was released in 2015.

SITES supports landscape architects, planners and others in implementing nature-based solutions. SITES projects enhance biodiversity and mitigate climate change, while conserving resources, improving public health and protecting critical ecosystems. The SITES Rating System is a 200-point system with four certification levels of achievement (projects must meet a minimum of 70 points to achieve certification). It has four levels: Certified, Silver, Gold and Platinum. Projects earn points based on their performance and the implementation of sustainable practices, with each level requiring a specific number of points. The 18 prerequisites within the rating system are required, plus the minimum number of credits to achieve the required certification level.

SITES certification is for projects located on sites with or without buildings – ranging from parks to corporate campuses, botanic gardens to universities, and more

- open spaces – local, state and national parks; botanic gardens; arboretums
- streetscapes and plazas
- commercial – retail and office areas; corporate campuses
- residential – neighbourhoods or individual yards
- educational/institutional – public and private campuses; museums; hospitals
- infrastructure
- government
- military
- industrial.

LEED integrated SITES principles and practices into the LEED rating system, effectively incorporating SITES into the LEED framework. For sites that include buildings, the SITES v2 Rating System focuses on the area from the building skin outwards. SITES promotes strategies such as protecting and restoring native habitats, restoring soil, and increasing key vegetation cover help to cool cities and communities, conserving water and promoting green infrastructure, while LEED applies to buildings and also considers the impact of the site on which the building is located. The SITES programme supports and encourages projects that plan to pursue simultaneous certification with LEED and the SITES v2 Rating System. SITES website (Sustainable SITES Initiative, n.d.) provides the synergies and potential equivalences between the LEED and SITES rating systems.

5.4.7 BREEAM infrastructure

BREEAM Infrastructure (BRE, n.d.a) is the evidence-based sustainability assessment, rating and awards scheme for civil engineering, infrastructure, landscaping and public realm projects. It was formerly called CEEQUAL, launched in 2003, originally developed by a team led by the Institution of Civil Engineers (ICE) with financial support from the UK Government. CEEQUAL became a world-leading benchmark for high performing infrastructure.

One of the CEEQUAL certified projects is Greening the 2012 Olympic Park, which was awarded CEEQUAL 2012 'Excellent Whole Project Award' – a prestigious civil engineering sustainability accolade (BRE, n.d.b). With London 2012 billed as the 'greenest Games ever', meeting sustainability targets was vital. Arup worked closely with the Olympic Delivery Authority to create the South Olympic Park Landscape and Public Realm (shown in Figure 5.14), transforming a derelict brownfield site into a stunning setting for Olympic competitors, spectators and Londoners.

Figure 5.14 London Olympic Park (photo credit to Mike Fleming licensed under CC BY-SA 2.0)

For the Olympic Park infrastructure and public realm, Arup pioneered solutions to achieve key sustainability targets, with design and specifications that included the reuse of 90% of demolition material and over 900 tonnes of pulverised fuel ash.

In 2015, BRE acquired CEEQUAL and it has become part of BREEAM suite of tools, bringing together all sustainability assessment methods for buildings, masterplanning and infrastructure. The scheme covers nine categories.

(*a*) Management.
(*b*) Resilience.
(*c*) Communities and stakeholders.
(*d*) Land use and ecology.
(*e*) Landscape and historic environment.
(*f*) Pollution.
(*g*) Resources.
(*h*) Transport.
(*i*) Innovation.

BREEAM Infrastructure is a self-assessment process, carried out by a trained BREEAM Infrastructure assessor. Using a manual and an online assessment platform, the assessor records the scores and evidence. It is then externally verified by a BREEAM Infrastructure-appointed verifier. The result is a ratified BREEAM Infrastructure score and rating, and the certificate is awarded to all project partners. There are two key differences from the BREEAM for buildings in that an assessor first determines assessment scope for a project and seeks agreement with a verifier. The independent verifier then verifies the assessment in accordance with the BREEAM Infrastructure procedures based on the assessor's submission before submitting the assessment to BRE for ratification.

BREEAM Infrastructure enables the achievement of high environmental, social and governance performance for infrastructure. The UK's new rail scheme, High Speed Rail 2, is one of the BREEAM Infrastructure pilot schemes currently ongoing, with strategy and design having been certified and construction ongoing (Watkins, 2022).

5.4.8 Sponge City

China has experienced extreme flooding catastrophes with significant economic damages in recent years. The Ministry of Housing and Urban-Rural Development launched the 'Sponge City Programme' in 2013 and issued a set of technical guidelines, aimed at ensuring that 70% of surface runoff be captured in place. The Sponge City guidelines provide green infrastructure strategies and calculation methods for city developments, including strategically implementing rain gardens, ponds, green roof/walls to achieve ecological sustainability. The programme promotes the wider adoption of measures such as permeable pavements, green infrastructure such as rain gardens, parks and wetlands, green roofs and rainwater reuse facilities.

The programme was developed in response to China's challenges on flood protection, access to clean water, and pollution of receiving water bodies as a result of China's rapid urban growth, exacerbated by climate change impacts. Professor Yu Kongjian, Dean of Beijing University's College of Architecture and Landscape, played an instrumental role to establish and promote the Sponge City concept of managing floods in China (Global Voices Brazil, 2025).

The Sponge City Programme has many similarities with equivalent international schemes including the following.

- UK: Sustainable Urban Drainage Systems (SuDS). In the UK, planning policy and Environment Agency requires the volume of attenuation storage to mitigate the 1 in 100-year storm event plus up to an approximately 45% allowance (depending on locations and conditions) for climate change in addition to a limit of discharge surface water runoff rate to a public sewer.
- USA: Best Management Practices (BMP), Green Infrastructure (GI) and Low Impact Development (LID).
- Australia: Water Sensitive Urban Design (WSUD).

Table 5.4 shows a comparison of different approaches based on the criteria provided by Shao *et al.* (2019)'s paper on '*Characteristics and effectiveness of water management methods for "sponge cities" in China*'. For more detailed comparison and information, Fletcher *et al.* (2014) described the difference in these urban drainage design concepts of SuDS, LID, BMPs, WSUD, GI and LIUDD and their evolution and applications.

The Sponge City Programme provides an implementation framework that China's national government has put in place that enables local government bodies to establish basic regulations, standards and targets, and funding/financing arrangements. There are many pilot cities where the Sponge City Programme has been made mandatory as planning requirements for designing drainage and attenuation systems. Sponge City guidelines provide specific design guidance and targets identified in relation to run off and pollution control based upon the regional conditions. Pilot cities adopting Sponge City programme cover cities and regions across China including Beijing, Tianjin, Xi'An, Chongqing, Naning, Shenzhen, Wuhan, Zhenjiang, Ningbo and Qindao (Shao *et al.*, 2019).

Table 5.4 Comparison of technical characteristics of major rainwater management systems (created using information from Shao *et al.* (2019))

Design concepts	Water quantity control	Water quality control	Small scale (<1 km^2)	Large scale (<1 km^2)	Top design	Source control	Process control	Terminal control	Compre-hensive manage-ment	Incor-porated into law
BMPs		Y	Y	Y	Y	Y	Y			Y
LID	Y	Y	Y			Y	Y			Y
GI	Y	Y	Y	Y	Y	Y	Y			Y
SuDs	Y			Y	Y		Y	Y		Y
WSUD	Y	Y		Y	Y	Y	Y		Y	Y
LIUDD	Y	Y		Y	Y	Y	Y			Y
Sponge City construction	Y	Y	Y	Y	Y	Y	Y	Y	Y	

The Sponge Cities Policy, in relation to stormwater capture, is based upon capturing a target percentage of the annual run-off. This target is set for each region based on the regional conditions – for example, if a development is in Sanya, this percentage target may be defined at approximately 65%, with a corresponding rainfall depth, which means 65% of the annual rainfall volume should be captured within the boundary of the development. The stormwater drainage solution for the development should therefore be capable of capturing and storing the first 65% of each rainfall event. During more extreme events, during which rainfall will exceed the rainfall depth, the stormwater may bypass the storage features by way of overflow systems.

LEED rainwater management credits use a different assessment methodology from the Sponge Cities – for example, 3 LEED points are available if the 90th percentile rainfall event is captured and retained on the site. However, the design measures promoted by the Sponge City would help attain the LEED points.

It should be noted that the Sponge City guidelines require measures that deal with 'low order' (high frequency) events and they include weighting of historic rainfall data to reflect rainfall experienced during recent years. A more extreme climate change scenario is not considered within the Sponge City design guidance. Designers may wish to apply a climate change uplift (e.g. 20%) to the Sponge City defined target rainfall depth, to provide extra resilience to flooding. This would result in an increased storage volume or more area of passive surface water storage.

Lingang Sponge City is an example, where the 79 km^2 Sponge City pilot project zone was implemented to capture, store, absorb and reuse rainwater within the neighbourhoods of the Pudong New Area of Shanghai, near the East China Sea (Yang, 2019). The 'renovated' sponge park included the following key measures.

- Replacing roads with permeable pavements.
- Building green roofs.
- Constructing new parks, wetlands and lakes.

- Planting within pools for water treatment to enable re-use of rainwater for car washing.
- A smart management system used to manage the stored rainfall and respond ahead of rainfall events to lower water levels.

Another prime example of a sponge city project is Tianjin's Qiaoyuan Park (shown in Figure 5.15), designed by Turenscape and completed in 2007/2008. Turenscape was founded by Professor Yu Kongjian, one of China's most prominent urban design thinkers and Dean of Beijing University's College of Architecture and Landscape. Tianjin's Qiaoyuan Park uses natural, green infrastructure to manage urban stormwater by absorbing, storing and purifying the water. The park was one of the early Sponge City projects and demonstrates the principles of using nature-based solutions to restore damaged urban environments, treat wastewater, mitigate flooding and enhance urban biodiversity (Global Voices Brazil, 2025).

Figure 5.15 Tianjin Qiaoyuan Park Sponge City (photo credit to Mydogistiaotiaohu licensed under CC BY-SA 4.0)

5.4.9 WiredScore, SmartScore

WiredScore is the global digital connectivity rating system for real estate. WiredScore works with landlords and developers to understand, benchmark, improve and promote the digital connectivity in their buildings and developments. It is estimated that 800+ million sq ft of real estate is certified by WiredScore. WiredScore can bring benefits of improved design, asset management, enhanced brand and attracting and retaining occupiers or residents.

WiredScore measures a building's resilience of critical tele-communication equipment, occupier experience of digital connectivity, mobile coverage and access, choice of providers in providing high speed fibre and competitive pricing, and future-readiness in their flexibility and capacity to adapt to new and advancing technologies.

The certification runs in six steps, from audit, pre-certify, improve by receiving expert guidance, to certify, under one of the four ratings ranging from WiredScore Certified, WiredScore Silver, WiredScore Gold and WiredScore Platinum. The final two steps are promotion, to communicate the WiredScore to the market, and utilising technical guidance and training throughout the certification period.

Launched by WiredScore in 2021, SmartScore is the global smart building certification that evaluates, improves, benchmarks and promotes cutting-edge smart technology in the built environment. SmartScore was created in collaboration with the owners and users of the world's most advanced properties. SmartScore certification provides clarity on what constitutes a smart building, guidance on how to achieve the status, and proof of the value it adds to a project. Providing a global standard, SmartScore identifies best in class smart buildings that deliver an exceptional user experience, drive cost efficiency, meet high standards of sustainability and are fully future-proof.

SmartScores measure a building's ability to deliver its user functionalities, such as individual and collaborative productivity, health and wellbeing, community and services, sustainability, maintenance and operations, and security. Criteria used for determining the SmartScore certification level in the technological foundations of a smart building include

- occupier digital connectivity
- building systems
- landlord integration network
- governance
- cybersecurity
- data sharing.

Similar to WiredScore, SmartScore certification includes four ratings: SmartScore Certified, SmartScore Silver, SmartScore Gold, SmartScore Platinum. There are 1000 WiredScore or SmartScore accredited professionals who are registered on the WiredScore site across 30 countries around the world (WiredScore, n.d.). WiredScore has also introduced an Accredited Solutions programme that validates technology providers whose systems are aligned with SmartScore certification standards, so that businesses can procure their solutions to achieve SmartScore certification goals.

The EQ building in Bristol, UK became the first building in the world to achieve WiredScore Platinum and SmartScore Gold certification in 2022 (WiredScore, 2021).

Owned and managed by Commercial Estates Group (CEG, n.d.), the 200,000ft^2 office building (shown in Figure 5.16) has achieved SmartScore Gold certification. EQ (EQ, n.d.) uses the latest integration of building systems and data rich technology to improve the user experience of everyone that comes into the building. This includes their health, wellbeing and productivity, improving operational efficiency while minimising environmental impact. The building includes a building app, contactless access, air quality monitoring and real-time energy usage.

5.4.10 UK Net Zero Carbon Building Standard

As described in Chapter 4, the UK Net Zero Carbon Buildings Standard (UKNZCBS, 2025) is the UK's first cross-industry initiative to bring together net zero carbon requirements for all major building types for the UK built environment. The implementation of this standard will have a

Figure 5.16 EQ Bristol (photo credit to CEG)

profound impact on the acceleration of net zero programmes across the world. The approach taken is the world's first balanced way of using top down and bottom up evidence-based assessment to set the science-based targets for a buildings standard. The first version of the standard is expected to be launched in 2026. Although the standard is an industry-led initiative and its adoption is currently anticipated to be voluntary, it is anticipated that it will be likely incorporated in some government policies such as planning or funding requirements.

The standard covers all of the UK's major building sectors, providing a clear set of requirements for buildings to be classified as Net Zero Carbon Aligned and Net Zero Carbon Aligned (plus offset), along with information on how to evidence and report this. It will cover the majority of common building typologies for both existing and new buildings (e.g. homes, offices, education, industry, retail, hotels, healthcare etc.). The Standard will not apply to infrastructure.

Figure 5.17 provides an overview of the three categories of the standard's requirement, where limits mean maximum numerical values that shall not be exceeded, targets mean a minimum numerical value that shall be met or exceeded, and reporting is for publicly reported data collection (UKNZCBS, 2025). It is the intention that the standard will evolve over time to update the limits currently set and include more aspects of limiting values and reporting requirements.

Figure 5.17 The UKNZCBS Standard requirement (reproduced with kind permission of UKNZCBS (UKNZCBS, 2025))

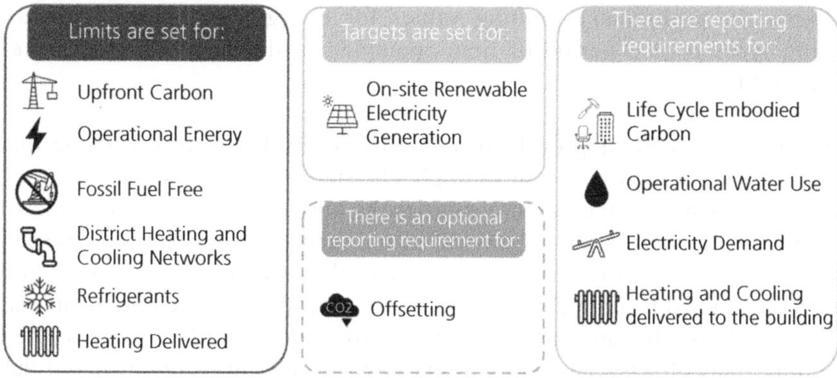

Among the 'limits' categories, upfront embodied carbon and operational energy are perceived as the most stringent requirements and will have the greatest impact on the design flow and process for the built environment.

5.4.10.1 Upfront embodied carbon

Life cycle total greenhouse gas (GHG) emissions and removals are associated with construction products and construction processes, over the whole life cycle of a building, including its disposal. In the NZCBS, these include the full life cycle modules: A0–A5, B1–B5 and C1–C4. The 'upfront' embodied carbon inputs are GHG emissions associated with construction products and construction processes, up to practical completion: life cycle modules: A0–A5.

Table 5.5 shows the preliminary limiting values for upfront carbon for a selection of the building types over a period of time contained in the pilot version of the standard.

The upfront carbon assessment is in line with RICS Professional Standard (RICS, 2024) methodology. A separate set of limiting values is provided for retrofit works.

Table 5.5 An extract of upfront carbon limits within the standard – pilot version (reproduced with kind permission of UKZNCBS (UKZNCBS, 2025))

Typologies	Upfront carbon limits kgCO2e/m2 GIA by year			
	2025	2028	2035	2050
Homes (Flats)	565	450	250	40
Offices	735	625	360	60
Schools (Secondary)	530	450	260	45

5.4.10.2 Operational energy

Energy Use Intensity reports the energy use of the whole building including renewable energy generated and consumed on site – that is, *all* energy used in the building from grid and renewables.

The standard prioritises energy efficiency and eliminating the performance gap by using measured performance data.

Table 5.6 shows the preliminary limiting values for Energy Use Intensity for a selection of the building types over a period of time contained in the pilot version of the standard (UKNZCBS, 2025)*.

The standard also covers the approach to procuring renewable energy and the treatment of residual emissions, including carbon 'offsetting'. The standard requires assessment of building performance based mostly on measured quantities, as opposed to predictive models. As such, conformity with this standard can only be assessed and verified for buildings whose construction has been completed and are fully in-use at the time of the assessment. To claim a building is a 'Net Zero Carbon Aligned Building' according to the standard, all its requirements *shall* be met and verified except those that are carbon offsetting. To claim a building is a 'Net Zero Carbon Aligned Building (plus offsets)' according to the standard, all the requirements of the standard *shall* be met and verified. This provides a level of independent assurance to the claims made at design stage and adds to the credibility of the scheme in the market place, befitting its current voluntary status.

It is envisaged that verification process would include the following, subject to be confirmed by the standard when it is finally published.

(a) Data submission: claimant will collate and submit to the third party verifier/verification team.
(b) Successful verification: claimant lodges verified performance data by way of NZCBS reporting format within the Building Environment Carbon Database (BECD).
(c) Public declaration allowed: accompanying communication materials under development. The Claimant will be able to publicly declare their building in accordance with the standard.
(d) Verification will be offered at the following points.

- First verification – 12 months of standard aligned operation.
- Ongoing verification – typically annually thereafter.

Table 5.6 An extract of operational energy limits within the standard – pilot version (reproduced with kind permission of UKNZCBS (UKNZCBS, 2025))

Typologies	Energy Use Intensity (EUI) kWh/m^2 GIA by year			
	2025	2028	2035	2050
Homes (Flats)	40	39	37	35
Offices	85	77	59	45
Schools (Secondary)	60	57	50	45

* After the editorial process was completed, UKNZCBS formally published the first edition of the Standard in March 2026 (https://8f2d86b0-7c72-4129-a02b-72f5adfae419.filesusr.com/ugd/790941_f53e6b4b6dc04fd3aeb4b122a1b95b15.pdf).

5.5. Design processes and tools

5.5.1 Soft Landings

Traditionally, design and construction projects follow a linear process, with a client management team setting up a project brief and design teams' and contractors' involvement ceasing at the handover stage. This often limits the transfer of knowledge to the management team as well as the opportunity for the design teams to learn from what has historically worked well or poorly. This shortcoming can emerge at any stage of a project's life cycle, from initial briefing, project design, construction, all the way through to the operational stage. The problem is often exacerbated by the separation of construction and management teams, shortened commissioning periods, inadequate training at handover and a lack of feedback throughout the design and construction process.

To address this shortcoming in the construction industry, BSRIA introduced the Soft Landings Framework in 2018 (BSRIA, 2018). This framework is a six-phase approach, as indicated in Figure 5.18, to help the project team focus more on the client's needs throughout the project, including better commissioning to smooth the transition into use and to address issues during post-occupancy evaluation (POE). The Soft Landings approach is designed to give clients and their project teams a delivery process to follow that will lead to a better performing product.

Figure 5.18 Soft Landings' six-phase approach, reproduced with kind permission of BSRIA (BSRIA, 2018)

The Soft Landings Framework aims to bring a change of culture as much as a change of process (BSRIA, n.d.a).

Given the significant value of Soft Landings to a development project, Government Soft Landings (GSL) was published to specifically help government projects enabling a smooth transition from construction to operation. This approach, and its focus on early end-user engagement, helps to realise better public-sector value from the construction process, a key objective of the UK Government. This guidance has the aim of aligning it with the UK BIM Framework (including ISO 19650 and BS 8536) and industry best practice, providing further specific guidance around the 'GSL Champion's role' to offer specific assistance to clients (Philp *et al.*, 2019).

Consideration of the operation of a building in-use must start from the outset of a project. This consideration involves extensive consultation and interaction with the stakeholder working groups. Table 5.7 gives an indication of the key activities recommended under Soft Landings framework using a university development project as an example.

In summary, Soft Landings promotes a whole-life approach to buildings that is embedded throughout the development process. This approach enables performance outcomes, not a set of pre-defined project specifications; it is a 'how to do' as opposed to a 'what to do' framework (BSRIA, n.d.a). Soft Landings comes into its own in closing the loop between design, construction, operation, and feedback into design again.

Table 5.7 An example of key activities at Stage 0/1 strategic brief and feasibility stage

Client	Architect / Design Team
Be an active client. Identify a Soft Landings champion on the client side who will be involved all the way through.	Identify a matching Soft Landings champion on the project team side. Consultation and workshops with the Estates and Facilities when developing the design. Use samples, mock ups, prototypes and exemplar projects to help individuals understand key aspects of the design.
Agree stakeholders groups, ensuring the Estates and Facilities individuals who will take over the installed systems is represented in the relevant groups and the ownership is dear.	Set building environmental and other performance targets including suitable predictions and method for measurement post occupation. Obtain and contribute insights from the performance in use of comparable projects.
The terms and reference of each stakeholder group is dear and fully committed to engaging with the project.	Ensure agreed environmental, sustainability, manageability, buildability, usability and maintainability themes are embedded with every element of the project design.
Committed to lessons learned, usability and manageability.	Champions best practice ideas. Ensure that whole life costs are considered as part of the design development, materials and product selection.

Soft Landings helps to solve the performance gap between design intentions and operational outcomes. Better Buildings Partnerships (BBP) in partnership with BSRIA published Soft Landings: The Benefits to Commercial Property Owners (BBP, 2020), demonstrating that the commercial property sector is waking up to the fact that there is a broad recognition that a significant gap exists between the design expectations and operational performance of buildings. To address the performance gap and deliver better buildings, clients and project teams should adopt a collaborative mindset focused on operational outcomes. Soft Landings supports this process by enabling closer collaboration between the various stakeholders involved in a project, helping clients and project teams deliver buildings that meet their design intent and, ultimately, creating better buildings.

Soft Landings provides a foundation for environmental assessments like BREEAM and LEED, all forms of energy performance certification, post-occupancy evaluation, building logbooks, green leases, and industry KPIs.

5.5.1.1 Post-occupancy evaluation (POE)

POE is a process of assessing a building's performance after it has been occupied and used for a period, typically 1 year after hand over. It involves gathering feedback from occupants on various aspects, including building performance, user satisfaction and energy consumption, to identify areas for improvement and inform future design decisions. POE provides data to help understand how buildings are performing compared with their design intention. Without this data, the construction industry will unknowingly continue to make the same mistakes, wasting time and money. The Royal Institute for British Architects have highlighted in their publication on Post Occupancy Evaluation (RIBA, 2020) that despite many benefits of carrying out POE, it is still not embedded in all projects. Ensuring the widespread adoption of POE will require systemic cultural changes in the construction sector. This should be led through a top-down approach from Government, who should require POE as a condition when public money is spent, but all clients have a responsibility to understand the benefits of the exercise.

There are three broad levels of POE.

- Light touch POE: a simple but meaningful rapid evaluation undertaken post occupancy, before the building contract concludes.
- Diagnostic POE: feedback from a light touch POE may highlight the need for a more detailed review of a building.
- Detailed (forensic) POE: a comprehensive investigation, by independent evaluators, to identify and resolve any significant and persistent performance issues.

POE provides actionable insight into whether the building meets the desired requirements for the people who inhabit it. Undertaking a POE can, therefore, determine whether the initial brief was achieved and can rectify any weaknesses. This ensures that the user's intentions are met and that they are satisfied with the project. This can also strengthen the relationship with client as they can provide feedback on operational improvements.

Typically, a POE exercise consists of five core components.

User surveys: gathering feedback either online or on paper from building occupants about their experiences, satisfaction and suggestions for improvement.

Interviews: talking with building managers, maintenance staff and other stakeholders to gather their perspectives.

Building visits: conducting site visits to assess the building's physical condition, layout and functionality, and recording how occupants use the building and systems. This could include on-the-spot data collection using sensors.

Building performance monitoring: collecting data on energy and water consumption, indoor air quality, temperature, humidity and other relevant factors.

Reporting: analysing the collected data and reporting key findings, recommendations and lessons learned, and sharing these findings with key stakeholders.

POE is an integral part of the Soft Landings Framework. To complement the operational building performance on energy and water consumption, BSRIA Occupant Wellbeing (BOW) Survey (BSRIA, n.d.b) was introduced to measure end user satisfaction, as occupant wellbeing has come to the focus of building owners as an important factor in the assessment of a building's overall performance, impacting its green credentials as well as its value.

The BOW survey is a tool that helps assess occupant satisfaction and wellbeing in buildings, covering the building's physical environment, indoor facilities, functionality and accessibility. It provides qualitative information that allows building owners and facilities managers to measure the impact of building services operation on occupant perception of wellbeing. The BOW survey can also be used as a tool to gather evidence to support building assessment methodologies including BREEAM, WELL, LEED and POE requirements.

The BOW survey can be conducted on-line or on-paper and supplemented with a forensic walk-through of the building. The data are independently collected and analysed by BSRIA engineers, presenting findings within a collection of statistical results, charts, trends of factors and benchmarks.

5.5.1.2 Building performance diagnostics

Building performance diagnostics involves identifying and analysing issues associated with a building's performance, aiming to improve efficiency, comfort and occupant satisfaction. This process uses data and analysis to understand how a building is functioning and identify areas for optimisation. Diagnostics is often applied as part of a detailed forensic POE programme as it helps ensure that buildings are performing as intended and meeting the needs of their occupants.

Building performance diagnostics start with data collection, including gathering information from various sources, including building management systems, sensors, and occupant feedback, from which trends, patterns and areas for improvement can be derived. A range of diagnostics tools are available, such as infrared thermography survey to visualise heat patterns and identify areas of heat loss, airtightness testing to measures the amount of air leakage in a building, acoustics testing to test sound insulation quality of a building as well as energy modelling to simulate building energy use and test and identify potential areas for optimisation.

Technology advancement such as drones and artificial intelligence (AI) are transforming building diagnostics; drones can safely conduct visual inspections of hard-to-reach areas, while AI can analyse vast amounts of data to predict potential issues. Integrating these technologies enhances the accuracy and efficiency of building diagnostics, providing more comprehensive assessments with less human labour cost. By leveraging new techniques in data management, machine learning and AI, building performance diagnostics specialists are able to interrogate the growing amount of raw data and help building owners and asset managers to identify and quantify optimisation measures for energy, comfort and facilities management improvement (Hulatt and Studysmarter, 2024). Building diagnostics will see more integration of Internet of Things (IoT) devices and blockchain technology in the future. IoT devices will provide a seamless flow of real-time data, while blockchain ensures data security and transparency. Such advancements will lead to smarter, safer and more efficient buildings, revolutionising the construction and maintenance industries.

Funded from the European Union's Horizon 2020, the SmartSPIN project researched a methodology of enabling accessibility of early building performance diagnostics, aiming to promote greater uptake of smart energy services deployed by way of performance-based contracting (Eguiartea *et al.*, 2023). The tool has a web dashboard with user-friendly API facilitating communication with the algorithms, simplifying the underlying calculations and enabling users to upload specific building information and datasets to obtain early building energy diagnostics. This is one of the examples of tools using data-driven energy diagnostic approach to identify the most significant energy saving streams.

Overall, building performance diagnostics enhance building performance by helping ensure that buildings are performing as intended and meeting the needs of their occupants, leading to a more efficient, comfortable and sustainable building.

5.5.2 Design for Performance (DfP)

For any given development project, a performance gap between anticipated and actual building performance may result from a combination of factors, including an unclear client brief regarding the project's aims and success criteria, a lack of focus on operational outcomes at the design stage, cost-cutting exercises using poor quality construction materials, a tightened commissioning period, insufficient training for the facilities management and lack of feedback about the performance of the building in operation. Figure 5.19 shows the vast range and difference between the declared Energy Performance Certificate (EPC) ratings for commercial buildings in the UK and actual building's energy intensity (kWh/m^2 per year).

The UK has a design-for-compliance culture, which has led to the well-known 'performance gap' that exists between original design intent and how a building truly performs in-use. This design-for-compliance culture has been exacerbated since Building Regulations and EPCs were enforced in 2008. A new approach to delivering energy-efficiency in UK offices, the DfP initiative was created and promoted by BBP, as an industry backed project established to tackle the performance gap and provide an approach, based on measurable performance outcomes, to ensure new office developments deliver on their design intent. The BBP is a collaboration of leading property owners who are working together to improve the sustainability of commercial buildings.

Given that both Soft Landings and Design for Performance (DfP) share the aim of providing the built environment industry with clear operational performance goals for each stage of project cycle, Figure 5.20 (BSRIA, 2019: BG 76/2019) outlines diagrammatically the specific DfP activities required at each Soft Landings Phase in projects where both Soft Landings and DfP can be deployed, and how the activities fall within each stage of the *RIBA Plan of Work* (RIBA, 2013).

Figure 5.19 Comparing whole building energy intensity for buildings with different EPC grades, reproduced with kind permission of BSRIA and Better Buildings Partnership (BSRIA, 2019: BG 76/2019; BBP, 2021)

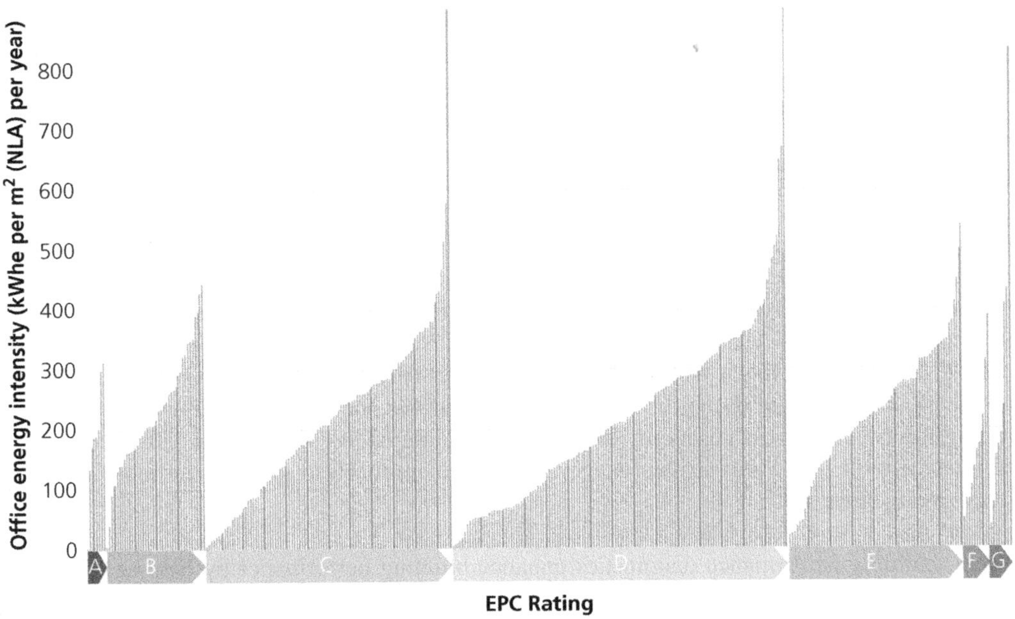

Each grey column represents a single office building's energy intensity over the course of a year

Source: Real Estate Environmental Benchmark 2017, Better Buildings Partnership

Figure 5.20 Design for Performance, RIBA plan of work and Soft Landings (reproduced with kind permission of BSRIA (BSRIA, 2019: BG 76/2019))

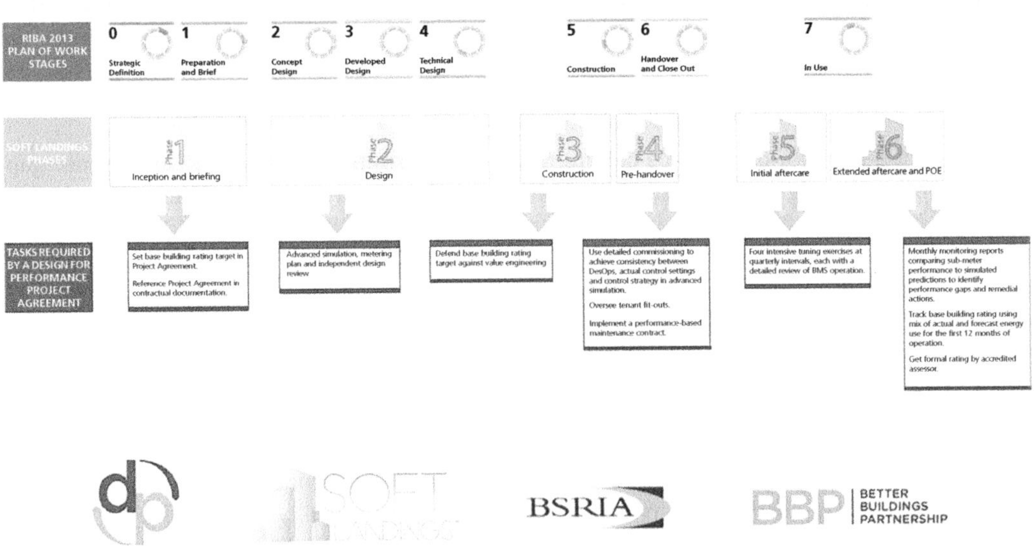

In contrast, Australia has had a system to measure and rate the operational efficiency of its commercial offices since 1999 – NABERS (National Australian Built Environment Rating System) (NABERS, n.d.). The scheme now covers 86% of the office market and the energy intensity of landlord services has improved by 36% since the scheme started. Australia has learnt to deliver far better energy performing office buildings than those in the UK, driven by the transparency of the NABERS ratings.

Figure 5.21 illustrates the key difference in the UK design compliance led culture and the Australian DfP approach.

Unlike design-based energy ratings, NABERS measures and rates the actual energy use of offices, helping building owners to accurately track and communicate the energy performance of their buildings. It also helps identify areas for savings and improvements. DfP in new buildings (or major renovations) allows building owners or developers to pledge to achieve a specific NABERS star rating after the project is complete and operational for 12 months.

BBP worked with the DfP initiative to launch NABERS UK in October 2020, aiming to bring the knowledge and success of NABERS to the UK by applying the same principles in Australia for new UK office developments. NABERS UK is now administrated by CIBSE (CIBSE, n.d.). As of March 2021, NABERS UK focuses solely on the energy performance of offices, with different programmes for existing buildings and new buildings. The new building (or major renovation) scheme is NABERS Design for Performance and the existing building scheme is the NABERS Energy ratings programme.

DfP changes the entire process into a more energy-conscious design bringing closer engagement between the landlord and tenants considering how the building will one day be operated, what the owners or developers must do and what tenants moving into the building must do. NABERS

Figure 5.21 Comparing how the markets in the UK and Australia are informed about building energy performance at the moment of a sale or lease property transaction (created using information from BSRIA (2019: BG 76/2019))

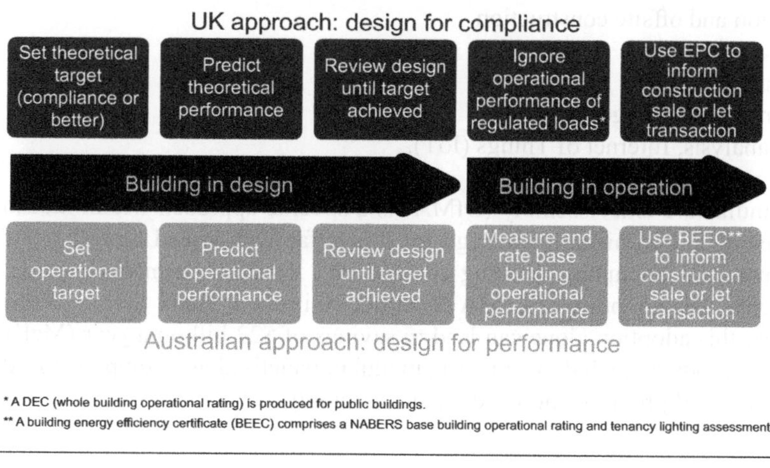

provides a rating from one to six stars. This helps building owners understand their building's performance when compared with similar buildings, providing an opportunity to set benchmarks and achieve new goals. In March 2024, Grosvenor's Toronto Square, Leeds, earned a NABERS UK Base Building Energy Rating of 4.5*, marking the first occasion in the UK where an existing office building has received such recognition. Grosvenor aims to achieve further NABERS UK ratings across their portfolio (BBP, 2024).

5.5.3 Design for Industrialised Construction (DfIC)

Despite the substantial advancements in digitalisation and automation we have seen in recent years, the construction industry's productivity has been trailing behind (Sawhney *et al.*, 2024). Unlike manufacturing and automotive industries, construction is a 'low volume – high variety' industry. The fluid and emergent designs that we see in construction typically call for supply chain flexibility and competitiveness, while manufacturing emphasises supply chain efficiency (Constructing Excellence, 2020).

To improve efficiency and productivity, while reducing costs and becoming more sustainable, the construction industry can learn several valuable lessons from the automotive industry, particularly in areas like standardised manufacturing, supply chain management and emissions reporting (Constructing Excellence, 2018). Design for Industrialised Construction (DfIC) is a focus area in the construction industry aiming to use manufacturing techniques to build buildings in order to improve efficiency, reduce costs and enhance the overall quality of construction projects. By incorporating design elements that support offsite manufacturing and assembly, DfIC allows for a more standardised, predictable and scalable construction process. This approach can lead to faster project delivery, reduced waste, and lower overall construction costs and safer construction process.

This initiative involves rethinking building design to optimise it for factory-based manufacturing and offsite assembly. It focuses on standardising components, using digital design tools, and embracing technologies like modular construction and 3D printing to reduce rework and errors, streamline the construction process and improve overall productivity. DfIC is a system that uses innovative and integrated techniques and processes such as building information modelling (BIM) and common data environments (CDE) to connect the design-to-make process by embracing five megatrends (Pothier and Sawhney, 2020)

- prefabrication and offsite construction
- additive manufacturing
- robotics and automation
- big data, artificial intelligence (AI)
- predictive analysis, Internet of Things (IoT).

Design for Manufacture and Assembly (DfMA) is a specific approach within industrialised construction, focusing on optimising the design for offsite fabrication and assembly. DfMA focuses on designing products, including building components, for efficient manufacturing of prefabricated components. The adoption of DfMA construction in the US and Europe is growing, and it is estimated that this adoption alone can lead to savings of $22 billion a year (McKinsey, 2019). Typically, DfMA products include volumetric modular, panelised and componentised assemblies, differing in the level of prefabrication and on-site assembly.

Volumetric modular is a method where buildings are built from large, pre-fabricated modules (or 'pods') that are manufactured off-site in a factory and then transported to the construction site to be assembled. Panelised modular construction involves the prefabrication of structural panels off-site, which are then transported to the construction site for assembly. This method offers efficient off-site manufacturing of standardised panels and streamlined on-site assembly processes. Componentised modular construction focuses on creating smaller individual building components and non-structural subassemblies, such as bathroom pods or electrical assemblies, window frames and doors, which are integrated into the main structure on-site.

Commonly perceived disadvantages of applying DfMA to a construction project includes initial investment, as implementing DfMA requires upfront investment in design time, tools and potentially software and training, although it can lead to long-term cost savings. Constructing Excellence (Constructing Excellence, 2021) looks to define KPIs for DfMA and benchmark these against existing building delivery models, to quantify the benefits. The KPIs identified by the Building Research Establishment UK include cost, time, quality, health and safety, labour requirements, environmental impacts and local disruption.

To fully realise the benefits of industrialised construction, it requires integration of all phases of the project life cycle (vertical integration), all members of the project team (horizontal integration), and inter-project learning and knowledge management (longitudinal integration). As illustrated in Figure 5.22, such full system thinking and integration driven by technology and a need to enhance efficiency, requires construction companies to re-construct the right teams and functional roles to enable these changes (Pothier and Sawhney, 2020), but the potential productivity gains would justify the shift.

Figure 5.22 Design for industrialised construction teams, skills and productivity gains (Pothier and Sawhney, 2020)

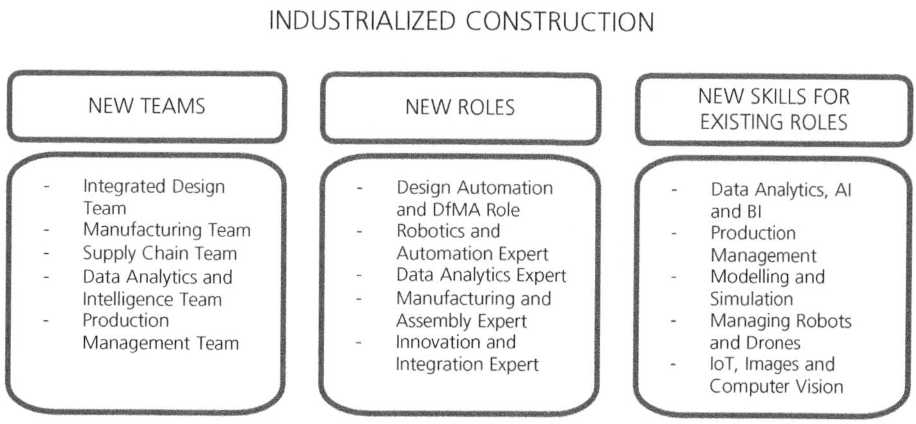

5.5.4 Environmental Product Declaration

The definition and calculation of the Whole Life Cycle Carbon following the BS EN 15978 (BSI, 2012a) framework for appraising the impacts of the built environment was introduced in Chapter 4. One of the key targets within the UK Net Zero Carbon Buildings Standard is embodied carbon materials and products of a building project.

An environmental product declaration, or EPD, is steadily growing in voluntary and mandatory engagements in the construction industry across the globe in line with awareness about environmental impacts. Both public and private stakeholders are increasingly demanding EPDs. All public procurement bodies in the EU and European Environment Agency (EEA) are required to use EPDs to assess the environmental footprint of products and Belgium requires the use of EPDs if a company is performing any environmental-related marketing. EPDs are required by specifiers, buyers, investors and regulators to demonstrate environmental impacts of construction products and building systems.

An EPD is a document that transparently reports the environmental impact of a product or material, based on a product life-cycle assessment (LCA). The LCA evaluates a product's environmental effects, commonly including data on resource consumption, air emissions, waste management, and other environmental aspects throughout the product's life cycle, from raw material extraction to disposal. EPDs, derived from these LCAs, support carbon emission reduction targets by enabling comparisons of the environmental impacts of different materials and products, helping to select the most sustainable option for buildings or infrastructure. An LCA is performed using a peer-reviewed Product Category Rules (PCR) document in line with the relevant European Environmental Declaration standard or other related international standards.

With EPDs for building materials, designers like architects and engineers are able to choose the most sustainable option for their project. The information can feed into whole building assessment schemes and other comparative tools. They can also inform BIM using computer-aided design (CAD) software to make comparisons and assess design options that improve a building's overall environmental performance. The use of EPDs can support innovative product development, such as improving efficiency for energy, materials and transport costs and help recognition for the product or material in building assessment schemes, such as BREEAM, LEED, Deutsche Gesellschaft fur Nachhaltiges Bauen (DGNB) and Green Star (BRE, n.d.c). EPDs are also an enabler for manufacturers to optimise the impact of their products and market their carbon transparency.

An EPD is usually valid for five years and is generated according to the relevant standards. EPDs for construction industry are based on the ISO 14040 (ISO, 2006a), ISO 14044 (ISO, 2006b), ISO 14025 (ISO, 2006c), BS EN 15804 (BSI, 2012b) or ISO 21930 (ISO, 2017) standards (One Click, n.d.). All European markets, and many other markets, accept EPDs created in compliance with BS EN 15804+A2 (a major amendment introduced that became mandatory in 2022) (BSI, 2020), which is now mandated under the revised Construction Products Regulation (CPR) for construction products in the EU. Many markets also accept EPDs created according to ISO 21930. US and Canada primarily use ISO 21930 for domestic EPDs and, in some cases, apply ISO 14040/44, which is the underlying standard for both the EN and ISO EPD standards (Pasanen, 2026).

The EPD Program Operator sets requirements for verifiers, writes up more detailed rules for creating EPDs (called Product Category Rules) and accepts and publishes the EPDs submitted by the manufacturers after their verification. Some EPD programmes charge annual fees to maintain the published EPDs, others only charge a lifetime fee to publish. There are dozens of EPD programmes worldwide. The globally operating ones include International EPD System (International EPD System, n.d.) and EPD Hub (EPD Hub, n.d.).

Product Category Rules (PCR) set more detail than the underlying standards. PCRs may be either product category specific or generic for all construction products. For EPDs produced following the European standards, the scope is often tightly defined because the standards are already detailed.

There is a five-step process involved in creating an EPD.

Step 1 – Collect data: including raw material, resource consumption and waste data for a specific product (type of data to be collected depends on EPD Program Operator and PCR).
Step 2 – Conduct an LCA: the LCA will need to conform to the PCR.
Step 3 – Prepare background report for EPD: known as an LCA report in the US, this report is a vital accompaniment to the public EPD, providing LCA methodology, assumptions and approach employed to support third party verification, as well as the standards adhered to.
Step 4 – Third party verification: every EPD needs to be verified by an independent third-party verifier before it can be published. This ensures accuracy, reliability and ensures that the EPD conforms to the requirements of the relevant PCR.
Step 5 – Publication: once the EPD has been verified by an independent third party, it is ready to be put into the public domain by way of publication by a programme operator, who will process, register and publish the EPD.

One Click, LCA's EPD Generator tool, has become a leading calculation tool for producing EPDs as it offers an end-to-end solution for manufacturers who need to create EPDs compliant with global ISO and EN standards. With this tool, users can perform unlimited LCAs and generate EPDs for any product category. Additionally, the EPD Generator provides the flexibility to create customer-specific private EPDs and TM65-compliant calculations.

For building services equipment, there are a limited amount of EPDs available for electrical, electronic and HVAC-R equipment as the process is not best suited to these products. Product Environmental Profiles (PEPs) are tailored specifically to electrical, electronic and HVAC equipment. A PEP is a comprehensive environmental assessment that provides detailed information about the environmental impact of a particular product or product range throughout its entire life cycle. The environmental information included in a PEP covers all the stages of the life cycle, typically described as 'cradle to grave', with the life cycle divided into specific stages ranging from the manufacturing stage through to the end-of-life stage. PEPs also include information relating to reuse, recovery and/or recycling; however, the inclusion of this information is optional. This means that they are third party verified in compliance with the ISO 14025 standard (ISO, 2006c), to ensure both consistency and reliability.

The PEP ecopassport programme is specific to electrical, electronic and HVAC-R equipment used in the construction industry.

While both EPDs and PEPs are growing in the awareness of design teams, their cost and time to implement within an organisation can be problematic. TM65 (CIBSE, 2021) (TM65LA; CIBSE, 2022) provides a methodology offering basic or mid-level calculations designed to aid consultants, researchers and manufacturers in evaluating and estimating the embodied carbon of building services equipment when an EPD (including PEP) is not available (BEAMA, n.d.).

TM65 assessments (CIBSE, 2021) can be considered as Type II environmental declarations that are classed as 'self-assessments' and are not third-party verified. As clearly indicated in the methodology text, asking manufacturers for EPD/PEPs should be the first step to establishing the embodied carbon of mechanical, electrical and plumbing (MEP) products. Requesting a TM65 should only occur when EPD/PEPs are not available from the manufacturer.

TM65 is more cost effective than typical EPD/PEP processes, and provides a useful stepping stone into embodied carbon evaluation. In the absence of an EPD or PEP, it provides an established methodology from an industry recognised body, albeit it is only designed to compare MEP products.

In summary, calculating embodied carbon for MEP products is a complex process. EPDs have broad applicability, but there are limited options available for electric or electronic products. PEPs are tailored specifically to electrical, electronic and HVAC equipment, but are primarily known within the EU. TM65 assessments were created by CIBSE to support the built environment, where a manufacturer does not already have embodied carbon data available, as clearly outlined in the TM65 guidance.

5.5.5 Universal Thermal Comfort Index

Chapter 3 introduced the concept of urban heat island as one of the design considerations to address the climate resilience for the built environment, as buildings and urbanised areas have a critical role to play in reducing vulnerability to climate risk, especially in a warmer climate.

One way to determine and measure the effectiveness of a design measure is to assess the outdoor thermal comfort people would feel with heat stress under an urban heat island effect. There is a growing need for valid assessment procedures of the outdoor thermal environment to assist public health systems in responding to climate change, urban planning and tourism and climate impact research. The Universal Thermal Comfort Index (UTCI) is a measurement index developed by a group of over 40 scientists from 23 countries collaborating within an EU-funded programme (COST 730 action) (Bröde et al., 2013). UTCI can be utilised to promote private and public health, develop appropriate urban planning, and study the effect of climatic conditions on heat stress.

The UTCI is the equivalent temperature for the environment derived from a reference environment. It is defined as the air temperature of the reference environment, which produces the same strain index value in comparison with the reference individual's response to the real environment. It is a measure of the human physiological response to the thermal environment.

There are ten UTCI thermal stress categories shown in Table 5.8 (Bröde *et al.*, 2013) that correspond to specific human physiological responses to the thermal environment, indicating the range within which occupants in an external environment would feel comfortable. Four factors are used to determine UTCI: air temperature, radiant temperature (solar and contextual), humidity and air movement.

Various efforts have been undertaken to compare different measurement indexes, including UTCI and other indoor and outdoor thermal comfort standards (Blazejczyk *et al.*, 2011; Zare *et al.*, 2018). UTCI has been widely adopted in the urban planning field and for reporting heat stress by the public health bodies (Hidalgo *et al.*, 2023) as the realisation grows of the profound human health impacts that can be caused by heat stress.

Table 5.8 UTCI assessment scale with comfort/stress categories, created using information from (Bröde *et al.*, 2013)

UTCI temperature (°C)	Comfort level	Stress category
> 46	Too Hot	Extreme heat stress
38 to 46		Very strong heat stress
32 to 38		Strong heat stress
26 to 32		Moderate heat stress
9 to 26	Comfortable	No thermal stress
0 to 9	Too Cold	Slight cold stress
−13 to 0		Moderate cold stress
−27 to −13		Strong cold stress
−40 to −27		Very strong cold stress
<−40		Extreme cold stress

The UHI effect means that the urbanised areas experience higher temperatures than rural areas or those with greater green infrastructure. This effect is profound in many densely occupied cities which exacerbates the impact of a warmer climate, especially as heat waves frequently occur. Broadly, the mitigation measures of reducing UHI include four categories.

- Reduce solar radiation by providing shading as much as possible.
- Increase wind flow.
- Material choices on nature-based solutions and use of high albedo, such as plant trees, green or cool roofs and cool pavements.
- Reduce heat rejection from buildings to the urban canyon (from air handling units, for example).

Another commonly applied design matrix is Physiological Equivalent Temperature (PET) for measuring outdoor thermal comfort. Both UTCI and PET are widely used for comfort indicators. PET is a thermal sensation index widely used in urban climate studies that takes into account mean radiant temperature, air temperature, relative humidity, wind speed and human factors. Wu *et al.*'s (2025) study on four distinct neighbourhood typologies with different configuration of street orientations and height-to-width ratios, trees and grass found that these configurations significantly affect the microclimate. Scenarios involving large trees consistently provide the most significant PET reduction across all typologies, primarily due to the extensive shading from large tree canopies. Small trees provide more PET reduction than all grass scenarios. Placing two rows of large trees on main streets is more cooling-efficient than a single row on both main and secondary streets on the high-radiation side.

Design mitigation measures should focus on these four impactful factors, especially for the warmer climates, due to the high air temperatures, providing solar shading and increasing air movement as well as evaporative cooling could be key to managing heat stress and enhancing external comfort. UTCI can be applied by the designers to assess the effect of the intervention measures including shading, green infrastructure, evaporative effect, materials and wind effect, in order to create a 'comfortable' thermal environment for outdoor spaces.

5.5.5.1 Solar shading

Providing shading to reduce solar radiations for users in outdoor spaces is one of the easiest and cost-effective ways of reducing heat stress. It often is done by buildings' self shading, planting tree canopies or other features or street furniture. Combinations of clustered shading types ensures shading at different times of the day and year, such as the concept shown in Figure 5.23.

One of challenges is designing in shading while making sure streets are accessible and walkable. A more sophistic shading provision could use a multi-layer strategy, as shown in Figure 5.24.

Figure 5.23 Clustered shading, concept

Figure 5.24 A multilayer strategy example (photo credit to combusean licensed under CC BY 2.0)

5.5.5.2 Evaporative cooling

In warm or hot climates, especially in conjunction with dry conditions, the ambient temperature can be significantly reduced by injecting water mist to the air to reduce the temperature closer to its wet bulb temperature. The scale of reduction varies with the relative humidity, but could offer up to 10 °C reduction. Street-located evaporative cooling installations are already prevalent in cities where excessive temperatures are common. With the changing global temperatures causing even countries in the temperate zones to experience unwelcome high summer temperatures, evaporative cooling installations, such as that shown in Figure 5.25, helping residents to take respite from the heat could become widespread.

Figure 5.25 Street-located evaporative cooling (photo credit to Strasbourg's 'surprise shower' by kewl licensed under CC BY 2.0)

5.5.5.3 Materials with low albedo

Albedo (from Latin albedo 'whiteness') is the fraction of sunlight that is diffusely reflected by a body. It is measured on a scale from 0 (corresponding to a black body that absorbs all incident radiation) to 1 (corresponding to a body that reflects all incident radiation).

The proportion reflected is not only determined by properties of the surface itself, but also by the spectral and angular distribution of solar radiation reaching the Earth's surface. A few simple examples for common construction materials are pavement and roof materials. Considering the albedo of the material has not been common for designers in most countries, where occasions of excessive heat have been rare in the past. With changing global temperatures, excessive heat is becoming more common and, therefore, the materials contained in buildings, structures and roads could have a much higher impact on the urban temperatures experienced. Figure 5.26 shows an illustration of different albedo in pavement materials.

Low albedo materials include limestone. Limestone slowly warms up, peaking by mid-afternoon before it cools to 0.8°C above the evening air temperature. Materials such as timber and marble peak before midday, yet remain only 1.2–2.4°C warmer than the ambient air temperature in the late afternoon.

Figure 5.26 An illustration of pavement materials with different albedo

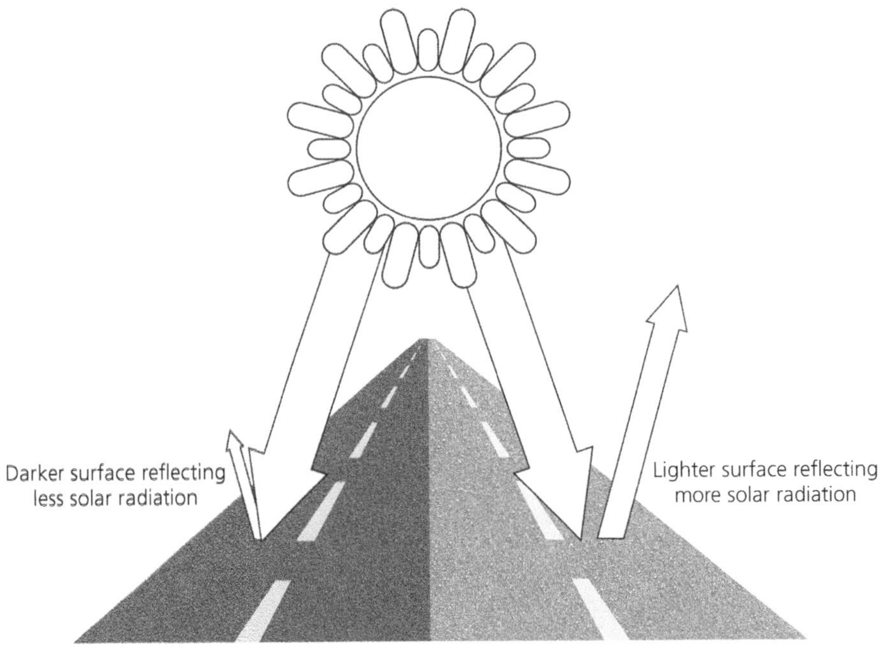

Granite and asphalt absorb significant heat very rapidly. Asphalt is the worst contributor, and remains 8°C hotter, even in the early evening. Adding a solar reflective coating can reduce ground temperatures by 17°C on hot days. Cooling pavement surface reflects UV and infrared rays and absorbs less solar radiation compared with dark asphalt. These considerations will need to be increasingly factored into design decisions in future.

5.5.5.4 Assessing the impact of UHI on building design

CIBSE has already recognised the effect of UHI in London's local microclimate. In CIBSE TM49 (CIBSE, 2014), weather data sets have three different locations: Central London (an urban environment), London Heathrow (a suburban environment) and London Gatwick (a more rural environment), so as to enable the most appropriate conditions for any development to be selected. This approach is in contrast to the previous CIBSE Design Summer Year weather data for London, which has only one data set taken from the Heathrow area and did not take account of the Urban Heat Island effect.

In order to assess the potential heat stress, the CIBSE weather files (DSY – Design Summer Year) contains three different sets to enable the simulation of different summer conditions and the impact each one could have on a development's peak summer time temperatures and cooling ability. These are

- DSY 1 – a moderately warm summer
- DSY 2 – a short, intense warm spell
- DSY 3 – a long but less intense warm spell.

Design for future climate has become a norm not only for new builds, but also any major refurbishment. Consideration of heat stress in buildings due to climate change is increasingly becoming standard practice. The intention is not to over-engineer cooling systems, or indeed to oversize chillers, rather, to design a building that can be adapted readily during a heat wave, and develop a plan in place for retrofitting in the coming 'warmer' years. This is one of the rapidly changing requirements that design teams and built environment professionals will need to address in the near future.

REFERENCES

Activescore (n.d.) *ActiveScore – Certify and Enhance Your Facilities for Active Travel*. https://activetravelscore.com/ (accessed 05/05/2025).

ADE (Association for Decentralised Energy) (2023) *London Legacy Development Corporation*. https://www.theade.co.uk/organisations/london-legacy-development-corporation#:~:text=The%20Concession%20was%20procured%20by,termination%20of%20the%20current%20appointment (accessed 15/05/2025).

AECOM (2012) *Marina Bay and Greater Southern Waterfront, Singapore*. https://aecom.com/projects/6sustainability-study-marina-bay-greater-southern-waterfront/ (accessed 10/10/2024).

APPG on Arts, Health and Wellbeing (All-Party Parliamentary Group on Arts, Health and Wellbeing) *et al.* (2017) *Creative Health: The Arts for Health and Wellbeing*, 2nd edn. https://ncch.org.uk/uploads/Creative_Health_Inquiry_Report_2017_-_Second_Edition.pdf (accessed 18/10/2024).

Arup (2017) *SPeAR Handbook 2017*. Arup, London, UK.

Barton H (2016) *City of Well-Being*. Routledge eBooks, 10.4324/9781315438689.

Barton H *et al.* (2015) *The Routledge Handbook of Planning for Health and Well-Being*. Routledge eBooks, 10.4324/9781315728261.

BBP (Better Buildings Partnership) (2020) *Soft Landings: The Benefits to Commercial Property Owners, Better Buildings Partnership*. https://www.betterbuildingspartnership.co.uk/sites/default/files/media/attachment/BBP_Soft%20Landings_Final.pdf (accessed 17/05/2025).

BBP (2021) *Real Estate Environmental Benchmark: 2020 Energy Snapshot, Real Estate Environmental Benchmark*. https://www.betterbuildingspartnership.co.uk/sites/default/files/media/attachment/REEB%202020%20Snapshot_Final_0.pdf (accessed 17/05/2025).

BBP (2024) *Grosvenor Achieves the UK's First NABERS Base Building Energy Rating*. https://www.betterbuildingspartnership.co.uk/grosvenor-achieves-uks-first-nabers-base-building-energy-rating (accessed 17/05/2025).

BEAMA (British Electrotechnical and Allied Manufacturers' Association) (n.d.) *An introduction to EPDS (Environmental Product Declaration)*. https://www.beama.org.uk/static/54b9649f-3015-4491-b22ab7e2aaddc22c/Explore-common-formats-of-Product-Embodied-Carbon.pdf (accessed 26/05/2025).

BGS (British Geological Survey) (2024) *New Report Assesses Deep Geothermal Energy in the UK*. https://www.bgs.ac.uk/news/new-report-assesses-deep-geothermal-energy-in-the-uk/ (accessed 15/05/2025).

Bikedock (2023) *Why Do We Need Active Travel?* https://www.bikedocksolutions.com/blog/why-do-we-need-active-travel/#:~:text=According%20to%20research%2C%20active%20commuters%20take%201.3,healthy%2C%20lower%20your%20risk%20of%20many%20diseases (accessed 18/05/2025).

Bioregional (2016) *The BedZED Story: The UK's First Large-Scale, Mixed-Use Eco-Village*. https://storage.googleapis.com/www.bioregional.com/downloads/The-BedZED-Story_Bioregional_2017.pdf (accessed 18/10/2024).

Blazejczyk K *et al.* (2011) Comparison of UTCI to selected thermal indices. *International Journal of Biometeorology* **56(3)**: 515–535,10.1007/s00484-011-0453-2.

BRE (Building Research Establishment) (n.d.a) *BREEAM Infrastructure*. https://breeam.com/breeam-infrastructure (accessed 09/05/2025).

BRE (n.d.b) *CEEQUAL Excellent Olympic Park Bridges*. https://bregroup.com/case-studies/ceequal-excellent-olympic-park-bridges (accessed 09/05/2025).

BRE (n.d.c) *What is an Environmental Product Declaration (EPD)?* https://bregroup.com/services/standards/environmental/en-15804-environmental-product-declaration-epd#:~:text=Declaration%20(EPD)?-,What%20is%20an%20Environmental%20Product%20Declaration%20(EPD)?,improve%20a%20building's%20environmental%20performance.&text=What%20is%20an%20EPD? (accessed 26/05/2025).

Brent Cross Town (n.d.) *Brent Cross Town Flourishing Index Putting Science Behind Health & Wellbeing*. https://brentcrosstown.co.uk/media/Brent-Cross-Town-Flourishing-Index.pdf (accessed 04/05/2025).

Bröde P *et al.* (2013) The Universal Thermal Climate Index UTCI compared to ergonomics standards for assessing the thermal environment. *Industrial Health* **51(1)**: 16–24, 10.2486/indhealth.2012-0098.

BSI (2012a) BS EN 15978: Sustainability of construction works – Assessment of environmental performance of buildings. BSI, London, UK. https://standardsdevelopment.bsigroup.com/projects/2023-01815, (accessed 25/04/2025).

BSI (2012b) BS EN 15804: Sustainability of construction works. Environmental product declarations. Core rules for the product category of construction products. BSI, London, UK. https://knowledge.bsigroup.com/products/sustainability-of-construction-works-environmental-product-declarations-core-rules-for-the-product-category-of-construction-products (accessed 25/08/2025).

BSI (2020) EN 15804+A2:2019: Sustainability of construction works. Environmental product declarations. Core rules for the product category of construction products, BSI, London, UK. https://knowledge.bsigroup.com/products/sustainability-of-construction-works-environmental-product-declarations-core-rules-for-the-product-category-of-construction-products-1 (accessed 25/08/2025).

BSRIA (Building Services Research and Intelligence Association) (2018) *Soft Landings Framework 2018 – Six Phases for Better Buildings*. BG 54/2018. https://www.bsria.com/uk/product/QnPd6n/soft_landings_framework_2018_bg_542018_a15d25e1/ (accessed 09/05/2025).

BSRIA (2019) *Soft Landings and Design for Performance*. BG 76/2019. https://www.bsria.com/uk/product/vBG24D/soft_landings_and_design_for_performance_bg_762019_a15d25e1/ (accessed 09/05/2025).

BSRIA (n.d.a) *Soft Landings Helps to Improve the Handover of New and Refurbished Buildings*. https://www.bsria.com/uk/consultancy/project-improvement/soft-landings/about-soft-landings/ (accessed 09/05/2025).

BSRIA (n.d.b) *BSRIA Occupant Wellbeing (BOW) Survey*. https://www.bsria.com/uk/consultancy/building-improvement/building-performance-evaluation/bsria_occupant_wellbeing_bow_survey/ (accessed 17/05/2025).

CEG (n.d.) *CEG. Commercial Estates Group*. https://www.ceg.co.uk/ (accessed 09/05/2025).

CIBSE (2013) (Chartered Institution of Building Services Engineers) *CIBSE TM51 Ground Source Heat Pumps*. https://www.cibse.org/knowledge-research/knowledge-portal/tm51-ground-source-heat-pumps/ (accessed 15/05/2025).

CIBSE (2014) *Technical Memorandum 49: Design Summer Years for London*. https://www.cibse.org/knowledge-research/knowledge-portal/technical-memorandum-49-design-summer-years-for-london-2014-pdf (accessed 30/08/2025).

CIBSE (2020) *CIBSE CP1 Heat networks: Code of Practice for the UK*. https://www.cibse.org/knowledge-research/knowledge-portal/cp1-heat-networks-code-of-practice-for-the-uk-2020-pdf (accessed 15/05/2025).

CIBSE (2021) *Embodied Carbon in Building Services: A Calculation Methodology (TM65)*. https://www.cibse.org/knowledge-research/knowledge-portal/embodied-carbon-in-building-services-a-calculation-methodology-tm65 (accessed 25/08/2025).

CIBSE (2022) *TM65LA: Using the TM65 Methodology Outside the UK*. https://www.cibse.org/knowledge-research/knowledge-portal/tm65la-using-the-tm65-methodology-outside-the-uk-pdf-2022 (accessed 30/08/2025).

CIBSE (n.d.) *NABERS UK. CIBSE Certification*. https://www.cibsecertification.co.uk/nabers-uk/ (accessed 09/05/2025).

Clements-Croome D, Turner B and Pallaris K (2019) Flourishing workplaces: a multisensory approach to design and POE. *Intelligent Buildings International* **11(3–4)**: 131–144, 10.1080/17508975.2019.1569491 (accessed 09/05/2025).

Collins A, Walker and UK Parliament (2023) *Local area energy planning: achieving net zero locally, UK Parliament Post*. POSTnote 703, 10.58248/pn703.

Constructing Excellence (2018) *5 Things Construction Can Learn from Automotive – Constructing Excellence*. https://constructingexcellence.org.uk/5-things-construction-can-learn-from-automotive/ (accessed 17/05/2025).

Constructing Excellence (2020) *Learning from Manufacturing to Improve Our Sector's Productivity – Transforming Construction Network Plus – Constructing Excellence*. https://constructingexcellence.org.uk/learning-from-manufacturing-to-improve-our-sectors-productivity-transforming-construction-network-plus/ (accessed 17/05/2025).

Constructing Excellence (2021) *Costs of MMC Set to Drop by a Third, Finds New Report – Constructing Excellence*. https://constructingexcellence.org.uk/costs-of-mmc-set-to-drop-by-a-third-finds-new-report/ (accessed 17/05/2025).

Cyclingscore (n.d.) *Cyclingscore*. https://cyclingscore.com/ (accessed 04/05/2025).

Department for Transport (2023) *Transport and Environment Statistics: 2023 (2021 data)*. https://www.gov.uk/government/statistics/transport-and-environment-statistics-2023/transport-and-environment-statistics-2023 (accessed 05/05/2025).

DESNZ (Department for Energy Security and Net Zero) (2023) *Energy Security Bill Factsheet: Heat Networks Regulation and Zoning (Updated 20 March 2023)*. https://www.gov.uk/government/publications/energy-security-bill-factsheets/energy-security-bill-factsheet-heat-networks-regulation-and-zoning (accessed 02/05/2025).

DoE (US Department of Energy) (n.d.a) *Enhanced Geothermal Systems*. https://www.energy.gov/eere/geothermal/enhanced-geothermal-systems#:~:text=GTO%20EGS%20Efforts&text=In%20the%20United%20States%2C%20the,geothermal%20drilling%20at%20The%20Geysers%20 (accessed 02/05/2025).

DoE (n.d.b) *FORGE – Frontier Observatory for Research in Geothermal Energy*. https://www.energy.gov/eere/geothermal/forge (accessed 02/05/2025).

Eavor (2023) Next-generation geothermal energy hits New Mexico, policy-makers want it to stay. *Eavor*, 10 March. https://eavor.com/blog/next-generation-geothermal-energy-hits-new-mexico-policy-makers-want-it-to-stay/ (accessed 02/05/2025).

EC (European Commission) (2022) *Policy Support for Heating and Cooling Decarbonisation: Roadmap*. Publications Office of the European Union, 10.2833/977806.

Eden Geothermal (2023) *The Drilling Rig – Eden Geothermal*. https://www.edengeothermal.com/the-project/phase-1/ (accessed 02/05/2025).

EEA (European Environment Agency) (2021) *Greenhouse Gas Emissions from Energy Use in Buildings in Europe*. https://www.eea.europa.eu/data-and-maps/indicators/greenhouse-gas-emissions-from-energy/assessment (accessed 02/05/2025).

EEA (2022) *Trends and Projections in Europe 2022*. EEA Report 10/2022. https://www.eea.europa.eu/en/analysis/publications/trends-and-projections-in-europe-2022 (accessed 02/05/2025).

EEA (2023) *Decarbonising Heating and Cooling – A Climate Imperative*. https://www.eea.europa.eu/en/analysis/publications/decarbonising-heating-and-cooling (accessed 02/05/2025).

Eguiarte O *et al.* (2023) Data-driven tool for early building energy performance diagnostic. *34th International Conference on Efficiency, Cost, Optimization, Simulation and Environmental Impact of Energy Systems (ECOS 2021)*, pp. 3307–3314, 10.52202/069564-0297.

Energy Transitions Commission (n.d.) *Demand Side Flexibility: Unleashing Untapped Potential Alongside Electricity Grids and Storage – Energy Transitions Commission*. https://www.energy-transitions.org/bitesize/demand-side-flexibility-an-important-route-alongside-electricity-grids-and-storage/#:~:text=Traditionally%2C%20the%20power%20system%20was,in%20response%20to%20system%20needs (accessed 04/05/2025).

EPD Hub (n.d.) *EPD Hub*. https://www.epdhub.com/ (accessed 26/05/2025).

EPG (Energy Policy Group) *et al.* (2022) *District Heating in National Long-Term Strategies*. Energy Policy Group, Bucharest. https://www.epg-thinktank.org/wp-content/uploads/2022/11/EPG_District_Heating_Report.pdf (accessed 02/05/2025)

EQ (n.d.) *Smart & Sustainable – EQ*. https://eqbristol.co.uk/smart-sustainable/ (accessed 09/05/2025).

Fervo (n.d.) *Technology – Fervo Energy*. https://fervoenergy.com/technology/ (accessed 02/05/2025).

Finnegan S, Jones C and Sharples S (2018) The embodied CO2e of sustainable energy technologies used in buildings: a review article. *Energy and Buildings* **181**: 50–61, 10.1016/j.enbuild.2018.09.037.

Fletcher TD, Shuster W, *et al.* (2014) SUDS, LID, BMPs, WSUD and more – the evolution and application of terminology surrounding urban drainage. *Urban Water Journal* **12(7)**: 525–542, 10.1080/1573062X.2014.916314.

Global Voices Brazil (2025) *Global Voice*. https://globalvoices.org/2025/09/29/the-legacy-of-sponge-cities-left-by-chinese-landscape-architect-yu-kongjian/#:~:text=Yu's%20legacy,CC%20BY%2DSA%204.0 (accessed 18/10/2025).

Greater London Authority (2023) *London Plan Guidance Urban Greening Factor*. https://www.london.gov.uk/sites/default/files/2023-02/London%20Plan%20Guidance%20-%20Urban%20Greening%20Factor.pdf (accessed 10/08/2025).

Hendriksen IJM *et al.* (2010) The association between commuter cycling and sickness absence. *Preventive Medicine* **51(2)**: 132–135, 10.1016/j.ypmed.2010.05.007.

Hidalgo J *et al.* (2023) Mapping the urban climate to address summer comfort management in French urban planning documents. *Cybergeo* [preprint], 10.4000/cybergeo.40319.

Hulatt and Studysmarter (2024) *Building Diagnostics: Techniques & Examples*. https://www.studysmarter.co.uk/explanations/architecture/building-performance/building-diagnostics/#:~:text=Definition,data%20and%20deliver%20accurate%20assessments (accessed 17/05/2025).

International EPD System (n.d.) *International EPD System*. https://www.environdec.com/home (accessed 26/05/2025).

Irfan U (2025) Grid-Scale battery storage is quietly revolutionizing the energy system. *WIRED*, 26 April. https://www.wired.com/story/grid-scale-battery-storage-is-quietly-revolutionizing-the-energy-system/ (accessed 30/05/2025).

ISO (International Organization for Standardization) (2006a) ISO 14040:2006: Environmental management — Life cycle assessment — Principles and framework https://www.iso.org/standard/37456.html (accessed 25/08/2025).

ISO (2006b) ISO 14044:2006 Environmental management — Life cycle assessment — Requirements and guidelines. https://www.iso.org/standard/38498.html (accessed 25/08/2025).

ISO (2006c) ISO 14025:2006: Environmental labels and declarations. https://www.iso.org/standard/38131.html (accessed 25/08/2025).

ISO (2017) ISO 21930:2017. Sustainability in buildings and civil engineering works — Core rules for environmental product declarations of construction products and services https://www.iso.org/standard/61694.html (accessed 25/08/2025).

IUCN (International Union for Conservation of Nature) (2020a) *Ensuring Effective Nature-based Solutions*. https://iucn.org/resources/issues-brief/ensuring-effective-nature-based-solutions (accessed 09/05/2025).

IUCN (2020b) *Guidance for Using the IUCN Global Standard for Nature-Based Solutions: First Editions*, 10.2305/iucn.ch.2020.09.en.

Living Future (n.d.) *Living Building Challenge Basics – Living Future*. https://living-future.org/lbc/basics/ (accessed 05/05/2025).

McCay AT, Feliks MEJ and Roberts JJ (2019) Life cycle assessment of the carbon intensity of deep geothermal heat systems: a case study from Scotland. *The Science of the Total Environment* **685**: 208–219, 10.1016/j.scitotenv.2019.05.311.

McKinsey (2019) *Modular Construction: From Projects to Products*. https://www.mckinsey.com/capabilities/operations/our-insights/modular-construction-from-projects-to-products (accessed 17/05/2025).

Merino-Garcia *et al.* (2024) *Britain's Buried Power: The Case for Next-Generation Geothermal*. https://britishprogress.org/uk-day-one/britains-buried-power-the-case-for-next-generation (accessed 04/05/2025).

MHCLG (Ministry of Housing, Communities and Local Government) (2024) *National Planning Policy Framework*. https://assets.publishing.service.gov.uk/media/67aafe8f3b41f783cca46251/NPPF_December_2024.pdf (accessed 08/10/2024).

Milman O (2024) US aiming to 'crack the code' on deploying geothermal energy at scale. *The Guardian*, 2 April. https://www.theguardian.com/environment/2024/apr/02/geothermal-energy-electricity (accessed 02/05/2025).

Modescore (n.d.) *Modescore*. https://www.modescore.com/ (accessed 04/05/2025).

NABERS (National Australian Built Environment Rating System) (n.d.) *What is NABERS?* https://www.nabers.gov.au/about/what-nabers (accessed 09/05/2025).

Natural England (2023) *Urban Greening Factor for England – Development and Technical Analysis (NERR132), Natural England – Access to Evidence*. NERR132. https://publications.naturalengland.org.uk/publication/5846537451339776 (accessed 25/01/2026).

Nijs W, Tarvydas D, Toleikyte A (2021) *EU Challenges of Reducing Fossil Fuel Use in Buildings*. Publications Office of the European Union, the Netherlands. https://data.europa.eu/doi/10.2760/85088 (accessed 12/12/2024).

One Click (n.d.) A simple guide to EPDs. https://oneclicklca.com/en-gb/resources/articles/guide-to-epds (accessed 26/05/2025).

Pasanen P (2026) *Global EPD Requirements by Country*. https://oneclicklca.com/en-gb/resources/articles/guide-to-global-epd-requirements-by-country#:~:text=What's%20involved%20in%20creating%20an,set%20a%20lot%20of%20detail (accessed 26/05/2025).

Passive House Institute (2023) *Criteria for buildings Passive House – ENERPHIT – PHI Low Energy building*. Version 10c, *Passive House Institute*. Version 10c. https://www.passivehouse.com (accessed 08/10/2024).

Passivhaus Trust (n.d.) *What Is Passivhaus?* https://www.passivhaustrust.org.uk/what_is_passivhaus.php (accessed 10/10/2024).

Pearson (2021) Case study: The House at Cornell Tech. *CIBSE Journal*, March. https://www.cibsejournal.com/case-studies/case-study-the-house-at-cornell-tech/ (accessed 10/10/2024).

Philp D, Churcher D and Davidson S (2019) *Government Soft Landings: UK BIM Framework, Government Soft Landings, Revised Guidance for the Public Sector on Applying BS8536 Parts 1 and 2. Updated for ISO 19650*, 10.17863/cam.45315.

Pothier and Sawhney (2020) *What will industrialised construction mean for the future of work?* www.rics.org. RICS. https://www.rics.org/news-insights/what-will-industrialised-construction-mean-for-the-future-of-work#:~:text=The%20benefits%20of%20industrializing%20construction,%2C%20Internet%20of%20Things%20(IoT) (accessed 20/08/2025).

Power Technology (2012) *Olympic Park Energy Centre, London – Power Technology*. https://www.power-technology.com/projects/olympic-park-energy-centre/ (accessed 04/05/2025).

Project LEO (2023) *Project LEO Final Report: a digest of key learnings, Project LEO*. https://project-leo.co.uk/wp-content/uploads/2023/02/LEO-Final-Report-v3a_Web_lr.pdf (accessed 04/05/2025).

Public Health England (2020) *Health Matters: Physical Activity – Prevention and Management of Long-Term Conditions*. https://www.gov.uk/government/publications/health-matters-physical-activity/health-matters-physical-activity-prevention-and-management-of-long-term-conditions#:~:text=The%20government's%20prevention%20green%20paper,for%20-individuals%2C%20communities%20and%20society (accessed 04/04/2025).

Quaise Energy (n.d.) https://www.quaise.com/ (accessed 02/05/2025).

RIBA (Royal Institute of British Architects) (2013) *RIBA Plan of Work*. https://www.architecture.com/knowledge-and-resources/resources-landing-page/riba-plan-of-work (accessed 09/05/2025).

RIBA (2020) *Post Occupancy Evaluation, an Essential Tool to Improve the Built Environment*. https://www.riba.org/work/business-tools/post-occupancy-evaluation-guidance/ (accessed 04/05/2025).

RICS (Royal Institution of Chartered Surveyors) (2024) *RICS Professional Standard: Whole Life Carbon Assessment for the Built Environment*. 2nd edn, Version 3. https://www.rics.org/profession-standards/rics-standards-and-guidance/sector-standards/construction-standards/whole-life-carbon-assessment (accessed 26/05/2025).

Sánchez-Carracedo F *et al*. (2020) The Sustainability Matrix: a tool for integrating and assessing sustainability in the Bachelor and Master Theses of Engineering degrees. *Sustainability* **12(14)**: 5755, 10.3390/su12145755.

Sawhney R and Luo D (2024) *RICS Construction Productivity Report 2024*. www.rics.org. https://www.rics.org/news-insights/rics-construction-productivity-report-2024 (accessed 30/08/2025).

Shao W *et al*. (2019) Characteristics and effectiveness of water management methods for 'sponge cities' in China. *Proceedings of the Institution of Civil Engineers – Water Management* **173(4)**: 163–171, 10.1680/jwama.19.00016.

Sofian S *et al*. (2015) Sustainable systems integration model-metrics in design process. *Procedia – Social and Behavioral Sciences* **184**: 297–309, 10.1016/j.sbspro.2015.05.094.

Srivastava A and Thompson C (2018) *Technology Innovations for Energy Masterplanning*. https://aecom.com/without-limits/article/technology-innovations-energy-masterplanning/ (accessed 10/10/2024).

Sustainable SITES Initiative (n.d.) *SITES rating system*. https://www.sustainablesites.org/ (accessed 10/08/2025).

Tellvik (2024) *The Future of Mobility in Real Estate – A Choice to Innovate*. https://www.esgrelab.com/. https://activetravelscore.com/wp-content/themes/jobify/theme-images/mobile_report.pdf (accessed 25/01/2026).

The House at Cornell Tech (2023) *Sustainability – The House at Cornell Tech*. https://thehouseat-cornelltech.com/sustainability/ (accessed 08/10/2024).

UKNZCBS (2025) *UK Net Zero Carbon Buildings Standard Pilot Version rev 2*. https://www.nzcbuildings.co.uk/_files/ugd/6ea7ba_1ef36b6835de46668f2ad8b589ff1b93.pdf (accessed 09/05/2025).

UKRI (UK Research and Innovation) (2022) *Smart Local Energy Systems: The Energy Revolution Takes Shape*. https://www.ukri.org/wp-content/uploads/2022/01/UKRI-250122-SmartLocalEnergySystemsEnergyRevolutionTakesShape.pdf (accessed 04/05/2025).

UKRI (n.d.) *Project LEO (Local Energy Oxfordshire)*. https://gtr.ukri.org/projects?ref=104781 (accessed 04/05/2025).

VanderWeele TJ (2017) On the promotion of human flourishing. *Proceedings of the National Academy of Sciences* **114(31)**: 8148–8156, 10.1073/pnas.1702996114.

Watkins Z (2022) *BREEAM Infrastructure – HS2 Learning Legacy*. https://learninglegacy.hs2.org.uk/document/breeam-infrastructure/ (accessed 25/08/2025).

WELL (n.d.a) *International WELL Building Institute*. https://www.wellcertified.com/certification/v2/ (accessed 05/05/2025).

WELL (n.d.b) *WELL v2: The WELL Building Standard Version 2*. https://v2.wellcertified.com/en/wellv2/overview/ (accessed 05/05/2025).

WiredScore (2021) *CEG's EQ Building Achieves First SmartScore Certification Globally – WiredScore*. https://wiredscore.com/blog/2021/06/29/cegs-eq-building-achieves-first-smart-score-certification-globally/ (accessed 09/05/2025).

WiredScore (n.d.) *WiredScore – Setting Buildings Up for Success*. https://wiredscore.com/ (accessed 09/05/2025).

World Economic Forum (2023) *How a New Digital Tool Could Help Cities Meet Key Sustainability and Mobility Goals*. https://www.weforum.org/stories/2023/05/new-digital-tool-cities-tackle-inclusivity-sustainability/ (accessed 08/10/2025).

Wu Y, Mashhoodi B, *et al.* (2025) Effective street tree and grass designs to cool European neighbourhoods. *Urban Climate* **61**: 102376, 10.1016/j.uclim.2025.102376.

Yang J (2019) Lingang sponge city nears completion. *City News Service Shanghai*, 22 August. https://www.shine.cn/news/metro/1908220642/ (accessed 08/05/2025).

Zare S *et al.* (2018) Comparing Universal Thermal Climate Index (UTCI) with selected thermal indices/environmental parameters during 12 months of the year. *Weather and Climate Extremes* **19**: 49–57, 10.1016/j.wace.2018.01.004.

Zargarian R *et al.* (2016) A new sustainability framework for urban underground space. *Proceedings of the Institution of Civil Engineers – Engineering Sustainability* **171(5)**: 238–253, 10.1680/jensu.15.00013.

Mei Ren and Rachel Waggett
ISBN 978-1-83662-853-8
https://doi.org/10.1108/978-1-83662-850-720261009

Chapter 6
Near term new thinking

6.1. Exploring emerging opportunities for sustainable transformation

In earlier chapters, the rapid application of sustainable development thinking to the built environment that occurred in the 1990s and early 2000s has been discussed. This period was remarkable not only for the speed with which new techniques and solutions were developed, but also for the creativity and enthusiasm shown by built environment professionals. Borrowing technologies and approaches developed from other sectors, working in interdisciplinary ways and open to learning and testing new methodologies paid dividends for the UK built environment sector. With the benefit of hindsight, the proliferation of opportunity that occurred at that time led to significant and substantial strides forward in sustainable development progress, at a speed since unmatched.

Globally, there is a sense that another such milestone has been reached, with the changes to global temperature and the impacts for planetary systems becoming evident to all; the only question is now how, not whether, action should be taken. Despite the difficulties associated with achieving any consensus on such a large, global scale, built environment professionals have an exciting opportunity to take another stride forward in mainstreaming beneficial action.

With so many potential avenues of opportunity, knowing where to focus can seem overwhelming; Chapter 4 discusses value-led frameworks for taking decisions in uncertain circumstances such as these, while Chapter 5 explores some of the design tools already available that allow built environment practitioners to work towards the strategic principles that a framework provides. Such is the effort now being expended across the world on developing practical solutions to achieving more sustainable buildings, that there are real opportunities to embed forward-thinking research into emerging designs and briefs.

This chapter focuses on some of the near-term opportunities that are already being used successfully for some pilot developments and could be integrated into mainstream projects rapidly. These are technologies and techniques that are already proven and although not widespread in use yet, at this important point in sustainable design development, the next few years could be transformative. Each use of innovative approaches moves the whole industry forward, improving performance and helping each successive building or development to expand upon the lessons learned. Importantly, it builds confidence and a community of practice that enables subsequent developments to leap forward, bypassing interim learning steps. Partial approaches and compromises are to be expected and should not be a reason for discouragement; this is a collective effort where small steps forward and hard-won progress, however minor, will help the whole sector to accelerate.

6.2. Circular economy and materials

Like many sustainability aspects considered in the built environment, circular economy principles (discussed in Chapter 1) are generally considered at the scale of a single building or development.

Important as this is, limiting the potential of a circular economy approach to single buildings restricts the potential benefits and opportunities. Increasingly globalised supply chains widen the impacts of manufacture and extraction to a huge extent, far beyond the footprint of an individual building. A similarly wide approach to considering how impacts can be reduced is therefore appropriate.

The UK Green Building Council (UKGBC) outlines a set of broad principles to promote the circular economy in construction (see Figure 6.1), however most buildings today are not designed according to circular economy principles.

Circular design of a new building or a retrofit requires a completely new process and mindset. Kanters (2020) carried out a series of semi-structured interviews with architects and consultants that have engaged in circular building design to identify the barriers and drivers of the transformation towards a circular building sector. The main barriers are due to the conservativeness of the building industry, the lack of political priority and dependencies throughout the building industry. However, there is a growing interest and commitment from supportive clients to drive this circular practice.

At a larger scale, beyond the individual building and even beyond the urban framework, it is easier to factor in the broadest possible interpretation of circularity, not just materials considerations but natural resources including water and energy, sustaining natural processes including natural

Figure 6.1 Circular economy principles for construction (adapted from UKGBC (2019) with kind permission of the UK Green Building Council)

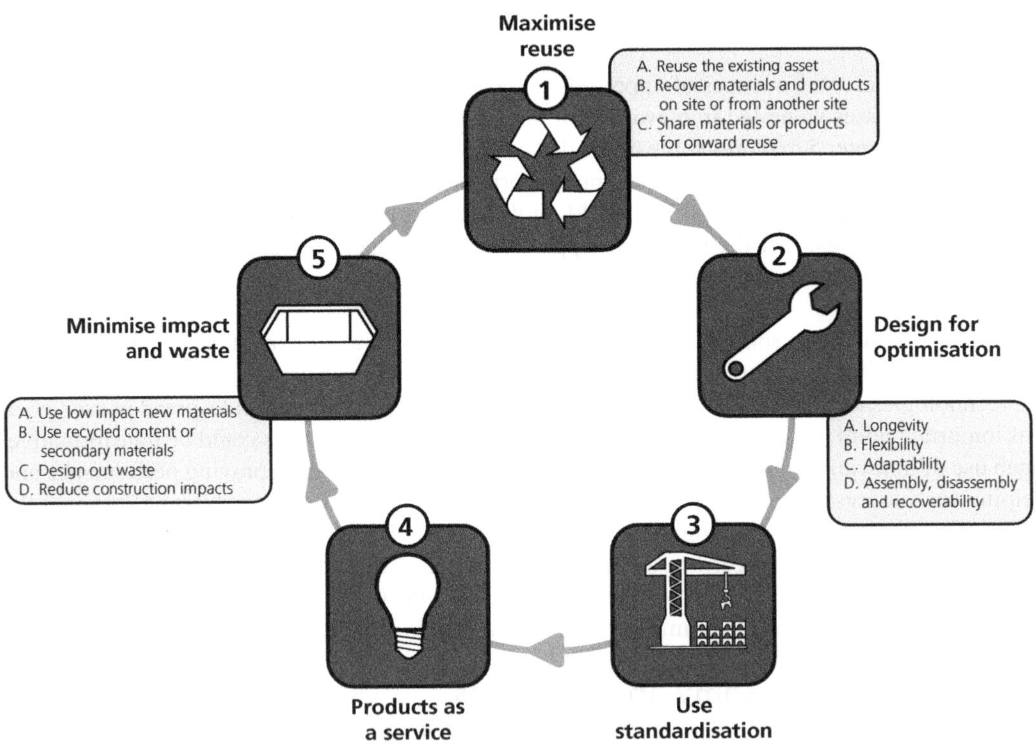

ecosystems and organisms. Taking account of a greater number of inputs to the circular process adds complexity, but it enables different combinations to be made to expand the range of useful outputs available.

6.2.1 City scale circular economy strategic thinking

Circular economy thinking at the scale of a building, piece of infrastructure or product can be scaled up to achieve a joined-up circular economy at the scale of a city or urban area. In practice, this would mean that all these different components of a city are durable, adaptable, modular, easy to maintain and repurpose and locally sourced, with everything capable of being reused, composted or recycled (Ellen MacArthur Foundation, n.d.). Because of the density and concentration within cities and urban areas, individual actions to embed circularity scale up to achieve considerably greater impacts. In addition, the concentration of opportunities could improve ability to innovate and create greater local benefit.

Achievement of a city scale circular economy requires each unit of the city and each component to play its part. Taking the Netherlands as an example, one of its circular economy initiatives is turning large infrastructure projects circular. Circular Copenhagen has been developed as an innovation platform run by the City of Copenhagen with the aim of developing tomorrow's circular economy solutions for today's challenges. Oslo is currently the first city worldwide to adopt a Climate Budget; the budget includes several measures quantifying changes and emission cuts needed by 2030. Circular economy is key to achieving the emission cuts to meet the budget. In Belgium, BC Materials was set up to solve this problem, with 100% of its operations committed to circular economy. BC Materials recovers surplus earth-mass from construction sites and transforms it into construction materials. Its ambition is to help the building sector lower its carbon emissions, reduce waste and create healthy and high quality environments for working and living. The 'Gent Waste Brick for DING' is such an example: it is a low carbon brick made from recycled municipal waste from the city of Gent. The brick has been developed for the construction of the new wing of the Design Museum Gent. This level of circular economy requires strategic leadership and commitment from all the clients, designers and constructors working together.

For built environment practitioners, understanding how the particular project or opportunity that is within their remit fits within the whole supply chain is essential. This will enable a broader context to be understood when considering the opportunities and decisions for each individual project. Coupled with a built environment practitioners' network that is committed to seeking opportunities together, major change could be achieved in the short term.

6.2.2 Maximising retrofit and renovation, adaptive reuse

There has been an ongoing dialogue in the built environment industry over several decades regarding the relative merits of renovation and demolition-rebuild. In the past, decisions have tended to be made on a purely financial basis, balancing the relative costs of upfront investment against life cycle running costs. Often, the complexity and investment of time and consideration into renovation, led to the balance being found in favour of demolition and re-build. More recently, the acknowledgement of a need to factor in the environmental consequences of demolition waste, the ecological and environmental footprint of new materials and the potential loss of social and historic capital have somewhat rebalanced the decision, even in cases where the existing building is in poor condition (Alba-Rodríguez *et al.*, 2017).

The concept of adaptive reuse of buildings and urban places is by no means new, with a burgeoning interest in architectural fields particularly in recent decades (Lanz and Pendlebury, 2022), often focused on the beneficial re-imagining of post-industrial buildings. In the context of the post-pandemic and technology-available world, urban centres are often struggling with reduced demand for office and retail space, leading to vacancies that could present an opportunity for repurposed uses. It is evident that many of these vacant buildings are no longer fit for purpose, being symbolic of a 'bygone era' (Construction News, 2024). Modifying these older, disused buildings as housing or for leisure use has been seen as a perfect solution in many parts of the world, incentivised in some cases by enacting (or removing) legislation, or by fiscal incentives. However, relaxation of legislation to drive such reuse projects has had some unfortunate consequences, where the conversions to a different purpose (such as, from office use to housing) were poorly executed, in some cases failing to deliver adequate daylighting or space (Clifford and Madeddu, 2024). This is a consequence of failing to understand that adaptive reuse requires careful consideration and design, in just the same way as construction of a new building.

The UKGBC (Hucks, 2022) highlight a series of steps that are essential when considering retrofit and renovation projects, the first of which are *understanding the building* and *assessing what is required*. These apparently simple steps are not necessarily straightforward, however, when considering buildings whose materials and components may be hidden within structural components with little information available. New thinking in the built environment industry is now beginning to allow time and resources to investigate the existing materials and structures, creatively considering different options rather than a default 'demolition' option. New techniques for retaining and repurposing materials, particularly the major components such as structural grid and facade elements, are being demonstrated and discussed. To help this process, sources of information to enable renovation many years in the future is being collected in some innovative projects (such as materials passports, discussed below).

Brand's building layers concept (Brand, 1995), shown in Figure 6.2, provides a clear illustration of building elements that helps designers and real estate developers in their design, construction, maintenance, re-use and decommissioning processes. The structure of a building tends to have

Figure 6.2 Brand's building layers and their lifespans (photo credit to Rosenfeld Media, licensed under CC BY 2.0 (adapted))

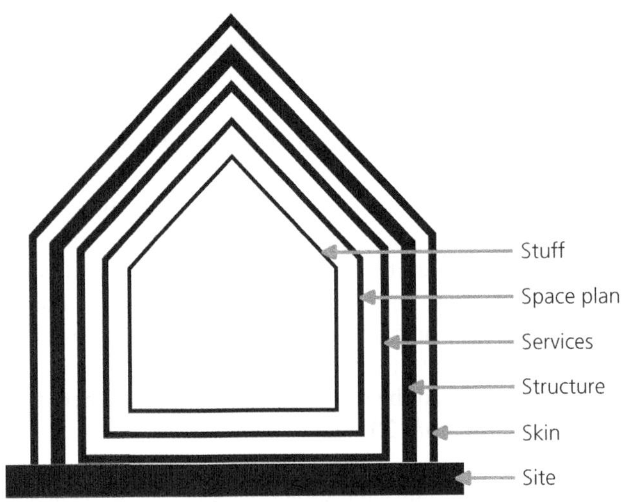

a lifespan of more than 60 years, whereas services inside the building typically last 15–30 years before a retrofit or an upgrade may be necessary. The building envelope, depending on construction materials, has typically 15–60 years. Conceptualising this building as a series of layers helps designers plan for easy refurbishment, providing easy access to short-lived components for regular maintenance, repair and upgrade. At refurbishment stages, layers requiring replacement can peel off without damaging the neighbouring layers. In order to improve adaptability, designs that allow a flexible floor plate and ensure that the primary structure is independent from others, would facilitate deconstruction in the future. When designing services, adopting a 'plug and play' approach enables future fitout requirement and reduces waste generation.

6.2.3 Design for disassembly and reuse

Design for deconstruction or disassembly is a way of ensuring that, at the end of a product's life, the different components can be easily taken apart for recycling or reuse, or easily modified into a new purpose. A simple concept, but to achieve it successfully it must be a target from the first stages of building design. Enabling buildings to be separated into structural and non-structural elements and enabling products to be separated from each other allows for easier maintenance as well as increasing the end of life options (Cheshire, 2019).

As one of the key challenges in applying circular principles to the built environment is the time to consider and discover the options for repurposing or reusing the building elements, it follows that a building designed to be easily separated from the rest of the components will minimise the time and resource needed for this process. Elements that have been assembled to enable them to be extracted will likely retain their value better, as they are less likely to have been damaged in the deconstruction process.

Many historic and traditionally constructed buildings can more easily be adapted and find new purpose for this reason – older buildings were constructed without the 'benefit' of modern composite materials, glues and bonding agents and without the many types of chemically complex components now commonly specified in building materials (Guy and Ciarimboli, n.d.). This made them simpler to modify, to repair and, ultimately, to sustain their usefulness for a longer period. A return to simplicity utilising fewer materials and less complex manufacturing techniques would benefit design for disassembly; this may also reduce the embodied environmental impacts of a material by a considerable amount (Roberts *et al.*, 2023). Connections have also been made to the use of modular construction and fabrication, enabling modular components to be replaced or reused in entirety in the future. Supporting circular economy principles, this enables the highest possible value of the material to be maintained for longer.

6.2.4 Materials passports and materials repositories

Materials passports provide information or data about a material or product in one trackable place, enabling easy access to required information. When collated together, materials passports can aggregate up to a central repository of data at the scale of a building or development (building passports/building logbooks are also used as terminology in this case).

Such passports provide the ability to contain a wide range of information including environmental characteristics, health and safety information and warranty or certification details in an easily accessible single repository. This type of approach can be used to record the materials and products used in a new building, for future use, or to investigate the materials used, and characteristics of an existing building. For the latter, this is a useful method of investigation prior to renovation or retrofit, to assess which materials can be reused, and how the different options compare in terms of environmental footprint (Owens, 2024). In the European Union (EU), digital product passports were highlighted as an important future opportunity in early iterations of their Circular Economy Action Plan (European Commission, 2020). In the latest version of this action plan, being drafted at time of writing, mandatory digital product passport (DPP) use for specific product categories is likely to be introduced.

Evidently, if such passports are to have maximum impact, it will be necessary for a range of actors along the value chain to be able to access the information easily, both now and in the future. A standardised approach is helpful because it will increase confidence that the information will be available when it is required. At this early stage of conceptual development of the materials passport concept, several organisations and government agencies are considering different ways of standardising the approach but no front-runner process has yet emerged. Naturally, it cannot be known at this stage what information future built environment professionals will find useful, but equally, it would be excessively bureaucratic and time-consuming to capture everything about a building project. Crowd-sourcing the views of the built environment industry on the most appropriate information to be captured is therefore a vital part of this work. The frameworks that have been developed to date (Costa and Hoolahan, 2024; Waterman, 2023) have taken this approach, working closely across the built environment and academic sectors to future-cast information needs and design appropriate repositories. A range of platforms for recording, storing and using the data to model options are now emerging (Madaster, n.d.; Upcyclea, 2024) and in use among pilot projects and vanguard organisations.

The gathering of a repository of information about materials contained within a building effectively allows buildings and urban areas to be seen as materials 'stores' that are a potential source of material in the future, once they have reached the end of their useful life. Such a change of mindset regarding the potential value of already-used materials is not unique to the built environment and is a result of realising the need to reduce or eliminate the damaging impacts of securing virgin materials.

Edenica is the first office development in London that systematically adopted a standardised approach to material passports. The 8400 m^2 speculative office development was designed by Fletcher Priest in collaboration with consultant Waterman. Four thousand, eight hundred and seventeen building elements representing 80% of the building's mass were passported, hosted on the Circuland online platform, from the structure and piling through to individual components such as the glossy blue bricks that contribute to the building's colourful identity. The hope is that this could feed into a future reuse market or recycling process for components, with the passport providing precise information on each element, such as the embodied carbon of a particular piece of steelwork, rather than generic information on the building's steelwork as a whole. This information can then be exported to external marketplaces when required (Buxton, 2025).

Recovering valuable materials from existing products and waste has gained significant attention, particularly in relation to electronic devices used in everything from renewable energy systems to consumer electronics. Considerable waste streams exist for products containing useful metals and other raw materials; demand for such products is still growing but the environmental and social (and economic) consequences of mining raw materials is becoming increasingly unacceptable. Therefore, it makes sense to seek to recover such materials from the existing waste stream.

In the case of consumer electronics that form a considerable element of municipal waste, recovery of valuable components is known as 'urban mining', demonstrating the value of the recovered materials. At present, such recovery is difficult and complex, due to the embedded nature of the components, but as methods of recovery are developed, recovery rates will be improved (Michielin, 2023). For buildings that often have materials that can be recovered more easily, the opportunity for 'mining' of a building's materials is a significant opportunity; this would be greatly aided by the existence of materials passports or other information repositories.

6.2.5 Building Information Modelling for the circular economy

Building Information Modelling (BIM) is a collaborative process to improve the specification and delivery of information throughout the whole asset life cycle (RIBA, 2024b), relating to the collection of specific information defined through the ISO 19650 series of standards (ISO, 2018). The information gathered and stored in the BIM process is used to create an asset information model, containing detailed validated and approved data that can be used during the operational phase of the building and during retrofit. To enable this model to function as it should and to enable it to continue to be a useful source of information in a future retrofit scenario, the data and information contained within it must be regularly updated, such as when any repairs or upgrades are completed. While BIM was not originally proposed with a primary aim to promote reduced life cycle impacts or circularity, it is easy to see how useful such information would be when considering any type of reuse or disassembly.

While use of BIM is by no means universal, use has been steadily growing and the most recent estimates (NBS, 2020) are that awareness of the BIM process in the UK is almost universal, with 73% of construction professionals using BIM. The application of BIM and the useful information it could contain is very much determined by the client and without specific direction to do so, the design team may not capture information that would be vital to consider in the reuse/disassembly phase. Essential information that might not be considered necessary for a maintenance purpose (such as the composition of partitions and structural grid components) would be vital to understanding the options for reuse, for example. Without capturing this information at the design and construction stages, it is much more costly and inconvenient to find this information later.

Existing materials in a building should be looked at as a bank for future projects. BIM can play an important role by not only providing spatial data, but also if developed properly, providing technical information for future re-use (Charef, 2022). The Edenica project is the first major pilot adoption of BIM with material passports hosted on the Circuland online platform providing materials bank information when required for future re-use (Buxton, 2025).

6.3. Design and construct for material efficiency and minimisation

As part of a desire to minimise the impact of the whole building life cycle, attention is increasingly being paid to minimising the construction impact of materials, utilising innovative technologies and methods to reduce the volume of raw materials used and divert waste materials from disposal to reuse. Innovative design methods and optimised retention is essential as part of a circular approach, but it is evident that despite this, new materials will continue to be needed in the built environment, even when reuse and refurbishment is maximised.

The life cycle impact of producing new materials is astonishingly high globally, with cement and concrete alone contributing around 7% of global carbon dioxide emissions (Global Cement and Concrete Association *et al.*, 2023). The high carbon dioxide footprint of these materials is because carbon dioxide is produced as a gas during the transformation of raw materials limestone and clay during the manufacturing process. High heat is also needed for manufacture, which is generally provided by fossil fuel sources. After first minimising the volume of new material that is needed to deliver the project objectives, attention is now turning to materials and building techniques that can minimise life cycle impacts, deliver carbon reduction or maximise other benefits. Some of these materials are new, a product of scientific and engineering innovation, while others are a re-discovery of older building materials and techniques that are being adopted by a new generation of built environment professionals.

6.3.1 Materials with reduced embodied impact

The embodied impact associated with concrete and cement has been known and understood for many years and even in the early 1990s products and techniques were becoming available that sought to reduce the use of virgin material. Use of demolition waste to provide sub-base and lower quality aggregate has been commonplace for decades, but for higher quality or structural uses, waste aggregate began to be used to replace a portion of the aggregate in a concrete mix. An early innovation was the production of concrete products such as blockwork and pre-cast units that incorporated waste materials, primarily from industrial waste streams such as from blast furnaces (known as GGBFS) and waste ash from power stations.

Increasing the proportion of waste material can be difficult due to the strict building codes and standards that structural concrete must meet. Recently, however, there have been experiments in replacing 100% of the aggregate with waste materials, for certain applications (Farooq *et al.*, 2023) and also experiments with novel waste materials such as plastic or glass as aggregate. Product development is also ongoing in creating concrete that actually absorbs carbon dioxide, locking carbon dioxide into the material (carbon sequestration). There are various methods including injecting captured carbon dioxide into the material during the curing process, through to manufacturing aggregate with a combination of carbon dioxide and industrial waste materials. There are also studies ongoing into methods of manufacture that would increase the amount of carbon that the concrete absorbs over its lifetime. Despite these very active research streams, the sheer volume of concrete and cement used across the globe makes this an area of very high environmental impact despite such advances.

6.3.1.1 Low carbon concrete and structural materials

Finding alternatives to materials such as concrete with high residual carbon footprints (even if the lower impact concrete is specified) is not always possible, where a particular structural load must be guaranteed, for example. In some cases, however, such as where no load-bearing is required, or in lower risk applications, replacement with other materials is certainly possible. Lime-hemp concrete (sometimes called 'hempcrete') mixes the woody plant hemp with lime and water to create a concrete-like material that can be used in some applications as an alternative to concrete block or slab materials, or even as a fibrous material in asphalt and surfacing materials (Muhit *et al.*, 2024). Like all plant materials, the hemp stores carbon in its structure as it grows; use in construction can therefore sequester carbon for the duration of the material life cycle. Other innovative biologically based materials also being actively explored and trialled (Alaneme *et al.*, 2023) include mycelium-based materials that are manufactured from specific fungal growth on a variety of possible materials (including agricultural waste products). When treated, the final material has been proposed for a range of construction products such as window and door frames, panel boards, insulation and others. Biodegradability at the end of life and the lower embodied carbon emissions are stimulating interest in these and other alternative materials that is likely to see their importance and availability grow in the near future.

Aside from these manufactured biologically based materials, use of timber as a building material has of course been commonplace for centuries. Use of manufactured materials such as steel and concrete has supplanted timber as a structural material in many parts of the world, because their replicability, consistency of product and, not least, the structural loads they can support are attractive. However, in an era where embodied carbon emissions are becoming of paramount importance, timber products are being considered mainstream once again. A report from the Biocomposites Centre at Bangor University published in 2019 under the auspices of the UK Climate Change Committee (Spear *et al.*, 2019) concluded that a reduction in embodied emissions of approximately 20% could be made by substituting timber frame for masonry in the UK (with even greater reductions where timber replaces concrete structures). An example of using such a timber-concrete-hybrid structure is showcased in Chapter 7, Section 7.3. The ability of timber to sequester (store) carbon within its structure means that atmospheric carbon emissions are locked inside the timber structure for the life cycle of the material. The report also considered engineered wood products such as cross laminated timber (CLT), which due to their greater structural properties, can replace concrete or steel frame in some applications. In that case, the reduction in embodied emissions was even greater, at around 60% compared with a concrete alternative.

CLT is manufactured from sawn timber laid crosswise and glued using structural adhesives, creating high stability suitable for wall and floor elements similar to concrete slab. Another engineered timber product, glulam, is manufactured by building up multiple layers of timber and bonded using adhesive, creating sizes that can be constructed in large lengths and, therefore, suitable as beams and columns. Use of such engineered products helps to overcome concerns about the risk of using timber in structural applications, as their properties can be demonstrated and certified to relevant building codes. Taking into account their comparable construction performance and their superior environmental performance, these materials provide an attractive opportunity whose use has been proven in a variety of real-world applications (Abed *et al.*, 2022).

These materials have wide applicability to northern hemisphere construction projects where timber has a long history of use. In southern hemisphere projects and particularly in the tropical regions, there is a longer tradition of using materials such as bamboo, which is exceptionally fast growing and likewise can be used effectively to replace high embodied carbon materials and sequester carbon. Bamboo's ability to be used as a structural and panel material and to be manufactured as board has great potential and is already being incorporated as a standard product in building codes in India (e.g. Krishna, 2020). Like softwood timber, much low-rise, vernacular housing is already constructed using bamboo. Applying building codes and standards to a non-uniform grown product is not a simple challenge to overcome for structural materials, but bamboo is already finding a ready market in non-structural applications such as flooring and panelling, even in areas where it is non-native, comparing favourably even to softwood timber. If uniformity challenges can be overcome then there are significant environmental benefits to using this material more widely (Harries, 2025).

At the other end of the technological spectrum, graphene concrete is a recent and very promising innovation. Prototypes have used the newly isolated material graphene fused with polymers and additives to develop graphene-enhanced concrete and production systems. It has been claimed that the manufacturing process can eliminate 88% of carbon dioxide emissions by avoiding the use of cement. The concrete compound also provides four times the compression strength of traditional concrete, despite being 30% lighter and faster in production. A similar process using graphene as an additive can also be applied to transform ground waste tyres and plastics as a sustainable waste upcycling platform. The company behind this, Graphene Innovations Manchester Ltd, won The Spectator's Innovator of the Year Award for Excellence in Sustainability in 2023 (University of Manchester, 2023).

6.3.1.2 Circular re-use of waste products

In line with circular economy principles, attention is increasingly turning to use of materials that include waste streams. Use of construction waste and demolition materials has been commonplace for years, as hardcore and aggregate, for example. But use of waste streams from other industries as a construction material has been much less common. The slow speed of adoption of innovation in the construction industry has meant that changes in waste stream components can take considerable time to gain acceptance as a mainstream material. Utilising waste products and specifying reduced impact new materials can considerably reduce the overall life cycle environmental impacts of a building and such materials will become commonplace as confidence in their quality and use grows.

Another example of potential sources of waste materials that could prove useful in construction projects is plastic manufacturing and waste. Plastic production has doubled since the start of this millennium – going from 200 m metric tonnes to over 400 m metric tonnes in just 24 years (Statista, 2025). More than 350 m metric tonnes joins waste streams each year with less than 10% recycled. Depending on the type of plastic, the waste can take more than 400 years to reach end of life and even then, the material does not fully decompose, breaking down into microscopic fragments of plastic ('microplastic' or even smaller 'nanoplastics') that can embed inside organisms causing harm in species of all kinds, including humans (Ghosh *et al.*, 2023). Of course, the ideal solution to this major pollution risk is to eliminate original plastic manufacture, replacing with more benign materials. Until such solutions are widespread, encapsulating plastic waste within long-lasting materials, such as those used in construction, is a sensible short-term potential solution.

Construction is a high consumer of materials in nations across the globe and, therefore, development of building products that can utilise plastic waste could reduce the volume of plastic material requiring disposal. Disposing of plastic is not a simple matter due to its robust nature and its long life. However, these are precisely the qualities that could make plastic waste attractive as a component of building materials. Certain plastics can be used as aggregate in concrete or asphalt mixes, rigid plastics in place of timber members or in insulation material, for example. By utilising wastes at the end of their useful life rather than raw materials and encapsulating them into a long-life construction product, the environmental benefits could be significant, reducing a problematic waste management issue, minimising material extraction and reducing the environmental contamination that can result from plastic waste disposal. New products of this type therefore have great potential in the short term (Awoyera and Adesina, 2020).

The construction industry is also a consumer of new plastic and therefore, in addition to utilising plastic waste streams, there is a great opportunity to specify products that avoid the use of plastic in favour of lower life cycle alternatives, such as bioplastics. Bioplastics are a range of materials that can be manufactured from various plant-based sources with a large spread of different characteristics, applications and impacts, depending on the type and composition of the bioplastic material. Their use in construction can range from additives in concrete mixtures, insulation products, facade cladding or surfaces or even structural elements. Strunge (2020) reported their research experiments through a collaboration between sectors on viability in using mycelium to leverage upcycling of local waste to produce valuable interior building components such as regenerative biomaterials grown from waste, with the overall aim of enabling regenerative building design.

6.3.1.3 Extending useful life cycle of materials

Extending the life cycle of construction materials is another way to minimise the environmental impacts of their use. Materials that remain in good condition longer will be less likely to require demolition and the environmental impact resulting from their manufacture is spread across more years of life, minimising the net impact. Recent research is attempting to find ways to further prolong the life of construction materials so that the investment of carbon made in their manufacture can be better justified.

Concrete infrastructure is prone to ageing deterioration over time; cracks can appear caused by climatic conditions including humidity, temperature changes or stresses in use. Concrete can to some extent 'heal' minor damage as water enters the concrete and binds to materials that remain un-hydrated from manufacture, filling the small crack. This only works in the presence of water and will only fill very small cracks. Self-healing concrete is now being produced that includes micro-capsules that would rupture in the event of a crack, allowing the chemical compounds to bind with the material and fill the crack.

There are now prototypes of 'bioconcrete' that incorporate bacteria or fungi in the manufacture of the concrete mix. In the event of a crack developing, the biological elements produce calcium carbonate in the presence of water, filling the crack and repairing the structural integrity of the concrete (at least, under test conditions) (Amran *et al.*, 2022). Evidently, such innovative methods will take some time to become market-ready, but this is a particularly promising area of opportunity for concrete structures that may be difficult to access, such as sea defences, offshore structures or more dangerous environments that cannot be easily inspected or repaired.

6.3.2 Materials that deliver additional in-use benefits

Building designers have always specified materials on the basis of their in-use benefits as a matter of course – for example, glazed panels and windows to let in natural light and reduce need for artificial lighting, thermally heavy materials to buffer excess temperatures, smooth or rough surfaces to improve acoustic performance. But a new suite of innovative materials is now becoming available that deliver enhanced performance over and above those already available techniques.

Thermal energy storage has been a particular focus of recent development, as a method to help management of indoor temperatures, particularly as the global temperature is rising rapidly. Use of the latent heat storage of phase-change materials is a potential opportunity that has resulted in several prototype and pilot installations now becoming ready to deploy more widely (Reddy *et al.*, 2024). Phase-change materials encapsulate a chemical mixture that can change phase from solid to liquid (or for some products, from a crystalline mixture to an amorphous mixture) when heated (absorbing heat) and then return to their original state as they cool. Applications include wall materials, ceiling and floor materials and, potentially, even structural materials such as block or brick. With the ability to reduce temperatures by a few degrees and without much controllability, these materials are not a single solution to overheating in buildings, but could become part of a solution when considering the building as a whole system. In a warming climate, with many homes and other buildings already constructed without the cooling systems that would help them to stay comfortable, simple retrofit materials that could absorb heat at peak times could be a very helpful tool.

The recent Covid pandemic placed a particular focus on indoor air quality, drawing attention to the lack of ventilation in many homes and places of work that can increase exposure to harmful substances. Emission of 'volatile organic compounds' (VOCs) from chemical applications, materials and furnishings has been known for many years and over time, some countries have introduced mandatory codes for such materials that have helped to reduce their sale and use. Quantified research into indoor air quality has been lacking and the great variation between different sources of pollution as well as the interaction with external sources have made indoor air quality an impossibility to regulate in any meaningful way (Lewis *et al.*, 2022). Nevertheless, it is recognised that air pollution from industry and fossil fuel combustion, together with off-gassing materials and chemical products used in buildings combine to make indoor air quality a considerable issue.

New materials are now being developed that not only reduce the amount of pollution they release, but also contain substances that can reduce existing air pollutants. Incorporation of titanium dioxide nanoparticles into materials ranging from paints and surface coverings to concrete mixtures, allows the material to react with pollutants such as nitrous oxides to degrade and remove them from the air (Jenima *et al.*, 2024). There are also promising developments in other materials that adsorb air pollutants onto their structure, such as metal-organic frameworks, that hold considerable promise for the longer-term future. Bio-materials that support moss growth and other bryophytes or plants can absorb and break down external air pollutants and can also have the benefit of reducing surface temperatures and absorbing surface water, slowing the flow of rainfall into urban sewers (Perini *et al.*, 2022).

These and other materials and technologies that deliver multiple benefits are illustrative of a growing realisation in construction and the built environment that buildings must deliver more than 'just' a structure. Delivery of multiple benefits for nature, for people and for environmental performance can now be demanded and innovators are focusing on this area in their research and development. In coming years, such materials will become more mainstream.

6.3.3 Minimising materials impacts with three-dimensional printing

Alongside innovations in new materials, new construction techniques and technologies are becoming more accepted. Three-dimensional (3D) printing (additive manufacturing) is a new and innovative construction technology; a system that can convert a digital design model into a physical object using different raw materials. 3D printing technology has been used for some decades in manufacturing and industrial prototyping, but is now making its way into the construction industry with large growth in use expected in the short to medium term, particularly in the Asia-Pacific market (Bidwai, 2024).

The largest uses to date have been to speed up and shorten supply chains for concrete components that are still often batched and constructed on site. The benefits to on-site manufacture using 3D printing could be considerable – lower waste, reduced materials use and reduced transportation impacts. The ability of 3D printed materials to be manufactured on site to high specification and assembled very quickly has meant that initial experiments tended to focus on rapid deployment for emergency housing and for residential low-rise applications, for prototype structures such as pedestrian footbridges, or for unique and complex forms that show the ability of the technology to deliver bespoke architectural visions (Guamán-Rivera, 2022). 3D printed materials can be used alongside or in place of traditional concrete or metal forms and have also been demonstrated utilising a range of post-consumer or post-industrial waste materials.

Use of sustainable materials and site-won materials such as local clay and soil or bio-materials such as straw, fibres and rice waste have been successfully incorporated in some of these experimental buildings (Di Salvo, 2024).

While there are undoubted benefits for sustainability and constructability, the lack of current codes and standards for 3D printed construction materials is likely to cause concern in a risk-averse construction industry (El-Sayegh *et al.*, 2020). Nevertheless, considerable research has focused on the specific applicability of particular materials, the environmental and economic benefits (cost- and time-effectiveness) of the solution. This research and successful case studies of its use has shown that the technology has enormous potential in construction. For specific uses and for complex non-structural solutions, 3D printing can be expected to be deployed much more widely in the near future and may become widespread if building codes and standards are updated to include such materials.

6.4. Harnessing new digital technologies

The built environment and construction industry has not been at the forefront of use in new digital technologies and techniques. Even the use of BIM has been relatively slow to be adopted and the industry still uses two-dimensional or even paper-based plans and records in many projects. This is now starting to change, albeit slowly, but the increased digital literacy of workforces and participants in built environment projects should mean an acceleration of take up. This means that the construction industry is in a good position to adopt techniques and technologies already tested and adopted in other industries, leading to rapid deployment.

6.4.1 Satellite data and technology

Satellite technology has become ubiquitous across much of the global community with the use of satellite generated mapping and visual information and location tracking using, for example, Google Maps and similar applications. Construction and built environment professionals do regularly use these technologies but often to a relatively simple level of integration into projects. On a day-to-day basis, the construction industry is finding satellite transportation management and locational tools to be highly accurate and helpful to manage the movement and storage of materials. However, there are many more opportunities for integration of the data that exists, which are now starting to be explored outside of research projects in real-world construction and built environment applications.

Modelling transformative energy retrofits across large areas is possible using satellite-derived data to estimate energy demand and opportunities in complex urban locations, combined with actual data from buildings where available to build an energy model across large areas (Anand and Deb, 2023). Similar applications include use of thermal scanning satellite data to identify likely targets for energy retrofit across large areas, enabling interventions to be better targeted, and identification of potential areas for biodiversity interventions and nature recovery.

Satellite data could also play a role in managing risks, factoring geospatial data into decision making for asset location or acquisition, taking into account risks of flooding, fire or other risks that have been exacerbated by climate change. Monitoring physical assets is becoming particularly important in dealing with risks of extreme weather and disruption, and there is good potential for future structures to be 'tagged' and monitored remotely using satellite data with high levels of accuracy – for example, embankment structures or reservoir walls, where slight movement that might indicate instability could allow early intervention preventing crises.

Satellite operators are working closely with innovation agencies to identify further appropriate uses for this valuable data, including inclusion in built environment digital twin models.

6.4.2 Digital twins and siblings

A digital twin is a virtual representation of an object or system that spans its life cycle. Importantly, a digital twin is connected to its real-world twin by data streams that constantly update the digital twin in real time. A digital twin can be used to simulate options, decisions and scenarios. It can also be integrated with artificial intelligence (AI) and machine learning to inform decision making based on predicted outcomes and learned responses. Examples in the built environment could include monitoring of air quality alongside traffic flows, monitoring water levels in drainage networks, or management of energy grids alongside decentralised energy generators. On the scale of a single building there are still many applications, from optimising the design process, through better construction management, to managing the completed building based on, for example, weather or floor-by-floor occupancy patterns and demands. It is easy to see how useful such a tool could be, if the data and usability of the model can be maintained and made available in a manageable way.

Sensat, an innovative software company in London has developed digital twin products focusing on civil infrastructure applications. Sensat digital twin architecture comprises five core components, serving as a valuable framework for the industry, enabling identification of current technical limitations, highlighting opportunities for innovation and development, and ultimately driving the evolution of more powerful and impactful digital twin technologies. A digital twin model of this type combines massive amounts of complex data – from surveys and 3D designs to GIS and programme information to empower decision making and drive efficiency across critical infrastructure projects.

Civil organisations are often managing complex interconnected networks that are relied upon to serve the health and wellbeing of communities. The risk of failure is high, but it is also vital to keep operational costs to a minimum, to minimise the cost to the public purse. An early example of digital twin ability to deliver these benefits has already been demonstrated in conjunction with a water supply and treatment company, showing significant time and cost savings (Sensat, 2025) see the illustrative process in Figure 6.3.

Figure 6.3 Digital twin in civil infrastructure plan/design/build/operation

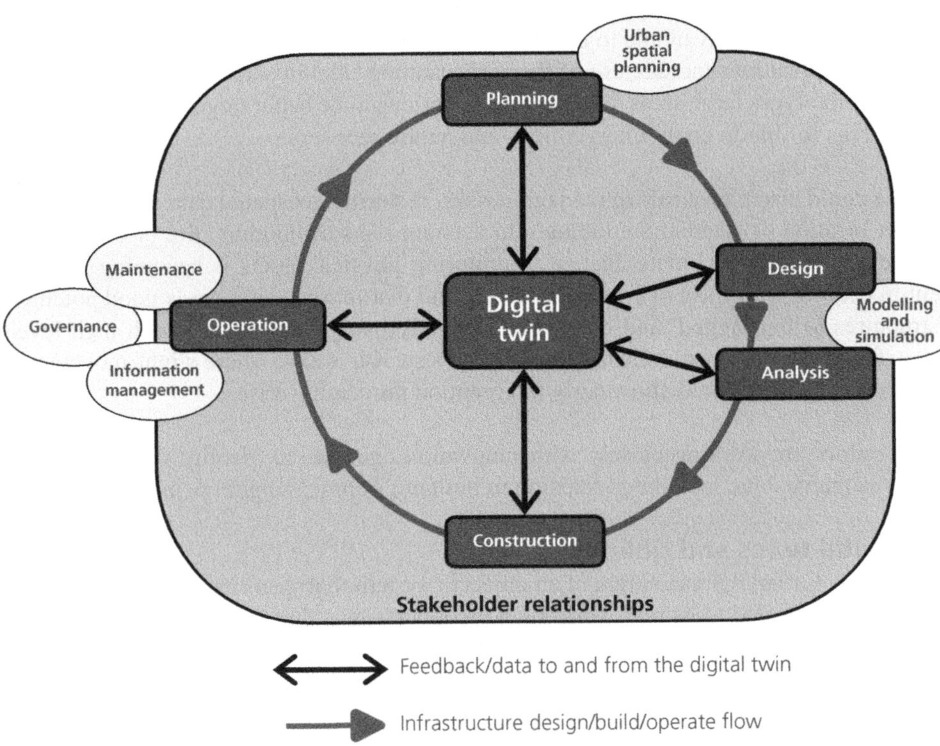

As of end 2023, the number of devices worldwide connected to the 'Internet of Things' is estimated at 16.6 billion (Sinha, 2024) and forecast to increase some 13% to 18.8 billion by the end of 2024. This includes devices connected to wi-fi, bluetooth and cellular internet sources, as well as wired connections. Just the number of devices already populating a single building and feeding information into connected data systems is sufficient to create a digital twin-type model, and the near future will see exponential growth as more and more products come with digital connectivity built-in. This connectivity provides evident potential benefits for control systems and for management of buildings as well as data systems that can benefit design and construction of buildings themselves.

In Sweden, a Digital Twin Cities Centre was established (Digital Twin Cities Centre, n.d.) as the foundation for digital planning, design, construction and management of sustainable, intelligent and liveable cities and regions throughout Sweden by 2030. The Centre supports the full integration of the digital twin concept in all parts of the planning process for the built environment. Utilising this technology in the right way has enormous potential for tackling some of the most pressing urban challenges, such as flooding, pollution, traffic congestion, spreading of fire and diseases. Gonzalez-Caceres *et al.* (2024) presented a case study demonstrating this potential in Gothenburg using an automated workflow in digital twinning for multi-domain environmental performance analysis of urban developments.

A recent systematic literature review of digital twin implementation in the construction industry (Omrany *et al.*, 2023) found that digital twin technology 'has the capacity to revolutionise the construction industry', while recognising that challenges with data management, privacy and safeguarding and governance need to be overcome to realise full potential.

6.4.3 Virtual and augmented reality

These linked technologies have been common in media, game and entertainment industries for many years but are now starting to make real impact in the built environment and construction sectors. Virtual reality presents a simulated experience, where the images, objects and surroundings appear to be real while being fully virtual. Augmented reality presents a combination of the real world overlain with computer-generated virtual imagery or information. For many years, the difficulty of presenting proposed or emerging designs for client or stakeholder approval has been overcome by using artist impressions or, more recently, photo-realistic images. Virtual reality is now relatively common in enabling people to experience a proposed building or project by 'walking through' a virtual model to see how the user will experience it when completed. Augmented reality can be used to 'scan tag' objects, materials or completed structures with information on maintenance, safety, embedded services and much more. This could be used during construction as well as post completion. As such, it is an extension of BIM processes and data collection that aims to present the information immediately, where and when it is required.

Augmented reality has also been used to interact with local communities and the public, before a building is constructed, during construction and after completion, to communicate information about community interaction with the completed building or surroundings.

6.4.4 Parametric modelling and generative (AI) design

Parametric modelling (design) is a method of designing products, components and structures whereby an algorithm-based modelling programme creates the design based on a set of programmed parameters. This could include, for example, the type of material, the site form, the required geometry or specific building requirements. This allows design options to be produced much more rapidly and more accurately even for exceptionally complex design challenges, such as the need to withstand extreme weather conditions or complicated geometries. Design professionals can then explore the design options and make adjustments to the evolving model to produce the final selected design.

Parameters and requirements also form the basis of generative design tools, but in general, the process of option selection is also built into the generative design model, so that the AI within the model selects the generated designs that most closely match the objectives required.

In both of these model types, the eventual design will only be as good as the initial parameters and data input at the start of the design process. Incomplete or inaccurate data or misleading weighting attached to certain parameters can potentially lead to unforeseen and unwelcome outcomes. In the generative design model, the AI embedded into the process will have a set of pre-learned 'parameters' based on previous learning experiences – for example, use of geometries used elsewhere that might suit the new parameters with which the model is presented. Evolutionary algorithms build on the inputs provided by designers in past projects to learn from their mistakes and to improve the parameters and how they are applied through the algorithm – over time, adverse outcomes will be minimised, to continually improve the accuracy of the result.

There is evidently huge potential for integration in the built environment industry using these approaches and the potential benefits for optimisation of components, materials and design could be immense. In the field of sustainability, many problems and constraints are now being faced by designers that have never been encountered (such as, rapidly changing weather patterns that are not allowed for in existing codes and standards). In such circumstances, even considerable experience on the designer's part may not equip them to assess the options fully. Integration and optimisation of multiple components are now often being demanded of designers (such as, shading, roof configuration to accommodate maximised solar PV energy generation, energy performance and more). These multiple competing demands add huge amounts of complexity that such design tools can help the designer to manage. Utilising parametric and generative design tools can help the designer to narrow down the options and arrive at an optimised design more quickly (Gradišar, 2022). Importantly, the use of design tools that can optimise designs to account for these complex demands reduces the risk of 'overdesign' and the risk of excessive material consumption that would result.

A recent report (Arup, 2025) found, in a survey of built environment professionals in ten countries, that 87% of respondents used AI or AI tools in their role at least weekly. Waste reduction and management and sustainable materials development were two of the main areas that respondents thought would be positively impacted by generative AI in the future.

6.5. Embedding wider-scale benefits

6.5.1 Embedding current and future social, cultural and community benefits

As well as environmental innovations presenting near-term opportunities, there are many social changes that are now starting to become important. In the near term, these changes will start to impact all aspects of built environment design and construction, influencing the type of development, layouts and designs.

One of the most notable changes is the older average age of the global population, a trend that is accelerating. The World Health Organisation (WHO, 2024) estimates that between 2015 and 2050, the proportion of the world's population over 60 years will nearly double, from 12% to 22% (2.1 billion). The number of people over 80 years will triple in this time. Higher-income countries have already started to experience this ageing population impact but this is a global trend that will impact all nations and countries and, therefore, all built environment projects will need to adapt to this change. In countries where the infrastructure and built environment has already been constructed, there are many potential barriers for older people – for example, in the London Underground system, only one-third of the stations has step-free access, effectively rendering the system significantly limited for people with mobility issues. The historic nature of much infrastructure in high-income countries means that modifying existing infrastructure and buildings to accommodate the needs of an older population is a great challenge, and costly to achieve.

Built environment professionals will need to ensure that the needs of an older population and those with potential health considerations are built into all future designs, to ensure that the development remains functional and useful to an older demographic.

Equally important is the function of the built environment in helping people to age well and more healthily. Physical accessibility will play a large part, enabling older people to work, socialise and participate in community activities for longer, with immeasurable benefits to their health and well-being. Provision of community spaces, both inside and outside of buildings is necessary, and buildings and homes that can be adapted to the specific needs of older people will become commonplace in the near term. Constructing the built environment to prevent exclusion is becoming more than just physical accessibility – for example, the consideration of sensory and cognitive aspects. In this respect, the built environment at the moment is lagging behind product designers and retailers who are realising the benefits of designing more inclusively, working with focused user groups during the design process to ensure that their needs are understood – a process that could be beneficially adopted in built environment designs (Zallio and Clarkson, 2021).

It is also the case that in most countries, birth rates are falling sharply. Across all Organisation for Economic Co-operation and Development (OECD) countries, total fertility rates (the average number of children that a woman would have during her reproductive years) fell to an average of 1.5 children per woman in 2022, well below the 'replacement level' of 2.1 children per woman (OECD, 2024). In some countries, including the UK, total fertility rates are at the lowest levels on record; the UK Office for National Statistics (ONS, 2023) records a fertility rate of just 1.44 children per woman in 2023. Combined with people living longer, the split of the population according to age is predicted to change considerably over the next few decades; within 50 years, almost one-third of the UK population will be over 65 years.

ONS confirm that net migration is projected to be the only source of population growth in the UK over the next 25 years, primarily working-age people who are likely to be recruited to replace older people who are no longer part of the workforce. Higher numbers of working people with different ethnic and cultural legacies brings an additional need for consideration in the built environment, again emphasising the need for inclusive design processes to be developed and embraced. Embedding cultural considerations into the built environment will become more than looking to the local vernacular for architectural inspiration and for locally derived materials (important though these are).

While cultural heritage has been a consideration in best-practice built environment design for many years, increased cultural diversity provides new opportunities for creativity – for example, a more ethnically diverse population may have different housing traditions, or different uses for outdoor social and community spaces. A vibrant and sustainable built environment will consider embedding these considerations now and providing adaptable spaces that can be changed to accommodate the population of the future, not solely the current cohort of users.

6.5.2 Embedding wellbeing and health benefits

Even in the early days of sustainable consideration in the built environment, human health and wellbeing was very much a key element. The early BREEAM schemes in the late 1980s and early 1990s included indoor and outdoor air quality, natural light and daylight, wind effects around buildings and other criteria whose main purpose was to minimise any negative impact of the built environment on occupant and community health.

That there are real and tangible links between building quality and human health has been clear for many years, since the days when insanitary housing was a focus of attention for Victorian campaigners and philanthropists. The scientific evidence is now becoming clearly documented and synthesised for the first time, however, allowing direct attribution of health impacts to be associated with the built environment. A recent systematic review of the evidence (Ige *et al.*, 2018) found that 'the built environment exerts one of the strongest directly measurable effects on physical and mental health', yet the consideration of physical and mental health has rarely been identified as a key outcome of building design.

Given the economic costs of ill health for occupants as well as for nations (both in terms of healthcare provision and impacts on workforce productivity), planning policy has been enhanced in recent years aiming to rectify the lack of health consideration of a proposed development. In the UK, London Healthy Urban Development Unit (HUDU) and NHS developed a HUDU method (HUDU, 2019), to enable designers to rapidly assess the likely health impacts of development plans and proposals, including planning frameworks and masterplans for large areas, regeneration and estate renewal programmes. Health impact assessment, which focuses on evaluating health impacts, is in addition to an environmental impact assessment, commonly required to be submitted with outline and detailed planning applications. Europe has put in place a wide range of policies and measures to reduce health impacts to overcome the slow progress in many areas of environmental hazards affecting public health (European Environment Agency, 2025).

Many studies comment on the high proportion of time that people now spend inside buildings compared with past populations, which will only increase in a rapidly urbanising global population. Naturally, this tendency to be enclosed within buildings means that any deficiencies in the design or maintenance of the building will have a proportionately higher impact on occupants. Much of the research to date has focused on the condition of homes and residential buildings, considering how improvements to housing can have a beneficial impact on physical or mental health. This is an understandable focus in countries where there has been a long history of urbanisation, leading to older homes whose condition does not meet modern standards. The impact is exacerbated in nations experiencing a rapidly ageing population (as discussed above), since older adults are more likely to spend high percentages of their time in their homes.

Somewhat surprisingly, given the focus on housing and residential buildings in the research literature, the earliest moves to include and codify health into building design were in the commercial building space. Starting with the embedding of specific issues into building standards, such as the Building Research Establishment Environmental Assessment Method (BREEAM) and the Leadership in Energy and Environmental Design (LEED), it soon became clear that the expanding nature of the issues backed with the latest science evidence required separate consideration. As a result, a standalone 'building wellness' standard, the WELL building standard, was launched in 2014, with a version 2 update published in 2020 (International WELL Building Institute, 2020).

The standard covers a very wide range of interventions that have been assembled by building practitioners and health professionals and refined through pilot testing. The concepts range from internal criteria such as fresh air and light, to spatial movement, sound quality and external considerations. The benefits to building owners and commercial property developers are primarily promoted as delivering added value and an advantage in a competitive commercial marketplace. The usefulness of the standard as an adjunct to broader sustainability considerations is demonstrated by the alignment of the WELL standard to the LEED standard, providing a more rounded suite of criteria that covers environmental and social aspects more fully.

It was not until 2024 that the WELL residential standard was launched, utilising the same concepts as the commercial standard but with additional focus on issues such as night-time occupation (affecting sleep quality) and a greater focus on individuals – necessary since management of residential buildings is primarily the remit of the occupants rather than professional building managers. The higher awareness (and marketability) of healthy homes as a result of the Covid pandemic is likely to have been the main reason why residential wellbeing became more of a focus at this time. Increasing awareness of the costs and impacts of buildings on health is likely to grow and these concepts, while currently the preserve of higher-value vanguard developments, are likely to become mainstream in the near term.

6.5.3 Embedding natural environmental systems to benefit wider populations

Embedding benefits to the natural environment into buildings and cities has never been more important. The 2021 Dasgupta Review (Dasgupta, 2021) of the economics of biodiversity articulated the dire state of global biodiversity and natural capital; while human and produced capital has grown immensely in recent decades, the stock of natural capital per person has declined by nearly 40%, with biodiversity declining faster than at any time in human history. A significant driver of terrestrial biodiversity loss is food production at a scale sufficient to provide food for the large populations of towns and cities who are unable to grow their own food stocks. Pressures on land use for more sustainable food growing programmes in competition with development and other land uses exacerbate the problem. In the face of this global emergency, built environment designers and planners have begun to find ways to ameliorate the impact and to embed a 'nature positive' approach into development designs and management.

Inclusion of space for nature and wildlife in built environments has often been relatively minimal in the past, comprising small areas of token planting more for aesthetic purposes than for any nature benefit. In their 2020 report on the future of nature and business, the World Economic Forum (World Economic Forum, 2020) emphasises the need for the built environment to be more hospitable to nature and wildlife, while reversing its negative impact on the surrounding natural environment. The creation of nature networks that connect areas of habitat enabling species to move between them, enabling their populations to grow and to be more resilient to disruption is recognised as an effective response to environments fragmented by human development (Lawton *et al.*, 2010).

Creating space for such nature networks in new development has been recognised as a key opportunity that built environment designers can promote, identifying the key network connectors between surrounding habitats and planning supportive green and blue spaces that enable species movement. Such networks can be retro-fitted to existing infrastructure and developments when buildings are refurbished or as part of re-thinking the space around them. Nature positivity is now being embedded into design codes and, over the coming years, will become more mainstream; but to reap the potential benefits for nature, designers need to think bigger than the scale of a single building to consider 'landscape scale' connections and benefits.

Use of nature-based solutions to common built environment concerns such as air pollution, water management (flood alleviation) and urban heat and cooling has been shown across the world in best practice developments. These solutions are now recognised as having very tangible economic and social benefits as well as environmental, particularly in existing urban environments where they can alleviate infrastructure challenges while also delivering great benefit to the physical and mental wellbeing of citizens. As with nature positive developments, built environment designers will need to consider interactions beyond the immediate boundary of their site to integrate nature-based infrastructure benefits effectively, considering such matters as surface and sub-surface water flows, local microclimates and wind patterns.

6.6. Design for future adaptability

Designing for future adaptability has often been considered by built environment designers to protect investments and to minimise risks of 'stranded assets' particularly for commercial developers and investors. In designing new homes, the concept of 'lifetime homes,' where the home could accommodate changing needs of their occupants over time throughout the life of the resident, has been considered important for some time. The recent pandemic brought into even sharper focus how consideration of potential changes in use for a building can extend its useful life (benefitting the life cycle carbon emissions) and protect its economic value. With a sharp change in work practices driven by the pandemic, resulting in more people working from their home, demand for commercial space was reduced, while homes that had ability to accommodate desks and superfast internet connectivity became more valued. In addition, city centre retail space suffered from reduced footfall as commuter workforce diminished and had to consider re-purposing or changing to service a more leisure-based market. This experience placed a higher premium on those buildings that could more easily be modified or re-purposed.

The UK has a relatively old existing building stock – in England 15% of homes were built before 1900 and in Wales is even higher at 23% (ONS, 2022). Over more than a century of occupation, it is not surprising that buildings will change in response to the needs of their occupants and community. A study of building stock in Liverpool, England, found that more than 60% of buildings had changed their original use over their life cycle. More surprisingly, the researchers also found that the frequency of change of use is accelerating in more recent years, with around 10% of the buildings studied changing use more frequently over the last few decades, even before the global pandemic brought this issue into even greater focus (Manewa *et al.*, 2016).

Designing the built environment to be adaptable is not as simple as imagining a range of possible futures and designing for those. Even if this were commercially and economically possible, the changes that will be experienced during the life cycle of a typical built structure cannot be accurately predicted. Building codes are regularly updated but the speed of change can be such that such codes struggle to keep pace. Looking beyond existing building codes to a range of possible futures for the built environment and for society can allow adaptability to be factored into the design. This does not necessarily require that the building is currently able to deal with future challenges in its current form, but making such changes easier, simpler or cheaper to achieve can extend the useful lifetime of a building – for example, most residential buildings in the UK do not have any mechanical cooling as standard, and in most areas of the UK such systems are rarely needed at present.

However, in the near future, with a warmer climate and an older population more vulnerable to heat stress, it could be anticipated that mechanical cooling and greater shading will be required to keep the home comfortable. With more people spending a high proportion of their day in their own homes, such impacts could have an important influence on population health. Considering how this could be achieved as a retrofit would be helpful in protecting future occupants and in extending the valuable life of the building.

To provide focus, considering a range of possible future conditions and how likely it is that the building or development will be exposed to those risks can be beneficial – for example, a project proposed for a coastal location will need to consider the risks associated with sea level rise, extreme storms and waves to a greater extent than a project located further inland, calculating and considering the ways in which buildings and infrastructure can be protected in under a range of future scenarios. This is already being done in coastal developments across the world to different extents, depending on national standards and attitudes to risk that prevail. In Germany, a new redevelopment of the dock to the south of Hamburg, known as HafenCity, ensured that all buildings were elevated 8.5 m above current sea level to protect the building structures and occupants, while allowing a range of public spaces at lower levels to allow connection to the water edge, where risk of inundation was not so critical (Huang-Lachmann and Lovett, 2015).

Many commentators have highlighted the importance of focusing on the longest life elements of a building to maximise the opportunity for adaptive reuse. Others have also commented on the concept of providing a 'loose fit' functional development, that is not designed too tightly to the original intended purpose, but considers how needs and uses might change and provides opportunity to accommodate them more easily, considering future compartmentalisation, ability to open up atria or glazing and install new service ducts, for example (Conejos and Langston, 2010). A comprehensive analysis of the research into building adaptability (Heidrich *et al.*, 2017) found a large range of different suggestions for adaptability characteristics, with the most cited being those that enable future occupants to easily change layouts and services, such as accessible floors and modularisation. This wide range of approaches would indicate, however, that there is no fixed methodology that suits all building types and structures; moreover, the changing societies within which the built environment functions add another layer of complexity to designing for future adaptation.

The annual UK architectural competition for 'house of the year' provides an interesting snapshot of the latest trends and issues of importance over time. In 2024, the winning design was a family home designed to allow change and modification over time as the needs of the occupants change and the world in which they live places new, as-yet unknown demands on the home (RIBA, 2024a). This ability to adapt to a range of possible futures requires creativity and a collaborative, multidisciplinary approach to the built environment, considering impacts beyond the scope of an immediate development boundary, to be successful.

6.7. Autonomous and regenerative, closed-loop systems

This large and wide-ranging area has many different aspects but all share the same underlying principle – that the built environment, even when impacts are limited by design, still consumes resources and damages the natural environment.

In the 1970s and 1980s, the concept of self-sufficiency was very popular, with increased interest in homes and buildings providing all the requirements to live successfully within the boundary of the site. Various proponents considered different system aspects, ranging from food self-sufficiency to energy, water and wastewater autonomy. Generation of all needs and disposal of all waste streams within the boundary enables a clear and very immediate understanding of the systems on which human settlements depend and brings into sharp focus the extent to which such systems are taken for granted in most built developments. Pioneers of this movement began to propose architectural responses in the early 1970s as the environmental movement was gaining significant momentum. One such response resulted in the construction of a residential home by two architects aiming to put the theories into practice, documented in their vanguard book, *The New Autonomous House* and subsequent updates (Vale and Vale, 1976). The home was the first use of grid-connected solar and battery to power a dwelling, using passive architecture and thermal storage to minimise energy consumption to near-zero.

Over the decades since, this concept has evolved and become more nuanced, taking into account other factors including extractive resources, life cycle impacts and natural ecosystem impacts. Thinking is now moving from autonomy, the ability to function independently, to the ability to *add value* as a fundamental principle of development. The concept has evolved from seeking a neutral impact to a net-positive impact on the natural systems that underpin the built environment, from energy, to nature, to water and so on. These concepts are now generally termed 'regenerative' practices but naturally have overlaps with other associated aspects such as closed-loop and zero-waste design.

A regenerative building would be one where natural systems are reinstated and enhanced, resulting in a positive benefit to those systems on which all life depends. Various different aspects of this concept have been explored by researchers and pioneering designers. More recently, design frameworks that can help to articulate and structure these complex matters are now starting to be published and made accessible to built environment professionals to assist with embedding these concepts into their practice on the scale of individual buildings (International Living Future Institute, 2024) and at a larger scale (Cheshire, 2024; Naboni *et al.*, 2019). As a very high-level summary, this concept requires practitioners to consider the state of natural systems before any development, taking into account the site itself and also the systems that support and interact with it. Their challenge is then to envision a development that mimics and enhances those systems – for example, this might include estimating the amount of water stored by the site in its pre-development condition, understanding the flow of water through, upstream and downstream of the site and aiming to replicate or increase the volume of storage and beneficial interactions within the development.

Vertical farms where crops can be grown in efficient vertical layers have been explored in many applications. It was found that the integration of vertical farms with buildings decreases the use of energy, water and nutrients from external sources, and offers great potential to reduce the environmental impacts, while producing food in urban environments. Applications of the synergetic integration of vertical farms and buildings in apartments, offices, restaurants, swimming pools and supermarkets located in the Netherlands were investigated (Blom *et al.*, 2023) and the results showed that the integration reduced the year-round energy use of the climate systems of both buildings and vertical farms collectively by between 12–51% in comparison with a non-integrated approach.

Across the world, many planning and development control legislative and policy frameworks are being strengthened to support a more regenerative approach to urban development. This is particularly the case with provision of green (and blue) infrastructure and biodiversity networks, which is being addressed as a response to the increasing densification of cities and past lack of consideration of such matters. In England, the Green Infrastructure Framework has set out the need for improved consideration of these matters in new development, building on experiences from other pioneering countries (Natural England, 2023a, 2023b). The overall aspirations have been somewhat codified by the introduction of an 'urban green factor', which aims to quantify the minimum proportion or percentage of green space that should be provided for a proposed development. The suite of standards covers size, accessibility, biodiversity (nature) and proximity to homes. New architectural and structural approaches to integration of vegetation that go beyond just site landscaping are encouraged by this process (such as green roofs and walls, or vertical food growing), as well as the consideration of sustainable drainage and water infrastructure as part of a nature-positive approach.

Further legislation and planning frameworks are being introduced to reinstate more natural water systems such as sustainable drainage systems that act to retain surface water on the site, promoting infiltration at a similar rate to the pre-development state. Increasingly, onsite renewable energy generation and storage is being promoted to decrease the burden of electrical infrastructure provision, reducing energy demand and generating as much as possible on site. While use of site-won materials and site-based recycling of waste streams are not yet becoming mainstream, some vanguard developments are showing that this is indeed viable and can have significant benefits. This more strategic, regenerative approach to development is gaining traction and while autonomy is unlikely for the majority of urban sites, the insights that can be gathered by utilising this approach have the potential to deliver a genuine environmental and social step-change for the built environment.

REFERENCES

Abed J *et al.* (2022) A review of the performance and benefits of mass timber as an alternative to concrete and steel for improving the sustainability of structures. *Sustainability* **14(9)**: 5570, 10.3390/su14095570.

Alaneme KK *et al.* (2023) Mycelium based composites: a review of their bio-fabrication procedures, material properties and potential for green building and construction applications. *Alexandria Engineering Journal* **83**: 234–250, 10.1016/j.aej.2023.10.012.

Alba-Rodríguez MD *et al.* (2017) Building rehabilitation versus demolition and new construction: Economic and environmental assessment. *Environmental Impact Assessment Review* **66**: 115–126, 10.1016/j.eiar.2017.06.002.

Amran M *et al.* (2022) Self-healing concrete as a prospective construction material: a review. *Materials* **15(9)**: 3214, 10.3390/ma15093214.

Anand A and Deb C (2023) The potential of remote sensing and GIS in urban building energy modelling. *Energy and Built Environment* **5(6)**: 957–969, 10.1016/j.enbenv.2023.07.008.

Arup (2025) *Embracing AI: reshaping today's cities and built environment: a global survey*. Arup.com. https://www.arup.com/insights/embracing-ai-reshaping-todays-cities-and-built-environment/ (accessed 10/05/2025).

Awoyera PO and Adesina A (2020) Plastic wastes to construction products: status, limitations and future perspective. *Case Studies in Construction Materials* **12**: e00330, 10.1016/j.cscm.2020.e00330.

Bidwai S (2024) *3D Printing Construction Market Size, Report by 2034*. https://www.precedenceresearch.com/3d-printing-construction-market#:~:text=The%20global%203d%20printing%20construction%20market%20size%20is%20expected%20to,period%20from%202024%20to%202034 (accessed 03/01/2025).

Blom, T, Jenkins A and Van Den Dobbelsteen A (2023) Synergetic integration of vertical farms and buildings: reducing the use of energy, water, and nutrients. *Frontiers in Sustainable Food Systems* **7**, 10.3389/fsufs.2023.1227672.

Brand S (1995) *How Buildings Learn: What Happens After They're Built*. Penguin, NY, USA.

Buxton P (2025) *Edenica: A Pioneering Use of Materials Passports in the City of London*. https://www.ribaj.com/intelligence/materials-passport-first-in-city-of-london-edenica-fletcher-priest (accessed 23/01/2026).

Charef R (2022) The use of Building Information Modelling in the circular economy context: several models and a new dimension of BIM (8D). *Cleaner Engineering and Technology* **7**: 100414, 10.1016/j.clet.2022.100414.

Cheshire D (2019) *Building Revolutions: Applying the Circular Economy to the Built Environment*. Routledge, Newcastle upon Tyne, UK.

Cheshire D (2024) *Regenerative by design: Creating Living Buildings and Cities*. RIBA Publishing, London, UK.

Clifford B and Madeddu M (2024) Planning, adaptive reuse, housing quality and health in a post-pandemic England. *Town Planning Review* **95(4)**: 379–389, 10.3828/tpr.2024.9.

Conejos S and Langston CA (2010) Designing for future building adaptive reuse using AdaptSTAR. *International Conference on Sustainable Urbanization: ICSU 2010*, pp. 1–10. https://research.bond.edu.au/en/publications/designing-for-future-building-adaptive-reuse-using-adaptstar (accessed 23/01/2026).

Construction News (2024) *New or Renew? The Global Debate on Building Reuse*. https://www.constructionnews.co.uk/buildings/new-or-renew-the-global-debate-on-building-reuse-07-10-2024/#:~:text=%E2%80%9CAdaptive%20reuse%20projects%20are%20essential,challenges%20of%20high%2Ddensity%20environments (accessed 23/01/2026).

Costa AR and Hoolahan R (2024) *Materials Passports: Accelerating Material Reuse in Construction*. Lancaster University and Orms Designers & Architects Ltd., 10.5281/zenodo.10472214.

Dasgupta P (2021) *The Economics of Biodiversity: The Dasgupta Review: Full Report*. HM Treasury, https://www.gov.uk/government/publications/final-report-the-economics-of-biodiversity-the-dasgupta-review (accessed 23/01/2026).

Di Salvo S (2024) Insights into new trends and contemporary challenges in 3D printing in architectural sector. *Engineering Innovations* **8**: 43–61, 10.4028/p-vpn0le.

Digital Twin Cities Centre (n.d.) *About – Digital Twin Cities Centre*. https://dtcc.chalmers.se/about/ (accessed 25/09/2025).

Ellen MacArthur Foundation (n.d.) *Cities and the Circular Economy*. https://www.ellenmacarthurfoundation.org/topics/cities/overview#:~:text=A%20circular%20city%20has%20embedded,value)%2C%20and%20regenerate%20nature (accessed 23/01/2026).

El-Sayegh, S, Romdhane L and Manjikian S (2020) A critical review of 3D printing in construction: benefits, challenges, and risks. *Archives of Civil and Mechanical Engineering* **20(2)**, 10.1007/s43452-020-00038-w.

European Commission (2020) *A New Circular Economy Action Plan: For a Cleaner and More Competitive Europe, European Commission*. https://eur-lex.europa.eu/legal-content/EN/TXT/?qid=1583933814386&uri=COM:2020:98:FIN (accessed 25/09/2025).

European Environment Agency (2025) *Environmental Health Impacts*. https://www.eea.europa.eu/en/topics/in-depth/environmental-health-impacts?activeTab=e3e6b879-fef4-4a88-9436-5f0064698270 (accessed 25/09/2025)

Farooq MU *et al.* (2023) Mechanical and durability performance of 100% recycled aggregate concrete pavers made by compression casting. *Journal of Building Engineering* **73**: 106729, 10.1016/j.jobe.2023.106729.

Ghosh S *et al.* (2023) Microplastics as an emerging threat to the global environment and human health. *Sustainability* **15(14)**: 10821, 10.3390/su151410821.

Global Cement and Concrete Association *et al.* (2023) *Making Net-Zero Concrete and Cement Possible*. https://3stepsolutions.s3-accelerate.amazonaws.com/assets/custom/010856/downloads/Making-Net-Zero-Concrete-and-Cement-Possible-Executive-Summary.pdf (accessed 23/01/2026).

Gonzalez-Caceres A *et al.* (2024) Towards digital twinning for multi-domain simulation workflows in urban design: a case study in Gothenburg. *Journal of Building Performance Simulation* **18**: 1–22, 10.1080/19401493.2024.2320112.

Gradišar L *et al.* (2022) Generative design methodology and framework exploiting designer-algorithm synergies. *Buildings* **12(12)**: 2194, 10.3390/buildings12122194.

Guamán-Rivera R *et al.* (2022) Recent developments and challenges of 3D-printed construction: a review of research fronts. *Buildings* **12(2)**: 229, 10.3390/buildings12020229.

Guy B and Ciarimboli N (n.d.) *Design for Disassembly in the Built Environment: A Guide to Closed-Loop Design and Building*. Hames Center for Community Design, Pennsylvania State University. https://www.lifecyclebuilding.org/docs/DfDseattle.pdf (accessed 05/12/2024).

Harries KA (2025) Adoption of full-culm bamboo as a mainstream structural material. *NPJ Materials Sustainability* **3(1)**, 10.1038/s44296-025-00050-2.

Heidrich O *et al.* (2017) A critical review of the developments in building adaptability. *International Journal of Building Pathology and Adaptation* **35(4)**: 284–303, 10.1108/ijbpa-03-2017-0018.

Huang-Lachmann J-T and Lovett JC (2015) How cities prepare for climate change: comparing Hamburg and Rotterdam. *Cities* **54**: 36–44, 10.1016/j.cities.2015.11.001.

Hucks N, Advancing Net Zero and UKGBC (2022) *Delivering Net Zero: Key Considerations for Commercial Retrofit*. https://ukgbc.org/wp-content/uploads/2022/05/Commercial-Retrofit-Report.pdf (accessed 05/12/2024).

HUDU (London Healthy Urban Development Unit) (2019) *HUDU Planning for Health Rapid Health Impact Assessment Tool*, 4th edn. https://www.healthyurbandevelopment.nhs.uk/wp-content/uploads/2019/10/HUDU-Rapid-HIA-Tool-October-2019.pdf (accessed 23/01/2026).

Ige J *et al.* (2018) The relationship between buildings and health: a systematic review. *Journal of Public Health* **41(2)**: e121–e132, 10.1093/pubmed/fdy138.

International Living Future Institute (2024) *Living Building Challenge 4.1 Program Manual: A Visionary Path to a Regenerative Future*. https://living-future.org/lbc/ (accessed 23/01/2026).

International WELL Building Institute (2020) *WELL Building Standard V2*. https://v2.wellcertified.com/en/wellv2/overview (accessed 21/04/2025).

ISO (International Organization for Standardization) (2018) ISO 19650-1: Organisation and digitization of information about buildings and civil engineering works, including Building Information Modelling (BIM): Part 1: Concepts and principles. https://www.iso.org/standard/68078.html (accessed 23/01/2026).

Jenima J *et al.* (2024) A comprehensive review of titanium dioxide nanoparticles in cementitious composites. *Heliyon* **10(20)**: e39238, 10.1016/j.heliyon.2024.e39238.

Kanters J (2020) Circular building design: an analysis of barriers and drivers for a circular building sector. *Buildings* **10(4)**: 77, 10.3390/buildings10040077.

Krishna S (2020) Review on bamboo as mainstream construction material. *International Journal for Research in Applied Science and Engineering Technology* **8(12)**: 794–799, 10.22214/ijraset.2020.32617.

Lanz F and Pendlebury J (2022) Adaptive reuse: a critical review. *The Journal of Architecture* **27(2–3)**: 441–462, 10.1080/13602365.2022.2105381.

Lawton *et al.* (2010) *Making Space for Nature: A Review of England's Wildlife Sites and Ecological Network*. Report to DEFRA. https://webarchive.nationalarchives.gov.uk/ukgwa/20130402170324mp_/http://archive.defra.gov.uk/environment/biodiversity/documents/201009space-for-nature.pdf (accessed 10/03/2025).

Lewis AC *et al.* (2022) *Indoor Air Quality*. Zenodo (CERN European Organization for Nuclear Research), 10.5281/zenodo.6523605.

Madaster (n.d.) *Madaster Global – Our Platform*. https://madaster.com/platform/ (accessed 17/11/2024).

Manewa A *et al.* (2016) Adaptable buildings for sustainable built environment. *Built Environment Project and Asset Management* **6(2)**: 139–158, 10.1108/bepam-10-2014-0053.

Michielin D (2023) *From Waste to Resource: The Rise of Urban Mining. Foresight: The CMCC Observatory on Climate Policies and Futures* [preprint]. https://www.climateforesight.eu/articles/raw-materials-urban-mining/ (accessed 23/01/2026).

Muhit IB, Omairey EL and Pashakolaie VG (2024) A holistic sustainability overview of hemp as building and highway construction materials. *Building and Environment* **256**: 111470, 10.1016/j.buildenv.2024.111470.

Naboni E *et al.* (2019) A digital workflow to quantify regenerative urban design in the context of a changing climate. *Renewable and Sustainable Energy Reviews* **113**: 109255, 10.1016/j.rser.2019.109255.

Natural England (2023a) *Green Infrastructure Standards for England – Summary: Green Infrastructure Framework – Principles and standards for England*. https://designatedsites.naturalengland.org.uk/GreenInfrastructure/downloads/Green%20Infrastructure%20Standards%20for%20England%20Summary%20v1.1.pdf (accessed 23/01/2026).

Natural England (2023b) *Urban Greening Factor for England User Guide: Green Infrastructure Standards Framework – Principles and Standards for England*. https://designatedsites.naturalengland.org.uk/GreenInfrastructure/downloads/Urban%20Greening%20Factor%20for%20England%20User%20Guide.pdf (accessed 23/01/2026).

NBS (2020) *10th Annual BIM Report*. https://www.thenbs.com/knowledge/national-bim-report-2020 (accessed 23/01/2026).

OECD (Organisation for Economic Co-operation and Development) (2024) *Society at a Glance*, 10.1787/918d8db3-en.

ONS (Office for National Statistics) (2022) *Age of the Property is the Biggest Single Factor in Energy Efficiency of Homes*. https://www.ons.gov.uk/peoplepopulationandcommunity/housing/articles/ageofthepropertyisthebiggestsinglefactorinenergyefficiencyofhomes/2021-11-01 (accessed 23/01/2026).

ONS (2023) *Births in England and Wales*. https://www.ons.gov.uk/peoplepopulationandcommunity/birthsdeathsandmarriages/livebirths/bulletins/birthsummarytablesenglandandwales/2023 (accessed 23/01/2026).

Omrany H *et al.* (2023) Digital twins in the construction industry: A comprehensive review of current implementations, enabling technologies, and future directions. *Sustainability* **15(14)**: 10908, 10.3390/su151410908.

Owens R (2024) Transforming homes, transforming lives: the role of property logbooks in scaling retrofit. *Journal of Building Survey, Appraisal & Valuation* **13(2)**: 185–193, 10.69554/vtwg9101.

Perini K *et al.* (2022) MosSkin: a moss-based lightweight building system. *Building and Environment* **221**: 109283, 10.1016/j.buildenv.2022.109283.

Reddy VJ, Ghazali MF and Kumarasamy S (2024) Advancements in phase change materials for energy-efficient building construction: a comprehensive review. *Journal of Energy Storage* **81**: 110494, 10.1016/j.est.2024.110494.

RIBA (Royal Institute of British Architects) (2024a) *Six Columns, Southwark*. https://www.ribaj.com/buildings/riba-awards-2024-london-south-east-six-columns-31-44-architects-house-southwark (accessed 23/01/2026).

RIBA (2024b) *Smart Building Overlay to the RIBA Plan of Work*. https://www.architecture.com/knowledge-and-resources/resources-landing-page/smart-building-overlay-to-riba-plan-of-work (accessed 23/01/2026).

Roberts M *et al.* (2023) Understanding the global warming potential of circular design strategies: life cycle assessment of a design-for-disassembly building. *Sustainable Production and Consumption* **37**: 331–343, 10.1016/j.spc.2023.03.001.

Sensat (2025) *Severn Trent Wins Water Industry Award with Sensat's Innovation*. https://www.sensat.co/news/severn-trent-crowned-winners-at-water-industry-awards-for-revolutionising-cso-planning-with-digital-innovation#:~:text=%E2%80%8DSensat%20is%20a%20leading%20digital,significant%20time%20and%20cost%20savings (accessed 25/09/2025).

Sinha S (2024) *State of IoT 2024: Number of Connected IoT Devices Growing 13% to 18.8 Billion Globally*. https://iot-analytics.com/number-connected-iot-devices/ (accessed 21/04/2025).

Spear M *et al.* (2019) *Wood in Construction in the UK: An Analysis of Carbon Abatement Potential*. https://www.theccc.org.uk/wp-content/uploads/2019/07/Wood-in-Construction-in-the-UK-An-Analysis-of-Carbon-Abatement-Potential-BioComposites-Centre-1.pdf (accessed 26/01/2025).

Statista (2025) *Global Plastic Production 1950–2023*. https://www.statista.com/statistics/282732/global-production-of-plastics-since-1950/ (accessed 23/01/2026).

Strunge JW (2020) Enabling regenerative building design through biohacking. *IOP Conference Series Earth and Environmental Science* **588(4)**: 042057, 10.1088/1755-1315/588/4/042057.

UKGBC (UK Green Building Council) (2019) *Circular Economy Guidance for Construction Clients: How to Practically Apply Circular Economy Principles at the Project Brief Stage*. https://ukgbc.org/wp-content/uploads/2019/04/Circular-Economy-Report.pdf (accessed 23/01/2026).

University of Manchester (2023) *Graphene Spin-Out Wins The Spectator's Innovator of the Year Award for Excellence in Sustainability*. https://www.manchester.ac.uk/about/news/graphene-spin-out-wins-the-spectators-innovator-of-the-year-award-for-excellence-in-sustainability/ (accessed 25/09/2025).

Upcyclea (2024) *Upcyclea – Platform*. https://upcyclea.com/en/platform/ (accessed 17/11/2024).

Vale B and Vale RJD (1976) *The New Autonomous House: Design and Planning for Sustainability*. Thames & Hudson, London, UK.

Waterman (2023) *Waterman Materials Passports Framework – Waterman*. https://www.watermangroup.com/watermanmaterialspassportsframework/ (accessed 23/01/2026).

World Economic Forum (2020) *New Nature Economy Report II – The Future of Nature and Business*. https://www3.weforum.org/docs/WEF_The_Future_Of_Nature_And_Business_2020.pdf (accessed 10/03/2025).

WHO (World Health Organisation) (2024) *Ageing and Health*. https://www.who.int/news-room/fact-sheets/detail/ageing-and-health (accessed 23/01/2026).

Zallio M and Clarkson PJ (2021) Inclusion, diversity, equity and accessibility in the built environment: A study of architectural design practice. *Building and Environment* **206**: 108352, 10.1016/j.buildenv.2021.108352.

Mei Ren and Rachel Waggett
ISBN 978-1-83662-853-8
https://doi.org/10.1108/978-1-83662-850-720261010

Chapter 7
International case studies of net zero and climate resilience

7.1. Introduction

In today's world, societies strive to adopt sustainable development models that ensure continuous improvements in the wellbeing of their citizens and protection of the natural environment societies live in, despite the ongoing changes in culture, politics, economics, technology and climate risks they face (Clark and Harley, 2020). To successfully navigate these transformations, societies must first envision what a sustainable future looks like and employ reliable tools that can guide them and support the formulation, testing, implementation, and evaluation of integrated policies and practices to steer their development towards more sustainable pathways (Clark and Harley, 2020).

Different regions around the world takes different actions, initiatives and programmes to respond to climate change and to conserve natural resources, with efforts spanning four dimensions: environmental, physical, social and economics. The actions and programmes vary due to differences in values and contexts influenced by land area, population size and topographic heterogeneity (Silva *et al.*, 2024). These sustainability transformation efforts require societies to shift to maintain ecological integrity and supply basic human needs over generations.

Clark and Harley (2020) identified six capacities in multiple contexts to support nature–society interactions in guiding development pathways toward sustainability. These are the capacities to

(*a*) measure sustainable development
(*b*) promote equity
(*c*) adapt to shocks and surprises
(*d*) transform the system into more sustainable development pathways
(*e*) link knowledge with action
(*f*) governance arrangements that allow people to work together in exercising the other capacities.

Previous chapters have explored past and present responses and the practical tools that exist to move forward on this journey, including a consideration of the exciting near-future opportunities that are beginning to open up to those working in the built environment. Many of these solutions are already being demonstrated in vanguard projects across the world.

This chapter examines sustainability transformation in each region, using a selected number of case studies. These case studies have been chosen based on several criteria, including availability of the information, representativeness of the region, and diversity of projects that illustrate the wide range of interactions and interventions between nature and society. Together, these examples reflect the different capacities identified by Clark and Harley (2020) and demonstrate how they contribute to advancing sustainability transformations.

Each case study has been described using three capacities to showcase their impact on sustainability transformation: Vision and Technology advancement, leading in Action and Governance.

In the earlier discussion of framework planning (Chapter 4), the concept of using a systematic series of steps to help users to develop a framework with a stronger purpose to guide their development was described. Those steps include: identification and alignment of core value and establishing a clear vision, defining indicators and clear goals to achieve the vision and values, setting out the governance models, planning a roadmap that will guide progress and collaboration, and following through on implementation while capturing lessons learned to inform future projects.

The capacities used to demonstrate transformative impact are broadly aligned with the systematic steps outlined in framework planning.

- Vision and Technology advancement – adopt innovative technologies, processes to pilot and drive transformation.
- Leading in Action – link knowledge with action, measure sustainable development and transform the system into more sustainable development pathways.
- Governance – devise governance arrangements that allow people to work together in exercising the other capacities as well as promote equity.

7.2. The UK

Since Building Research Establishment Environmental Assessment Method (BREEAM) was devised and launched in 1990s, the UK has been at the forefront of developing approaches to the sustainable built environment, especially in policy frameworks (e.g. planning framework to encourage adoption of sustainability practices) and methodologies to measure and benchmark the sustainability performance of developments. As a developed economy, the region has well established both ecological infrastructure and socio-economic infrastructure, supported by mature policies, research and development efforts, and implementation instruments that have enabled sustainability practices to achieve a notable degree of success.

Four case studies in the UK have been chosen to indicate sustainability transformation of the built environment.

- Major London urban developments have played a significant transformative role in London's economy in the past 10 years. Initiatives to rejuvenate areas such as Battersea and King's Cross have transformed them into vibrant, sought-after neighbourhoods. In addition, transport and mobility investment promotes a cycling culture, evident from the increase in cycling superhighways and bike-sharing. The London Plan (2021) has played a key role in driving sustainability in London by introducing several initiatives to promote sustainability including an increased emphasis on green infrastructure and air quality improvement, as well as promoting net zero practices and a circular economy. Three case studies in London have been chosen, each reflecting different initiatives.
    - King's Cross Re-development. This case study shows how the public and private sector partnership supported by major transport led investment can successfully create vibrant places, underpinned by a governance structure capable of delivering the long-term vision for the local area.
    - Post occupancy evaluation (POE) of housing development at Chobham Manor, London Legacy Development Corporation (LLDC). The London Olympic Development developed

Chobham Manor as part of its efforts to regenerate the area using the 2012 Olympic Games as a catalyst not only to speed up development in east London, but also create a sustainable neighbourhood for local residents. A few years on since residents moved in, a POE was carried out to assess how well the sustainable design measures perform in reality.
- Battersea Power Station is a leading example of adaptive re-use of existing assets. Battersea Power Station is a decommissioned coal-fired power station located on the south bank of the River Thames in Nine Elms. It was built by the London Power Company (LPC) as one of the world's largest brick buildings and used to provide a majority of power supply to London.

- One of the key design issues in many developments is resource efficiency. Designers are often challenged by how far minimisation of energy, water and waste for a development can be pushed within the budget constraints and amid the uncertainties associated with new technologies. Achieving a well thought through balance is essential. NOMA Manchester provides an example of an urban development in the North of England where such an approach has been successfully implemented.

7.2.1 Major London re-developments

These case studies, including King's Cross Regeneration for its place making and governance, POE of housing by London Olympic Legacy and Battersea power station as a best in class major adaptive re-use example, have been selected to showcase the transformation across these capacities.

7.2.1.1 King's Cross Regeneration London

The King's Cross Regeneration Programme entails the transformation of a 27-hectare area in central London on former rail land to the north of King's Cross and St Pancras stations into a mixed-use urban regeneration project, with up to 739 690 m^2 of floor space. This regeneration programme, one of the largest in Europe, has managed to attract a number of lead tenants, including Google, and has turned a disused area of London into a new vibrant urban space. It is also expected to provide substantial returns to its developers in the medium to long term.

King's Cross is a unique development, involving many years from design and delivery of the initial phases. The process included four rounds of public consultation, involving around 30 000 people, leading to a well-tailored solution for urban regeneration.

This case study illustrates the process of leveraging high connectivity and public private partnerships (PPPs) in increasing real estate market value of areas around a major railway hub. Such an approach exemplifies a strategic approach to urban transformation through Transit Oriented Development (Salat and Ollivier, 2017).

Leverage infrastructure investment

The 2004 London Plan aimed to make King's Cross the most accessible location in Greater London, with the completion of the Channel Tunnel Rail Link – High Speed 1 (HS1), Thameslink 2000 and the Cross River Tram, as well as being the biggest inner-city transit interchange in London, linking six metro lines at one venue. The King's Cross re-development involved major infrastructure investments in the St Pancras station and its surroundings, with reconstruction of interchanges to the metro links. The area was identified as one of the five 'Central Area Margin Key Opportunities' in the strategic planning guidance for London. Over £2 billion has been spent on the local transport infrastructure, including St Pancras station and public areas. Encouraged by the restoration of the underground stations and national mainline stations on the site completed by 2007, the landowners – London & Continental Railways (LCR) and DHL – realised that the site had immense untapped potential, particularly in its ability to respond to and accommodate the large number of people who would use the new international interchange. The re-development site of about 27 hectares of land has been planned to contain more than 1900 homes, 50 new and refurbished office buildings, 500 000 ft^2 (about 47 000 m^2) of shops and restaurants, 20 new streets, and ten major new public spaces for a projected 50 000 people.

Partnership approach

The King's Cross regeneration programme aims at delivering an accessible, high-quality mixed-use environment, with a strong focus on art, culture and heritage. In 2001, LCR selected property developer Argent, one of the UK's most respected property developers, as its private development partner for the programme, recognising the company's extensive experiences in regeneration and mixed-use developments. The partnership arrangement is illustrated in Figure 7.1, including partnership with the central government, local government, private sector companies as well as local communities. The process included active public consultation, two master planning teams and four independent design review panels to advise and inform the development of a tailored solution for the site over six years of planning.

Argent brought the essential financial backing from a large pension fund (BTPS managed by Hermes Investment Management), which was critical to enabling the private development of the site. By early 2015, LCR's primary focus had shifted to the area of property development and land regeneration. It had become a joint venture partner in two major regeneration programmes, including King's Cross in partnership with Argent and DHL, and the International Quarter, Stratford City, in partnership with Lend Lease.

While the majority of private floor space is intended to generate business profits, more than 40% of the redeveloped former 'brown-field' site has been designated for public purposes. In addition, 20 historic buildings across the redevelopment site are being restored for modern use (PPIAF, 2017; Salat and Ollivier, 2017).

Figure 7.1 King's Cross partnership arrangements (created from information in financing transit – oriented development with land values (Suzuki *et al.*, 2015))

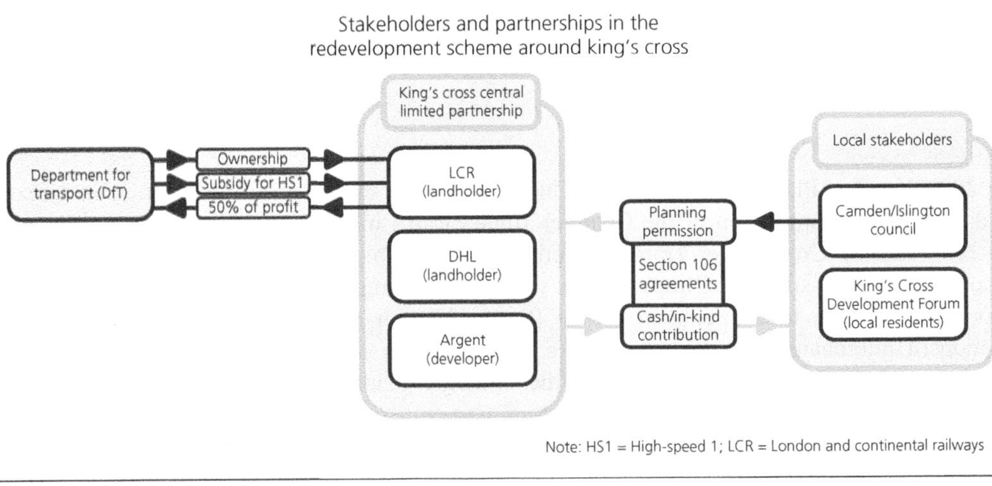

Note: HS1 = High-speed 1; LCR = London and continental railways

Joined up with local communities

The joint developer was required to contribute to the provision of local infrastructure and community services by the local government, Camden Council. This contribution in the form of cash and in kind contributions was provided by way of the Section 106 agreement for the King's Cross project, including (Camden Council, 2006) the following.

- £2.1 million to create 24 000–27 000 local jobs through a Construction Training Centre and Skills and Recruitment Centre.
- 1900 homes, more than 40% of which will be affordable housing.
- Community, sports and leisure facilities.
- New green public spaces, plus new landscaped squares.
- Well-designed and accessible streets, accounting for about 40% of the entire site.
- A new visitor centre.
- Education facilities.
- A bridge across the canal to link streets; and cash contributions to improve adjacent streets, transit stops and bus services.

The development of the site is anchored by a vision under which it transforms from a slightly decentralised derelict place to a core part of central London, which would work 24 hours a day and 7 days a week, and would be open, democratic and accessible. That meant dividing the area into development zones, with maximum building heights, and environmental specifications, but also producing hundreds of urban design options, to define possible use and leverage the existing historical heritage in the area such as Granary Square. The historic fabric was embedded in the plan in a sophisticated manner, rather than simply preserved. All these options have been consulted extensively and, to a large extent, co-created with local communities.

The provision of good quality public space enhances the image and appeal of the location. The master plan presented a network of public open spaces, streets, lanes, squares and parks that permeated the urban blocks and made connections beyond the site into the wider city.

Social and economic benefits

The regeneration scheme around King's Cross is expected to directly deliver significant social and economic benefits and to fundamentally improve the physical environment, throughout the construction period and beyond. The construction programme required a large amount of infrastructure to be built before any buildings work could begin. This included improvements to local roads, phased opening of some routes through the site, improved drainage and storm water run off (preventing flooding) and enhanced local electricity capacity. The length of the King's Cross Central construction period also means that there is the potential for the construction sector to become a long-term stable employment base within the local economy.

The joint developers are committed to the creation of a high-quality public realm maintained by a comprehensive management regime. King's Cross has rapidly become one of the most attractive places to live, work and visit in London, and there has been strong demand for both the offices and residential units, together with restaurants and retail shops opening. New public streets, squares and gardens have opened, including Granary Square with its striking fountains, Lewis Cubitt Park, Lewis Cubitt Square and Gasholder Park. King's Cross is now a much-loved part of London and a popular destination for locals and visitors from further afield (Figure 7.2).

Figure 7.2 King's Cross (photo credit to © John Sturrock, courtesy of The King's Cross Group LP)

Sustainability summary

King's Cross has been shaped explicitly by the compact city principles set out in the GLA 2010 strategic planning policy, the 2004 and 2008 London Plan (Adelfio et al., 2020). Several sustainability features are incorporated in King's Cross.

- King's Cross is built around a green framework. More than 40% of the 67-acre development is open space.
- A low carbon strategy that is expected to achieve a 60% carbon reduction between 2000 and 2050.
- A district heating system that accounts for 99% of heating and hot water needs.
- More than 80% of recycled public waste.
- An extensive use of green roofs and walls integrated in a living-landscape approach.
- Some of the buildings (namely One, Two and Five Pancras Square) have been certified with a BREEAM 'Outstanding' rating (BRE, n.d.; Kings Cross, n.d.).
- The combination of using 100% renewable energy, having energy-efficient buildings and offsetting carbon emissions enabled King's Cross to become carbon neutral in November 2021. King's Cross set up their journey to become net zero as soon as possible.
- Established the King's Cross Central Carbon Fund, to accelerate carbon reduction efforts across the King's Cross Estate and the surrounding community, supporting the estate's environmental, social and governance (ESG) priority areas. This fund is dedicated to empowering local projects that can demonstrate a tangible carbon reduction benefit.

Focus area (learning points) **GOVERNANCE**

King's Cross London is one of the largest development programmes in Europe; a transformational regeneration programme for a 27-hectare area in central London on former rail land at the north of King's Cross and St Pancras stations. The programme has created a mixed-use vibrant urban centre and has attracted a number of lead tenants, including Google.

This case study demonstrates that by leveraging infrastructure investment and building partnership among the public and private sectors, a successful urban regeneration programme is possible in not only realising real estate market value of areas around a major railway hub, but also bringing significant social economic benefits to local communities.

Given the extensive development programme, King's Cross has been shaped by policies in the Greater London Authority (GLA) 2010 strategic planning policy, the 2004 and 2008 London Plans and, less directly, the 2011 London Plan. In particular, sustainability features include green infrastructure, low carbon strategy, 100% renewable energy and the King's Cross Central Carbon Fund.

King's Cross London has won multiple awards including RIBA London Award 2024, RIBA National Award 2024 and was shortlisted for the RIBA Stirling Prize 2024, with some specific buildings such as Author King's Cross winning the 2024 New Homes Award.

7.2.1.2 Post-occupancy evaluation of housing development at Chobham Manor, London

London Legacy Development Corporation (LLDC), is the organisation responsible for Queen Elizabeth Olympic Park. Located at the Queen Elizabeth Olympic Park in Stratford, East London, Chobham Manor residential development contains 180 homes and is the first of four planned phases at Chobham Manor (Figure 7.3). The wider scheme is a multiple award-winning housing development, recognised by the New London Awards with the 2019 Mayor's Prize. LLDC's vision was to create a family oriented neighbourhood within an emerging part of London, designed as a child friendly, sustainable and inclusive place with inspiring landscapes and public realm. This case study demonstrated how the sustainable design practices at the time (2010s) for housing development performed over the time and LLDC's commitment to continuingly learn from lessons and improve their future developments.

Why POE?

The overall aim of the independent evaluation over a 12-month period is to understand if the scheme has achieved its design intent, as well as gain a deeper insight into how it is performing for residents who live there. LLDC have a total of 5000 homes to deliver on their landholdings. Understanding how the emerging built environment is performing is important as it will provide feedback to inform learning on future communities being created and to help developers and management companies respond to issues. In the longer term, this work can contribute significantly to LLDC's ambition to serve as a sustainability exemplar and demonstrate public sector leadership in promoting high standards of design, sustainability and inclusive development. By undertaking this evaluation, there is the opportunity to learn what works well in practice and to exceed existing standards in subsequent phases of development, grounded in a clear evidence base and a deeper understanding of actual performance (Buro Happold, 2022).

Figure 7.3 (a) Chobham Manor, London within the wider Queen Elizabeth Olympic Park (Buro Happold, 2022), Copyright © 1976–2022 Buro Happold); (b) a view of Chobham Manor from the Olympic Park, London (photo credit to sludgegulper licensed under CC BY-SA 2.0)

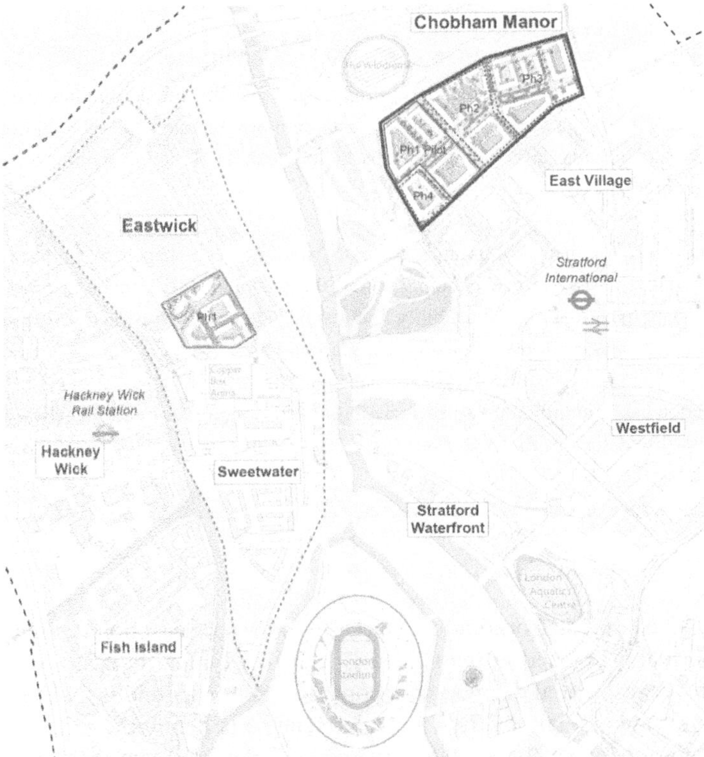

(a)

(b)

The scheme

Chobham Manor (all phases) is a 116 500 m^2, £270 million residential-led scheme designed as a family friendly neighbourhood and incorporates a wide range of houses and apartments. The overall scheme consists of four phases, with Phase 1 and Phase 2a now in occupation, and some residents having lived in the area for up to three years. As the first of five neighbourhoods being delivered as part of the Olympic Legacy, Chobham Manor plays a pivotal role in setting the benchmark for the quality, sustainability, and design standard for these future developments.

Phase 1 consists of 180 homes in total, comprising a mixture of apartments, maisonettes, townhouses, mews houses and houses. All properties are managed by L&Q and there is a mix of tenures (social rent, shared ownership, private rent), with new houses sold as freehold. Prices varied from around £600 000 - £1 million.

Sustainability sits at the heart of the development vision, alongside a strong focus on neighbourhood and community at Chobham Manor, beginning with a clearly defined framework that guided the design process (Figure 7.4).

Sustainability summary

- All homes were built to Code for Sustainable Homes (CfSH) Level 4 standard. A further 26 exemplar homes (phase 1 and 2) were designed to achieve 'zero carbon' as per the definitions set in the London Plan 2011 at the time (i.e. they have achieved at least 35% reduction in regulated carbon dioxide (CO_2) and a carbon offset payment has been made to offset 100% of remaining regulated carbon dioxide).
- As part of the design strategy, all buildings are also connected to the Stratford district energy network, which provides district heating from gas combined heat and power (CHP), supplemented by biomass heat delivered through a link from the Kings Yard Energy Centre to the Stratford energy centre.
- Ten Exemplar homes were also required to achieve the following requirements: CfSH Level 5 target, zero regulated emissions through the use of on-plot measures only, and potable water use at 80 l or less per person per day (CfSH Level 5–6 requirement).
- Commitment to long-term evaluation. London Plan (Policy S12 clauses 9.2.9 and 9.2.10) has a requirement for major development to monitor energy usage annually over a 5-year period in use.

Overview of methodology

LLDC places strong emphasis on post-occupancy evaluation (POE), with the work receiving high-level leadership support. To ensure robust oversight throughout the project, LLDC have also engaged with an external advisory panel with members from several local and national government bodies. In a parallel study, the LLDC also supported Catalina Morales Maya, of Oxford Brookes University, in undertaking a separate user survey focused on inclusive design at Chobham Manor and East Village.

The key stages of the assessment (Dowson, 2023) include: resident engagement, signing up residents to the POE, user surveys and stakeholder engagement, undertaking utility meter readings as part of 'Basic POE', and undertaking environmental monitoring as part of an 'Detailed POE' (Figure 7.5).

Heating and hot water for the properties is provided by way of a district heating network operated by ENGIE who are part of ELE (East London Energy). Each property has an individual Heat Interface Unit (HIU) supplying the space heating and hot water demands. The heat output of each HIU is individually metered and the data from these meters is collected centrally by ENGIE.

Figure 7.4 Chobham Manor sustainability framework at design stage (Buro Happold Chobham Manor Phase 1 Post Occupancy Evaluation for the LLDC, Copyright © 1976–2022 Buro Happold)

Neighbourhood and house design

A family friendly, characterful neighbourhood of high-quality homes comprising a choice of typologies, a varied architecture and a place where families can grow.

Sustainable community

A place with a sense of community, where people feel safe and welcome to participate in local activities and enjoy life satisfaction and wellbeing.

Inclusive design

A scheme that feels inclusive, and caters to the needs of disabled people, older people, children and families in housing and neighbourhood design.

Sustainable transport

A neighbourhood where sustainable transport options such as walking, cycling and public transport are readily available and used over cars.

Landscape design

A place close to the park, with high quality, safe urban spaces, and shared connective green spaces, where residents benefit from green infrastructure and enjoy private outdoor space.

Environmental sustainability

Housing that incorporates natural lighting, ventilation, water management, energy conservation/carbon reduction, waste management and biodiversity, with 26 'exemplar' sustainable units

District energy network

A district energy network (DEN) that provides cost effective, low-carbon heat, that is energy efficient to run and does not adversely affect resident comfort.

For both the basic and detailed POE, data were supplied by ENGIE for the required periods (i.e. monthly and for 2–3-month periods in summer and winter). No other additional equipment were required for collecting the data.

Results summary

The POE study concluded in September 2022 shows the following findings (Dowson, 2023).

- The majority of residents are satisfied with the overall urban structure and appearance of Chobham Manor.
- The team explored a well-designed neighbourhood featuring a wide range of housing typologies and a public realm that are well appreciated, well used and well liked. Chobham Manor has a thriving community that has developed a deep sense of place identity within a relatively short time.

- Energy use is low compared with published data for UK households. Site-wide measured space heating was 25 kWh/m^2/year compared with 50 kWh/m^2/year for the average new-build home. Measured domestic hot water and electricity usage was also low compared with UK new-build average at 30 kWh/m^2/year and 32 kWh/m^2/year, respectively.
- For electricity, in 13 properties with half-hourly electricity sensors fitted, peak household demand was 8.1 kW (although this home had an electric vehicle). For the remaining properties, peak was 1.7 kW to 3.5 kW.
- Residents were particularly satisfied with daylight (89% satisfaction) and temperature in winter (80% satisfaction).
- Anonymised district heating and hot water data for all 259 homes were provided by the energy provider for 26 months from September 2019. For 35 homes that consented to the POE, this data was analysed in more depth at a property level. Monthly electricity readings were provided by residents for 14 months from October 2020, covering 29 households (11% sample).
- In summertime, overheating was a dominant concern in the quantitative survey results.
- Low satisfaction levels (36% satisfaction) with district heating system, largely due to concerns over the associated fixed charges. Many residents question the economic justice of the heat network and the quality of landscape maintenance.

Figure 7.5 Process map for overall post occupancy evaluation methodology (recreated from Buro Happold Chobham Manor Phase 1 Post Occupancy Evaluation for the LLDC, Copyright © 1976–2022 Buro Happold)

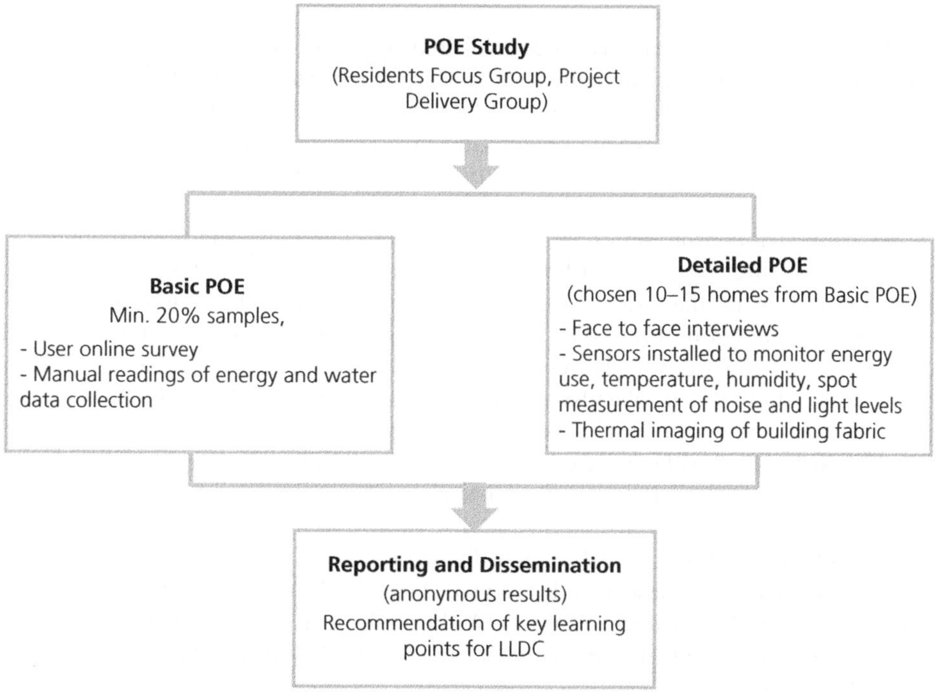

Focus area (learning points) **ACTION**

Chobham Manor residential development with 180 homes, is the first of four phases built at Chobham Manor, developed by LLDC. The scheme as a whole is a multiple award-winning housing development with LLDC's vision for a family friendly, sustainable and inclusive place with inspiring landscapes and public realm in Chobham Manor, Stratford.

This case study demonstrated how the sustainable design practices at the time (2010s) for housing development performed over the time and LLDC's commitment to continuingly learn from lessons and improve their future development. Since the completion of the POE of Chobham Manor development, several lessons have informed updates to the latest policies, including the following.

- The GLA has introduced new energy efficiency targets and daylight and overheating guidance within the London Plan 2021, specifically on managing overheating risk (London Plan 2021; OPDC, 2022).
- Heat Trust, is an independent, non-profit consumer champion for heat networks in Great Britain (England, Wales and Scotland) with a clear ambition to protect heat network consumers. Heat Trust sets the minimum standards for heat network operators, making sure the heat consumers are treated fairly and that suppliers deliver high-quality customer service.
- Heat networks had been unregulated in the UK until 2025. The UK Government introduced the Energy Act 2023 and named the Office of Gas and Electricity Markets (Ofgem) as the official regulator of heat networks across England, Scotland and Wales (Great Britain). The comprehensive new regulatory regime came into force in January 2026, with Ofgem officially acting as the regulator for heat networks..

7.2.1.3 Battersea Power Station, London

Battersea Power Station is a decommissioned coal-fired power station located on the south bank of the River Thames in Nine Elms, within the London Borough of Wandsworth. Battersea Power Station was built by the London Power Company (LPC) and, at its peak, produced a fifth of London's power, supplying electricity to some of London's most recognisable landmarks, such as the Houses of Parliament and Buckingham Palace from the 1930s to the 1980s.

In 2012, the Malaysian consortium of SP Setia, Sime Darby Property and the Employees Provident Fund (EPF) acquired the site and formed Battersea Power Station Development Company (BPSDC) to bring forward the vision for the new riverside neighbourhood and town centre. The redevelopment was guided by the Rafael Viñoly masterplan, which sets out the framework for a dynamic mixed-use town centre (Battersea Power Station, 2024a, 2024b, 2024c). The Viñoly masterplan approved in 2011 by Wandsworth Council, intended to position the power station as the central focus of the regenerated site, housing a blend of shops, cafes, restaurants, art and leisure facilities, office space and residential accommodation as well as the creation of a new High Street designed to link the entrance to Battersea Power Station tube station.

The overall development project of £9 billion is divided into eight phases, with each designed by a range of specialist architects. These include SimpsonHaugh and Partners and De Rijke Marsh Morgan (dRMM) in Circus West Village (Phase 1), WilkinsonEyre in Battersea Power Station (Phase 2), and Foster + Partners and Gehry Partners in The Electric Boulevard, Battersea Roof Gardens and Prospect Place (Phase 3).

It is anticipated that 25 000 people will be living and working here when the development is complete, creating one of London's largest office, retail, leisure and cultural quarters. Across the 42-acre site a new 24/7 community would be formed with over 250 shops, cafes and restaurants, theatre, hotel, medical centre and 19 acres of public space including 450 m of river frontage and a six-acre public park. Battersea Power Station, shown in Figure 7.6, is set to become a new office district with over 3 million square feet of commercial space alongside new private and affordable homes.

The first phase of the development, Circus West Village, was completed in 2017 and is now home to over 1800 residents, as well as a mix of bars, restaurants and leisure facilities, including a cinema and theatre. The power station was the second phase to open to the public on the 14 October 2022. This truly mixed-use building is home to Apple's London Campus, hundreds of new shops housed in the historic turbines halls, a state-of-the-art dolby cinema, a 24 000 ft^2 food hall, a glass chimney lift and hundreds of new homes. This case study focuses on the adaptive re-use of the power station building, one of the world's largest brick buildings.

Adaptive re-use of power station

WilkinsonEyre architect was appointed in 2013 to carry out the restoration of the Grade II* listed power station. Meticulously renovated over 9 years, the power station is the cornerstone of an entirely new neighbourhood. The work included the restoration of the art deco structure internally and externally, reconstruction of the chimneys, and refurbishment of the historic cranes and jetty as a new river taxi stop. Restoration work on the power station's chimneys was completed in 2017. In 2019, the jetty in front of the power station opened to the public for the first time in its history. In May 2021, the first residents moved into their new homes at the power station. As part of the development, a 200-seat theatre, the Turbine Theatre, was established in railway arches under the Grosvenor Bridge in September 2019.

The restoration of one of the world's largest brick buildings was by no means an easy task. Engineering challenges include the following.

Figure 7.6 Battersea Power Station, London (photo credit to Mei Ren)

- The power station is formed of a steel frame clad in brick. In order to maximise reuse of the building, the restoration necessitated careful consideration and interaction between the new frame and existing fabric.
- This prime riverside development has posed a number of significant challenges in the area of flood risk and water resource management. Significant amount of effort has been undertaken to minimise remedial works and optimise the raising of the existing listed river wall, as well as identifying future climate resilient sustainable drainage solutions.
- Below ground across the site, obstructions and deep geological scour hollows created a highly variable foundation system, with significant risk of differential settlement between new and existing structures.
- Repair and modification were necessary throughout the building. The power station's external elevations had become damaged in many places as the existing frame has moved and expanded as a result of long-term weathering. Given the scale and condition of the existing building, addressing the repairs to the external fabric was a major challenge in its own right. An innovative strategy that involved removing and retaining the historic bricks for reuse where possible, combined with installing new steel framework around the windows, enabled the restoration of the original façade's heritage value. This strategy also allowed slot windows on the sides of the building to be extended in height, improving daylight access for the new residential units.
- Some of the biggest challenges facing the design team revolved around the notion of creating a new, large-scale building within this historic shell. Controlling building movements and retaining construction tolerances between the new and existing structures was a key priority. This led to innovative solutions to maximise usable space within the building using steel ground beams. These measures optimised the sustainable reuse of the structure and materials, enabled careful sequencing temporary works and facilitated the installation of new structural elements designed to withstand exceptionally large forces.
- A conservation-led steering group was established to explore practical options to restore the chimneys. When it became apparent that these monoliths were beyond safe repair, the decision was made to dismantle them and create new geometric replicas on site around new stronger internal steel frameworks. Requiring 600 tonnes – or 25 000 wheelbarrow loads – of concrete and 1500 litres of paint to implement, this method upheld the very highest safety standards.
- Fire risk management is one of the challenges, as the existing constraints and complicated nature of the mixed-use building meant that bespoke solutions had to be developed. The team applied computational fluid dynamics (CFD) techniques to validate the design and to allow the removal of the smoke reservoir screens from the original design for these spaces. The intervention opened up the Turbine Hall, allowing it to fully meet the architectural vision. It also reduced the necessary smoke extraction rates significantly, enabling maximum flexibility for amenity space on the roof.
- To realise the client's ambition for the power station to become a destination in its own right and to deliver high quality green public space, it was necessary to construct the site-wide energy centre beneath the riverside park, between the listed power station and the listed river wall. A number of significant engineering constraints had to be overcome to realise this ambition while also ensuring that the construction programme did not stall around such a critical path activity. The below-ground structure spans the full 150 m frontage of the power station, and extends to over 10 m below ground to accommodate the CHP plant and gas boiler equipment.

Sustainability summary

Battersea Power Station project has been awarded with multiple awards including 2023 Building Awards – Project of the Year, 2023 Construction News Awards – Project of the Year (over £50 million). The project team has been focused on realising the client vision to reimagine this former power station as the centerpiece of a vibrant new urban community and destination. Their work has helped to bring this historic building back to life as a twenty-first century mixed-use complex, supporting the wider design team in creating the 'wow factor' throughout, while ensuring safety, maintaining strict tolerances and maximising commercial space.

Transforming Battersea Power Station was a complex undertaking that has secured a long-term future for this iconic building. Battersea Power Station was awarded Royal Academy of Engineering *Major Project Award for Sustainability 2023*. This award is for a team that has played a critical role in a major engineering project that has had a substantial impact on society and sustainability (Royal Academy of Engineering, 2023).

The key contribution to the adaptive re-use of the building and sustainability includes the extensive brick replacement programme, reconstruction of the four iconic chimneys, as well as the critical engineering work delivered across the former 42-acre brownfield site. Together, these interventions have preserved the architectural heritage of the power station while enabling its transformation into a modern, resilient mixed-use destination.

- Re-use of the existing building structures – foundation, columns, beams, slabs – led to saving more than 36 000 tonnes of carbon dioxide equivalent (CO_2e) of legacy carbon. This represents approximately 20% of the overall total embodied carbon within the primary structural frame.
- Over 1 750 000 replacement bricks were specially made by brickmakers in Gloucestershire and Shropshire that made the original bricks back in the 1930s and 1950s.
- Refurbishment of external elevations masonry resulted in a saving of approximately 4000 tonnes of CO_2e of legacy carbon.
- A tailored instrumentation and monitoring strategy to safeguard the listed structure throughout the works.
- Raised the existing listed Thames River wall and identified climate resilient sustainable drainage solutions.
- In designing for construction, the team devised clear, and cost-effective solutions that minimise risk and expedite programme delivery.

Focus area (learning points) **VISION AND TECHNOLOGY**

Battersea Power Station London has become a popular destination, attracting millions of visitors since its public opening in October 2022.

This case study presents a successful adaptive re-use example demonstrating that through careful repair and restoration of the existing building fabric and structure, the Grade II* listed building has been given new life. The project also acted as a catalyst for revitalising the nearby wider area. Significant carbon savings were made possible by overcoming engineering challenges and re-use of the existing building structures. Battersea Power Station was awarded the Royal Academy of Engineering Major Project Award for Sustainability 2023.

Battersea Power Station has received numerous awards for its restoration and redevelopment, including recognition for heritage, master planning and residential development. The neighbourhood was named one of the Top 20 'Best Cultural Spots' in National Geographic's Best of the World 2024 series and won the RIBA London Conservation Award 2024. Battersea Power Station was also awarded World Gold in the Master Plan category at the FIABCI World Prix d'Excellence Awards 2021 and World Silver in the Residential Mid-Rise Category for the first completed phase of Circus West Village in 2019 (Battersea Power Station, 2024a).

7.2.2 NOMA, Manchester

Outside London and the south-east regions, Manchester is often regarded as the 'capital of the North' in England. Some very large-scale regeneration projects have been successfully delivered in the North-West in the 2010s, including projects such as MediaCityUK in Salford and Spinningfields in Manchester. NOMA is another key large-scale redevelopment scheme that has completely transformed that part of the city, with sustainability at the heart of the redevelopment.

NOMA (a portmanteau of North and Manchester) is an £800 million, 20-acre (8-hectare) mixed-use redevelopment scheme in Manchester, as shown in Figure 7.7. The NOMA scheme has been one of the most significant development projects in the North-West. The name of the project is derived from the location of the project: North Manchester. A Strategic Regeneration Initiative and Masterplan for NOMA, a 20-acre piece of land centred on The Co-operative Group's existing Head Offices in Manchester City Centre, was endorsed in 2009 that set out the development vision, phasing and development framework. The emergence of NOMA was driven by The Co-operative Group seeking new landmark headquarters within Manchester. After an extensive site search, a £100 million investment was made by the group for the development of 1 Angel Square

Figure 7.7 NOMA Masterplan 2012 visual (photo credit to The Co-operative licensed under CC BY 2.0)

on Miller Street (shown in Figure 7.8). This formed the catalyst for the establishment of NOMA as a regeneration area, as new uses were sought for their vacated estate, including many listed buildings and surrounding vacant land. The 2009 masterplan and subsequent updates set out a commitment to create a commercially led, mixed-use destination underpinned by core development objectives of job creation, placemaking and supporting communities. The 20-acre site includes not only premium office real estate, but also more than 4 acres of high quality outdoor space for residents.

Focusing on the regeneration of existing disused areas of Manchester, covering an impressive 20 acres of land, the scheme is one of a kind, reshaping the heart of the city. Although the masterplan programme is expected to complete in 2029, it has already welcomed over 5000 jobs into the city. There has been a noticeable shift in business giants moving offices to the North and NOMA is now home to Amazon's first headquarters outside of London. In 2024, the global financial services company, the Bank of New York Mellon (BNY) decided to move into 4 Angel Square in the NOMA district of Manchester. The move is expected to take place by 2026.

One Angel Square

Located at the heart of the NOMA complex, One Angel Square is the iconic headquarter of the Co-operative Group. Part of the British high street since 1863, the Co-op is the UK's largest mutual retailer with a turnover of over £13 billion. The Co-op always has a strong ethical values, such as fair food production, democracy, honesty, openness and social responsibility, with doing sustainable business sitting at the heart of this. In 2009, the Co-operative commissioned a new 400 000ft^2, 16-storey head office building to provide a modern working environment and facilities that would allow greater freedom for flexible working by its staff.

Figure 7.8 One Angel Square, Manchester (photo credit to The Co-op Group licensed under CC BY 2.0)

Figure 7.9 Environmental design concept (photo credit to The Co-op Group (https://www.flickr.com/photos/39796699@N02/) licensed under CC BY 2.0)

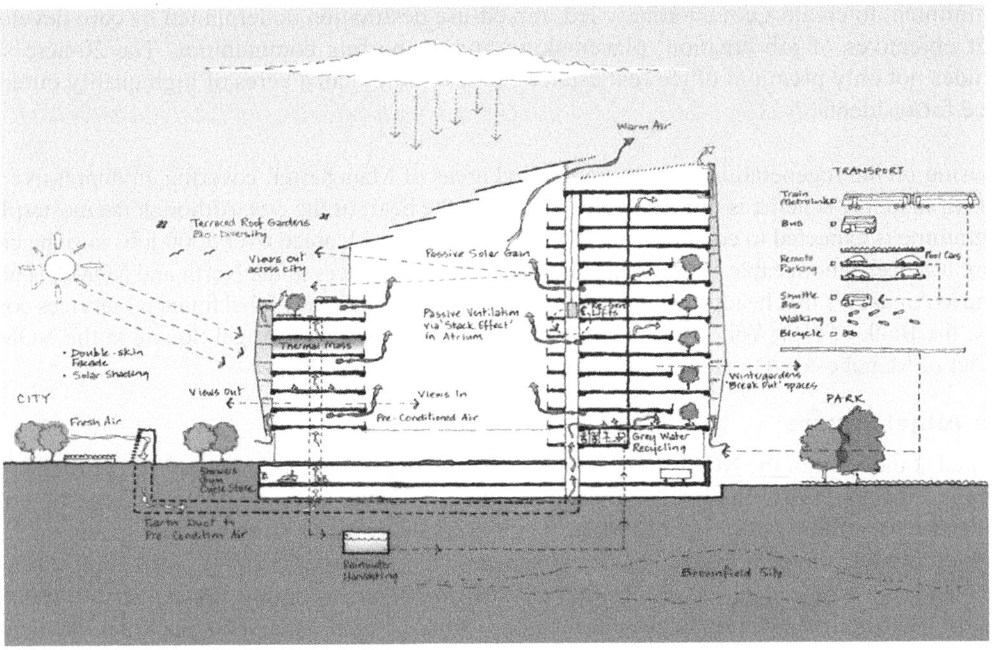

As the phase 1 development of NOMA masterplan, the 'bee hive' shaped building, with an emblem of the Co-operative ethos using the beehive as a symbol of working together for the common good, has exceptional sustainable design features, taking sustainability and innovation to the next level.

The brief

The new head office should consolidate a network of outdated offices into a commercially viable and modern workplace environment. The new building must respect their heritage, provide culture transformation, promoting community engagement and create a legacy. This ethos should shape the project right from design to construction to operation. The building should push boundaries and drive innovation, including re-setting British Council for Offices (BCO) standards; specifically, sustainability targets were set at day one and included the following.

- BREEAM Outstanding.
- EPC A (Energy Performance Certificate A).
- DEC A (Display Energy Certificate A).
- 'In use' total equivalent metered energy consumption of 150 kWh/m^2.
- The building must be designed to have the ability to 'plug-in' to, or provide, future innovative energy solutions.

The solution

One Angel Square is one of the most sustainable buildings in Europe. The building generates heat and power through an 'earth duct' heat recovery system. The system takes in the waste heat

Figure 7.10 Earth ducts and air intake drums at One Angel Square: (a) a view inside the earth duct; (b) the air intake drums as a feature located in the public realm square (photo credit to David Dixon, licensed under CC BY-SA 2.0)

(a) (b)

generated by the IT systems and uses it to heat the building. In addition to this, the double-skin exterior insulates the building during winter and reduces heat gain during warmer months. The environmental design concept diagram is illustrated in Figure 7.9.

Alongside this innovative heating system, the development includes low energy LED lights, rainwater recycling systems for toilets and high-efficiency passenger lifts. The development has been awarded a BREEAM 'Outstanding' environmental rating, achieving an exceptional score of 95.32%. The design has achieved a 40–60% reduction in energy consumption. The client brief has been fully met and exceeded in some areas.

Key design features include

- high performance double skin facade (Figures 7.11)
- integrated passive chilled beams and cooling towers, extensive free cooling (2600 kW heat rejection and 1100 kW free cooling)
- heat recovery from server rooms into fresh air supply
- variable speed and 2-port control, EC DC motor drives
- earth tube air intake system (Figures 7.10(a), 7.10(b))
- low energy lighting
- low pressure drop mechanical ventilation
- two 400 kW CHP Units – using rape seed oil, but capable of running on multiple fuels (fuel conditioning units)
- CHP is optimally matched to an absorption chiller to maximise overall efficiency
- IT energy use – reduce demand through 'thin client'
- seasonal commissioning
- fully integrated IT network building management system with all systems
- electrical management system (SCADA based)
- effective lighting control
- electric car charging zone
- an iPhone app: explaining the building's systems and operational performance as an engagement tool with the building users.

Figure 7.11 Ventilation concept and facade cavity at One Angel Square: (a) the concept of ventilation within the double skin facade; (b) a view of the facade cavity (reproduced with kind permission of Buro Happold)

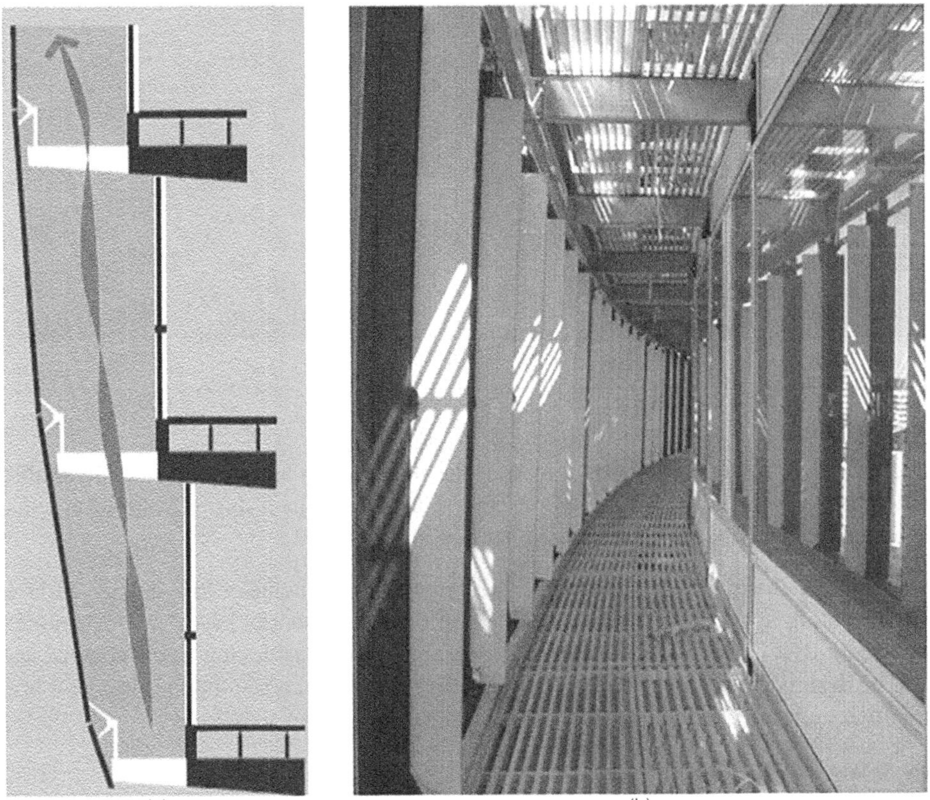

(a) (b)

Climate adaptation

A climate change resilience and adaptation strategy was developed for the head office. The assessment primarily focused on testing the cost and benefits of various intervention measures under two climate scenarios for the 2030s and 2050s, which represent the 'likely scenario' and an 'near extreme scenario' respectively. The innovative earth duct, as one of the key sustainability design features, has proved to be an effective and robust adaptation measure. It is effective now and its performance will improve further due to the ground temperature rising at a much slower rate than ambient temperatures in future climate models.

The ventilated double skin facade (Figures 7.11(a), 7.11(b)) has proved to be effective in reducing cooling energy due to its ability of removing hot air in the cavity after it absorbs solar gains. However, the study revealed that increasing the ventilation rate further within the cavity may not offer additional benefit, while increasing energy burden on fans.

In addition to the design adaptive measures, certain management adaptation measures including flexible seating and dress code during a heat wave period were evaluated in consultation with the

Figure 7.12 No. 4 Angel Square, Manchester (photo credit to Valienne licensed under CC BY 4.0)

Co-op management and human resource teams. Although the 'soft' intervention is cost effective and can offer significant benefits to staff, it would require a culture shift and a change programme. The pandemic in 2020 has accelerated this culture shift to flexible working, which inadvertently helps staff to adapt their working environment in response to changing climate.

4 Angel Square, Manchester

4 Angel Square provides over 200 000 ft^2 of commercial office space located within the NOMA mixed-use development. It was designed by SimpsonHaugh and completed on site in 2023. The building is one of the UK's most sustainable speculative office buildings, being Manchester's first building to receive a NABERS UK Design Reviewed Target Rating of 5*, having achieved BREEAM "Outstanding" (Figure 7.12).

The engineering strategy aspires to extremely high environmental standards including net zero carbon – by combining leading sustainability design and wellbeing knowledge to create a future-proofed working environment with low occupational costs. This pioneering project embraces an innovative Design for Performance (DfP) approach, which aims to close the performance gap.

Initially operated by the Building Research Establishment (BRE), National Australian Built Environment Rating System UK (NABERS UK) is a newly established system for rating the energy efficiency of office buildings across England, Wales, Scotland and Northern Ireland, and is fast becoming the gold standard for assessing operational efficiency across the UK. The 4 Angel Square site is the first in central Manchester to achieve the industry-leading NABERS DfP Target Rating of 5 Stars. Since 2024, NABERS UK is administrated by the Chartered Institution of Building Services Engineers (CIBSE).

The following are the key sustainability features.

- BREEAM Outstanding rating, ActiveScore Platinum, Fitwel 2-star Certification
- NABERS 5 stars and DfP applied throughout the design and construction stages.
- Fabric first approach applied to building envelop with U values and solar protection.
- Heat pumps to provide heating and cooling.
- Metering and controls to ensure energy wastage is minimised.

The Bank of New York signed into the building in 2024 and will be the sole occupier of the 200 000 ft^2 building when they are scheduled to move in by 2026.

VISION AND TECHNOLOGY

Focus area (learning points)

NOMA (a portmanteau of North and Manchester) has been one of the most significant development projects in the North-West of England, with the first phase being the Co-operative Group's Head Office in Manchester. The Co-operative, being part of the British high street since 1863, and the UK's largest mutual retailer, always has strong ethical values and social responsibility. The commercially-led, mixed-use destination NOMA masterplan has created a mixed-use vibrant urban centre and has attracted a number of lead tenants.

This case study outlines a series of innovative technologies and the processes that informed the climate change adaptation strategy. It demonstrates that how a strong vision, combined with pragmatic adoption of technologies led to a highly efficient building with the highest BREEAM 'Outstanding' scores at the time it was certified. The climate change adaptation measures and customer focused engagement approach contributed to high levels of occupant satisfaction. Further development at NOMA such as 4 Angel Square continues to pursue ambitious sustainability objectives, incorporating performance-led frameworks such as NABERS.

One Angel Square won the 2012 Property Week Award for Sustainable Achievement, and the 2013 British Council for Offices Award for Corporate Workplace.

7.3. Europe

In 2015, the European Union (EU) and its member states demonstrated their thought leadership by adopting the 2030 Agenda and the Sustainable Development Goals (SDGs), alongside the Paris Climate Agreement. Since then, more policy and technical agencies of the European Commission have been mobilised to develop integrated SDG actions and monitoring processes. As a developed economy, European countries have well established sustainability policies, supported by implementation tools. European regulations on sustainability are now frequently regarded as a global benchmark, influencing the behaviour of institutions, consumers, investors, businesses, farmers, non-governmental organisations (NGOs) and social organisations.

In 2023, the EU's first Europe Sustainable Development Report 2023/24 (Lafortune *et al.*, 2024) provided a voluntary review of the EU's progress towards implementing the 2030 Agenda. The report reviewed the current SDGs progress and called for decisive actions to be taken in the EU and globally before 2030 to avoid irreversible environmental and dangerous social tipping points and to preserve the possibility of achieving key global goals, including the 2030 Agenda with its 17 SDGs and the Paris Climate Agreement.

Three case studies in Europe have been chosen to indicate sustainability transformation of the built environment. The 2023/24 SDG Index (Europe Sustainable Development Report 2023/24, 2024) for Europe is topped by Northern and Western European countries. Two case studies were chosen from these regions, supplemented by a southern European Mediterranean project.

- EDGE Suedkreuz, Berlin. This case study shows how the innovative materials were adopted to reduce environmental impact while meeting regulations such as fire risk.
- Kristian Augusts gate 13 (KA13), Oslo. KA13, designed by Mad arkitekter, is a truly ingenious office refurbishment endeavour, by thinking materials first and building with the circular economy in mind. This new way of undertaking the design process provides a fundamental shift for implementing a circular economy in built environment.
- Valencia, Europe Green Capital 2024 and Valencia Conference centre 25 years on. The city has a proven commitment to improving air quality and restoring nature ecosystems. This case study is chosen for its demonstration of governance and continuous improvement.

7.3.1 Innovative use of materials: EDGE Suedkreuz, Berlin

Situated near the Südkreuz station in Berlin's Schöneberg district, EDGE Technologies developed a 10 000 m^2 property with the intention of creating 39 000 m^2 of office space for the German energy supplier Vattenfall. The architect is Tchoban Voss Architekten, with Buro Happold providing structural engineering, building services engineering (MEP), BIM management and sustainability consulting.

EDGE Suedkreuz is one of the largest buildings planned as a modular timber hybrid construction in Germany. It comprises two office buildings, each with six floors and a connecting underground car park, which is fully equipped with charging stations for electric vehicles.

In line with Vattenfall's corporate strategy as well as that of EDGE Technologies, sustainability has been imperative from the very beginning of the project. The aim was to reduce the building's carbon footprint, to use sustainable and healthy materials and to take into account the wellbeing of the future occupants. The modular timber hybrid construction method is an important part of achieving this and was integrated into the engineering concept.

The design features timber throughout and incorporates floor-to-ceiling windows. Oversailing the 1600 m^2 atrium, the roof construction comprises lightweight ETFE cushions, a visually appealing and cost-efficient design (Figure 7.13).

The atrium roof is constructed from timber, steel and lightweight ETFE cushions, enabling the G-value (G-value represents solar protection parameters of glazing) of the roof to be changed depending on the season. The ETFE cushions can be adjusted to regulate the level of daylight and

Figure 7.13 EDGE Suedkreuz, Berlin (Buro Happold, n.d.a) (photo credit to CREE BUILDINGS licensed under CC BY-NC-SA 2.0)

thereby control the temperature. A microshade glazing system offers a passive and highly efficient shading solution, which features a micro-structured film integrated into insulating glazing units that adapt to the summer and winter conditions (Figure 7.14). Smart controls automatically regulate air quality, light and temperature to enable optimal working conditions and save energy. All materials used were tested for their environmental impact.

Two sides of the atrium roof have been left open to create a natural ventilation system and allow fresh air to flow into the building (Figure 7.15). The atrium is a valuable source of daylight for the surrounding offices, minimising the need for artificial lighting while achieving the optimal balance between energy use, material efficiency, and occupant comfort.

The timber used for the structure, for instance, comes from certified forests, other materials are recycled or recyclable where possible and surface paint does not contain toxic substances (VOC free). Full life cycle carbon of materials was considered. All materials used have a life span of 50 years. The modular hybrid construction method means that dismantling at the end of life will be as easy and efficient as construction.

From the outset, the structural engineer worked closely with Cree by Rhomberg, an innovative construction system supplier, to achieve a fully modular timber-concrete-hybrid structure (Figure 7.16). The advantages of this approach include rapid construction, the sustainable nature of the material and, ultimately, the creation of a more desirable working environment. Additionally, the structure's light weight will significantly reduce costs associated with creating the building's foundation.

Figure 7.14 Microshade solution at EDGE Suedkreuz, Berlin (image reproduced with kind permission of Buro Happold)

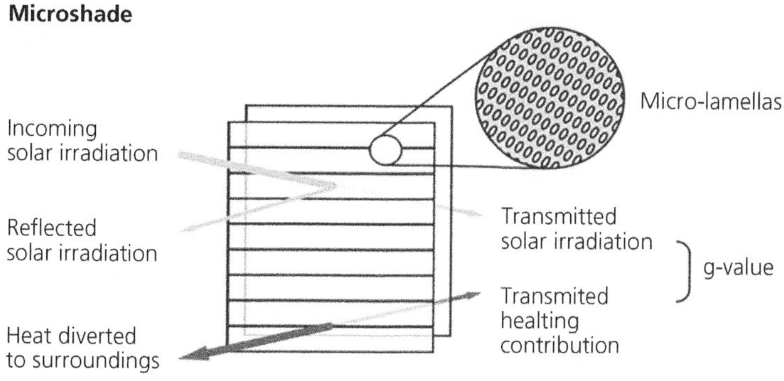

Figure 7.15 Cross section of ventilation solution at EDGE Suedkreuz, Berlin (image reproduced with kind permission of Buro Happold)

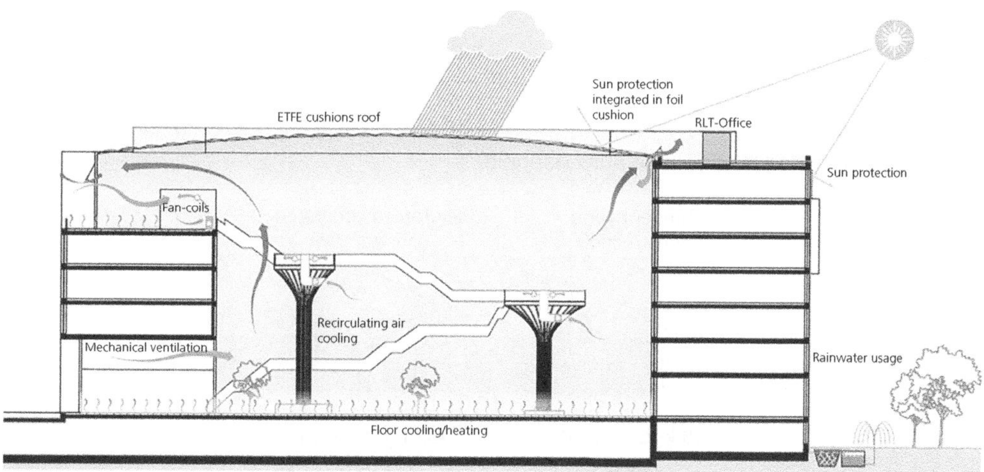

Sustainability and occupant wellbeing were important drivers for the EDGE Suedkreuz project right from the start. These imperatives continue to inform the entire development and have resulted in a DGNB (German Sustainable Building Council) pre-certificate in Platinum and WELL Gold. Vattenfall's employees will benefit from a healthy indoor working environment.

Summary

- By using the modular timber hybrid construction, embodied carbon has been reduced by 48% in comparison to a reinforced concrete structure (Figure 7.17).
- Reduced weight of hybrid ceilings allows material savings for the structure. The hybrid constructure saves time, waste and energy during construction, allows easier demolition and higher grade of recyclability of building.

Figure 7.16 A close view of the hybrid timber and steel modular construction (image reproduced with kind permission of Buro Happold)

Figure 7.17 Comparison of the embodied carbon of the two structure options (image reproduced with kind permission of Buro Happold)

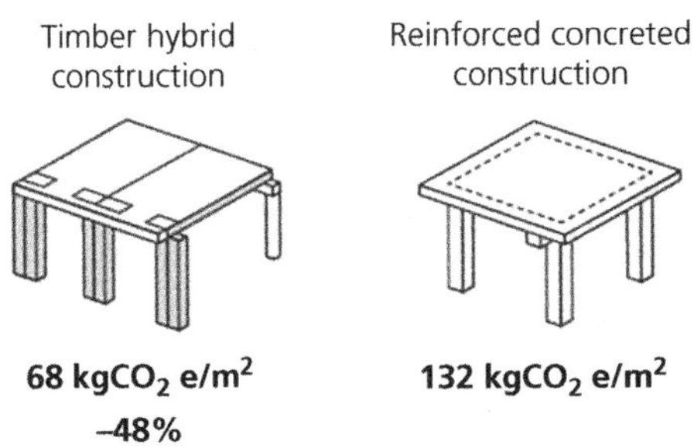

- EDGE Südkreuz building achieved DGNB Platinum and WELL Gold.
- The building's proximity to Südkreuz railway station and its underground car park for e-cars provides employees and visitors with access to alternative green mobility.
- The highly automated, wireless system monitors air quality, lighting and temperature to maintain optimum working conditions. The smart building design features include 12 000 sensors, in addition to digital ceiling, smart phone app, building IOT database, smart parking, robotic security and cleaning.
- The hybrid construction method challenged all the teams involved to come up with completely new and innovative approaches to the joint coordination of all services. Hybrid ventilation systems, for example, make use of the space between the wooden beams, bringing integrated lighting and sensor technology into the building at the same time. Design integration using BIM technology played a big part in managing the coordination effort to adapt working methods to needs.

Focus area (learning points) **VISION AND TECHNOLOGY**

EDGE Technologies developed a 39 000 m^2 of office space for German energy supplier, Vattenfall. The building is situated near the Südkreuz station in Berlin's Schöneberg district. Sustainability goals were imperative from the very beginning of the project in line with the EDGE Technologies' and Vattenfall's corporate strategies.

EDGE Suedkreuz is one of the largest buildings as a modular timber hybrid construction in Germany. This case study provides an example of innovative use of materials and smart building application with 12 000 sensors including digital ceiling, smart phone app, building IOT database, Smart parking and robotic security and cleaning. The highly automated, wireless system monitors air quality, lighting and temperature to maintain optimum working conditions. It demonstrates that the strong vision and technologies led to embodied carbon reduced by 48% in comparison to a reinforced concrete structure, achieving DGNB Platinum and WELL Gold.

EDGE Suedkreuz Berlin was awarded DGNB: Overall Compliance Rating: 95.4%, making it the most sustainable building in Germany, winning DGNB Diamond Award: for design quality (Schröder, 2022), 2023 Green GOOD DESIGN Award (Good Design Awards, 2023), International Architecture Award (Buro Happold Projects, n.d.a), Built Design Awards Winner in Architectural Design – Sustainable and Energy Saving (BLT Design Awards, 2023).

7.3.2 Kristian Augusts gate 13, Oslo, a circular example

Kristian Augusts gate 13 (KA13) at Tullinløkka is a groundbreaking reuse project where a 1950s building threatened with demolition was preserved and reused. The project transformed a 1950s office building in central Oslo and added its new eight-storey extension using a variety of reused building materials. The building is located in Ombruk, Oslo, with an area of 4300 m^2.

This innovative circular project, Kristian Augusts gate 13 (KA13), implemented the reuse of building materials on a large scale, reusing approximately 80% of materials. The project has been critically acclaimed both in Norway and abroad (Kristian Augusts gate 13 – Mad arkitekter, n.d.). The architect is Mad studio, the client is Entra with partners Asplan Viak, Futurebuilt, Haandverkerne, Insenti AS, Resirqel AS. KA13 continues to pave the way towards a more sustainable future for the construction industry.

Buildings built from the 1950s are prone to be demolished and can quickly disappear as they are reaching end of life and have low floor heights and are complicated to upgrade to today's technical requirements – for example, new technology means low room heights can be problematic. In KA13, this challenge is compensated for by creating internal vertical openings over several levels to create dynamic interior spaces and greater openness within the building. The vertical spaces are also used in connection with exhaust ventilation, and they provide more daylight into the office areas (Figure 7.18) The open access with an entrance from street level has been provided with the possibility of passing through to the courtyard, which is connected to a network of walkways within the block.

In connection with the reconstruction, the building was expanded with a small extension on nine floors. This extension is connected to the existing staircase and elevator in the building. Both hollow floors, steel structure and facades are recycled materials applied in the extension.

The design process is a complete shift away from a linear economy considering 'building form first and materials last'. Instead, knowing first what materials might be potentially available for reuse,

Figure 7.18 A close view of the building exterior with reused cladding tiles and green roof, Kristian Augusts gate 13, Oslo (photo credit to Mad arkitekter, Kyrre Sundal)

then considering physical and materials constraints, helped the design fall into place. Many of the 'new' elements in K13 are not new, but reused elements from other buildings. In this process, the building elements come either from those completely different projects that are being rebuilt, or from another place in the building. The recycled materials used in the renovation of the building include concrete floor slabs from a government building, grille panels from the Tøyenbadet swimming pool (Figures 7.19 and 7.20). The colourful facade panels (Figure 7.18) are recycled Steni composite stone rainscreen panels and cladding (Omnis, n.d.).

Part of FutureBuilt pilot projects that must reduce climate footprint by at least 50% with at least 50% re-used and reusable components, Kristian Augusts gate 13 achieved 80% re-use and 70% carbon reduction from the material components of the building (KoozArch, 2022). The project pioneered the reuse of structural materials overcoming challenges associated with precision, testing, manual work, logistics and responsibility.

Following the Kristian Augusts gate 13, a growing number of players across the value chain have come to the market including physical and digital platforms such as urban mining and mapping of buildings, along with components for potential re-use, marketplaces and upcycling (KoozArch, 2022).

Figure 7.19 Interior during reconstruction, Kristian Augusts gate 13, Oslo (photo credit to Mad arkitekter, Kyrre Sundal)

Figure 7.20 Wall enforcement, Kristian Augusts gate 13, Oslo (photo credit to Mad arkitekter, Kyrre Sundal)

Focus area (learning points) **ACTION**

Kristian Augusts gate 13 in Oslo is a groundbreaking reuse of 1950s building that has transformed the office building and added a new eight-story extension using a variety of reused building materials. The building has an area of 4300 m^2.

This case study demonstrates a pioneering large scale commercial application of circular economy principles, using approximately 80% reused materials including concrete slabs, steel structures and facade panels, demonstrating a commitment to circular construction practices. The project focused on circularity through strategies such as reuse of building components and design for disassembly.

Kristian Augusts gate 13 has provided an impactful catalyst towards increased circularity (KoozArch, 2022). Following the Kristian Augusts gate 13 example, a growing number of players across the value chain have come to the market including physical and digital platforms such as urban mining and mapping of buildings, along with components for potential re-use, marketplaces and upcycling. In 2021, the Norway Government initiated a regulatory change that would make it significantly easier to both reuse and recycle building materials.

KA13 project received the DOGA award from the Foundation for Design and Architecture (DOGA, 2021) in Norway for its future-oriented approach and the Norwegian State Award for Building Quality in 2021 (Omnis, n.d.), and EUmies Award 2022 (EUmies, n.d.).

7.3.3 Valencia, European Green Capital 2024

Valencia, located on the southeastern coast of Spain, has 837 000 inhabitants and is the capital of the province that has the same name. Valencia city earned the European Green Capital 2024 title due to its past and current achievements in the field of sustainable tourism and climate neutrality, as well as its commitment to a fair and inclusive green transition.

The city has a proven commitment to improving air quality and restoring nature ecosystems, such as the Devesa dune and wetland ecosystems. Valencia is home to the Albufera Natural Park, one of the most important wetlands in Europe. The measures taken by the city to protect this environment have also led to Valencia being designated the first Wetland City in Spain in 2022. Valencia also stimulates a healthier, sustainable and inclusive food production through its ‘Neighbourhood and Food Programme’.

Photovoltaic (PV) panels have been installed in different municipal buildings and the entire lighting system of the city has been technologically transformed, achieving greater efficiency and savings of €4 million per year, which has been reinvested in new energy efficiency projects (Garcia, 2022). By incorporating a mission-oriented innovation approach at the core of the Urban Strategy, the Valencia City Council has accelerated their decarbonisation journey in recent years, which has resulted in the city’s candidacy to be one of the 100 cities ‘Achieving 100 climate-neutral and smart cities in 2030’ chosen by the European Commission (Directorate-General for Research and Innovation, 2020).

From reducing carbon emissions by promoting sustainable transport to protecting local wildlife such as that of Albufera, five sustainable initiatives currently being implemented in Valencia, with the goal of ensuring the city’s green status beyond 2024 and of reaching its goal of climate neutrality by 2030.

- Use of AI to accelerate sustainability

Valencia has been chosen as one of three large European supernodes for the Citcom.ai project, funded by the European Union, to install an artificial intelligence (AI) experimentation and testing centre in the city. The aim is to apply AI to solve urban issues, such as mobility, sustainable transport and energy efficiency.

- Smart recycling

The city will soon have smart brown containers for the selective collection of organic waste. They will have an electronic closing system that requires the use of a card or app to open them, aiming to reduce the percentage of improper waste that is deposited.

- Wave power

The first installation of wave power in Valencia is already under construction in the city's Marina. Current figures estimate that it will generate about 130 000 kW/year from wave movements, which will contribute to the education of renewable energy and the city's carbon dioxide reduction.

- Pedestrianised public spaces

Fewer cars and increased numbers of people opting to walk has a positive environmental impact within cities. To ensure that locals and visitors alike feel safe and confident to walk in the city, so far 150 000 m^2 of land has been recovered for pedestrian use due to the pedestrianisation of Plaza de la Reina, Plaza del Ayuntamiento and Plaza de Brujas. As part of the 'Valencia, City of Plazas' scheme, the city aims to recover 207 000 m^2 of public space, by reducing roads to lanes with speed limits and expanding and re-vegetating pedestrian spaces (Figure 7.21). Additional measures, including installation of PV solar-panelled pergolas that generate electricity on a small scale, the creation of superblocks and the construction of more bike lanes, are also helping to enhance quality of life in Valencia.

- Renaturation

Since the creation of the Jardín del Turia in 1986, when the Turia riverbed was diverted and a new public garden built, Valencia has continued to expand its green spaces and renaturalise its urban areas.

Figure 7.21 Public spaces, Valencia, Spain (photo credit to Expedia: https://www.expedia.co.uk/Valencia-Valencia-City-Centre.dx6185713)

Figure 7.22 Green roof, Valencia, Spain (https://growgreenproject.eu/new-green-roof-installed-senior-centre-valencia/; photo credit to Las Naves)

In 2022, 1687 new trees were planted. Moreover, the city has a remarkable 600 hectares of parks and gardens, resulting in an estimated 90% of the population having access to a green space within 250 m. Figure 7.22 shows an example of green roof programme.

Extreme rain event 2024

Sustainable urban drainage systems (SUDS) are an alternative and supplementary approach to traditional urban drainage practices. SUDS were almost unknown in Spain two decades ago; however, urban drainage practices in the country are now transitioning towards a more sustainable and regenerative management within a global context where green policies are gaining prominence. Despite this progress, Spain still faces several challenges that hinder a full transition (Andrés-Doménech *et al.*, 2021), including governance-related barriers with limited stakeholder involvement, insufficient knowledge, and unclear organisational responsibilities.

Climate change is the key contributing factor to the more extreme and frequent storm rain events that the area is facing. For southern Europe, and especially Mediterranean countries such as Spain, Italy and Greece, the escalating impacts of climate change are becoming painfully evident. The region is a flashpoint for extreme weather due to its unique geographical and climatic conditions (Policy Circle Bureau, 2024). The Mediterranean has seen the warmest surface water temperature on record with a mean temperature of 28.47°C in mid-August. This facilitates a greater uptake of moisture in the air, resulting in more rain when the atmosphere starts to cool in the autumn.

Valencia suffered from a historic flood in October 1957, known as the Great Flood of Valencia, which resulted in significant damage to property and caused the deaths of 81 people. This flood was mainly caused by the overtopping of the River Turia which was then diverted in 1969 to avoid the centre of the city. During floods, most of the river water was diverted southwards along a new route that borders the city. Since then, Valencia city and the region have developed a plan in 2020 to reduce the amount of damage a flooding event such as this could cause to the region. The plan involves putting an investment in place to enable the nearby Poyo gorge to be upgraded. Unfortunately, this plan had not been implemented by the time the region was hit by the extreme storm event in 2024.

Due to the unprecedented volume of rainfall starting on 29 October 2024, enormous amount of water flowed through the gorge causing widespread flooding to the region. While suffering from nearly a full year's worth of rain in eight hours would cause a certain amount of flooding, regardless of how prepared a region may be, the situation in Valencia was exacerbated by the presence of the Rambla del Poyo, a 43.5 km gorge.

Although Valencia city itself largely escaped the worst effects of the floods, the towns and industrial belt surrounding the city were severely impacted, causing the tragic fatalities of 210 people and widespread destruction of property in the region. Valencia has now accelerated the implementation of the SUDS principles, models and measures to reduce the impact of future extreme events and improve climate emergency preparedness in order to stop such a catastrophe reoccurring (Johnson, 2024; Policy Circle Bureau, 2024).

Valencia Conference Centre

The Valencia Conference Centre building was a landmark architectural work designed by Norman Foster Architects. Figure 7.23 and Figure 7.24 show the centre covered with roof PV and a

Figure 7.23 Valencia Conference Centre is covered by the extensive PV arrays (photo credit to https://palcongres-vlc.com/es/

Figure 7.24 The centre entrance is shaded to provide passive cooling effect for the visitors. The centre is well utilised by communities for education purposes (photo credit to https://palcongres-vlc.com/es/)

view of the entrance area. It was awarded Best European Building by the Royal Institute of British Architects a year after its opening in 1999, due to its perfect fusion of aesthetics and usability. Since opening, the conference centre won the World's Best Convention Centre twice, in 2018 and 2010, the most prestigious international award from the International Association of Convention Centres (AIPC), which recognises excellence in management, facilities, equipment, services, accessibility and accommodation capacity. This centre was highlighted due to its continuing ESG effort.

Its sustainability plan covers four key axes: environmental, economic, social and governance. The plan details how the Valencia Conference Centre intends, within its area of responsibility and action, to generate value for the society in which it is integrated to carry out its activities. Supported by different agents who have contributed with their specific approach and vision, the plan also demonstrates the strategic thinking by the management team of the centre as a sign of the team's commitment to realising the organisation's main purpose.

This sustainability plan outlines a series of actions and objectives that will be monitored with data analysis tools. Through these tools, the status of progress across the different action plans developed as part of the sustainability strategy can be assessed and monitored. They also help ensure that all targets are met through ongoing oversight within the designated strategic forum. In addition, this strategic forum is responsible for defining the appropriate contingency plans and updating, if necessary, any of the plans set for the coming years.

For the environment part of the plan, the primary targets include: waste recovery, carbon neutrality and resource consumption.

- Waste recovery (% kg recovered or recycled), 50% by 2025, 100% by 2029.
- Minimise the amount of greenhouse gas (GHG) emitted into the atmosphere, achieving neutrality of the carbon footprint (t CO_2 generated – offset) is 100% by 2025.
- Resource consumption reduction for water, electricity and materials (% of the annual target for reducing resource consumption), 25% by 2025, 100% by 2029.

Focus area (learning points) **GOVERNANCE**

Valencia city earned the European Green Capital 2024 title due to its past and current achievements in the field of sustainable tourism, climate neutrality, as well as fair and inclusive green transition.

This case study showcases the southern European city's commitment and initiatives to improving air quality, health wellbeing, climate resilience and restoring nature ecosystems. Their mission-oriented innovation approach at the core of urban strategy, has accelerated their decarbonisation journey in recent years, which has resulted in the city's candidacy to be one of the 100 cities 'Achieving 100 climate-neutral and smart cities in 2030' chosen by the European Commission (Directorate-General for Research and Innovation, 2020).

Two particular examples were highlighted in this case study. Extreme rainfall events such as in October 2024 brought a renewed focus on climate resilience. Mediterranean countries, like Spain, Italy and Greece, will see the escalating impacts of climate change since the region is especially susceptible for extreme weather due to its unique geographical and climatic conditions (Policy Circle Bureau, 2024). This adds urgency for the city region to address those challenges and implement the climate resilience strategy to overcome obstacles that delay the transition (Andrés-Doménech *et al.*, 2021).

The second example is the Valencia Conference Centre. The award-winning centre was especially chosen to highlight its continuing ESG effort. Their sustainability plan covers four key axes: environmental, economic, social and governance. This sustainability plan defines a series of actions and objectives that are constantly monitored with data analysis tools.

7.4. North America

The United States has a variety of programmes and initiatives in place to promote sustainability but is not on track to meet all of the SDGs (EPA, 2024). The US Environmental Protection Agency (EPA)'s environmental regulations provide a framework for sustainability that can be applied to decision making across all EPA programmes, including decarbonising existing buildings.

The country has abundant natural resources, and it is capable of generating a large amount of renewable energy. The American built environment industry has a culture of seeking cutting-edge solutions, such as Leadership in Energy and Environmental Design (LEED), which was originated in the US and has since been influential across the globe. The Living Building Challenge emerged in the US recently from a growing movement within the American architectural and environmental community to push beyond standard green building practices and strive for a more regenerative built environment, where buildings not only minimise negative impacts but also actively contribute positively to the ecosystem.

Three case studies chosen in the US include the following.

- Santa Monica, an office building certified through International Living Future Institute (ILFI), aligns directly with long-term sustainability commitments made by the Santa Monica City Council. The project is committed to achieving community-wide carbon neutrality by 2050 or sooner, water self-sufficiency by 2023 and zero waste by 2030.
- Empire State Building decarbonisation. The Empire State Realty Trust launched a bold programme to decarbonise this world famous tower. This case study reflects on a key challenging dilemma facing the construction industry today: to build new buildings, or to renew the existing ones?
- Following the aftermath of Hurricane Sandy 2012, a new vision for the BIG U emerged in 2024 located in Lower East Side of New York, initiating the redevelopment to protect Lower Manhattan from floodwater, storms and other climate related impacts affecting the low-lying topography of Manhattan. This is one of the most ambitious storm resiliency projects in the world.

In Canada, as the city experiences intensified climate changes, the City of Toronto wishes to protect their citizens against negative climate impact. The City Planning Division commissioned a thermal comfort study to serve as one of the foundational climate action resources to inform their Heat Relief Strategies.

This case study is chosen to show the City of Toronto's climate action to maximise thermal comfort in public spaces.

7.4.1 Santa Monica, CA, Living Building Challenge

The Living Building Challenge is an attempt 'to raise the bar for building standards from doing less harm to contributing positively to the environment'. It 'acts to rapidly diminish the gap between current limits and the end-game positive solutions we seek' by challenging architects, contractors and building owners. The Living Building Challenge is administered by the International Living Future Institute (ILFI) and consists of seven performance categories, including net zero energy, net zero water, net zero waste (Living Future, n.d.a)

The Santa Monica City Hall East (Figure 7.25) is a three-storey office building. It was designed to meet the rigorous standards of the Living Building Challenge. Recognised as one of America's greenest city hall buildings (Stamp, 2020), the project reflects the Santa Monica council's ambition to pursue a certified Living Building Challenge as a way to inspire others to improve their green credentials while showcasing how to do so.

The 50 200 ft^2 Santa Monica City Hall East is a modern extension to the existing City Hall and houses state-of-the-art, energy efficient civic offices. Commissioned by the City of Santa Monica, Architecture Services Division, the design team includes Frederick Fisher and Partners Architects and Buro Happold as an integrated engineering and consulting team. The design ensured the new office building meets the Living Building Challenge criteria and stands as a leading example of sustainable design. Completed in 2020, the project cost approximately US $57 million (Creighton, 2020).

Living Building Challenge standards are the most rigorous performance standards in the built environment, surpassing the highest LEED certification requirements, as they demand that a building

Figure 7.25 Santa Monica City Hall, CA (photo credit to Joseph Coriaty is licensed under CC BY-SA 4.0)

actually makes positive contributions to its environment as opposed to merely lessening its negative impact. An inspiring model aimed at efficient operations, a healthy and productive workplace, and positive contributions to one of America's most successful communities, this new municipal operations center supports the city's commitment to sustainability with the pursuit of full Living Building Challenge certification (Dean, 2022).

Energy Petal

A passive approach to design was adopted to make the most of all available natural resources. The building uses natural ventilation on a diurnal cycle, applied with phase change materials incorporated within internal ceilings, walls and the perimeter spandrel glass areas to absorb the heat gains to reduce peak cooling impact. Heating and cooling delivered to internal spaces are through radiant means by incorporating pipes within the slabs. The building's electricity requirements were reduced significantly so that they can all be met through a PV array on the roof. In addition, heating and cooling loads were minimised by making strategic use of both daylight and natural ventilation (BNIM, n.d.).

The predicted Energy Use Intensity is 24.9 kBtu/ft^2 (78.5 kWh/m^2) per year. On-site PV installations of 292 kW generate 399 430 kWh per year, 27.1 kBtu/ft^2 (85.5 kWh/m^2) in 2021 (Dean, 2022).

Water Petal

The Water Petal presented among the most challenging goals, given the drought-prone location of the project. In response to the broader global climate crisis, the design team grounded their approach in several key principles such as rethinking how water is used onsite, that water be respected as a precious resource, and that the building's water needs can be fully met on site (Creighton, 2020). Figure 7.26 illustrates the water strategy.

Three separate water strategies collectively enable all the building's needs to be met by minimising use and through water harvested on site.

- Firstly, the projected water and sewer demand for the building was reduced by half with the introduction of a foam-based composting system for all the lavatories. The system units are located in the basement of the structure and require periodic maintenance.
- Secondly, for potable water uses, rainwater from the roof is captured in a 40 000-gallon (182 m^3) cistern located beneath the building; the rainwater system is supplemented by groundwater pumped from a well drilled on site with reverse osmosis water treatment for top-up during low rainfall conditions. All rainwater and groundwater is treated with a combination of cartridge filters and granulated active-carbon filters, also located on site.
- Lastly, for non-potable water uses, the system captures greywater and condensate from the showers and the cooling system, treats it through a moving bed-membrane bioreactor and reuses it for onsite irrigation.

The composting toilets save about 270 000 gallons (1022 m^3) of water per year and they also produce nutrient-rich organic materials, called biosolids (Creighton, 2020). All sanitary wastewater is treated on the building site, and biosolids are exported offsite for composting. Leachate – the liquid that passes through the matter and extracts soluble or suspended solids – is pumped out and

Figure 7.26 A technical schematic diagram of the water system (Creighton, 2020) (image reproduced with kind permission of Frederick Fisher and Partners)

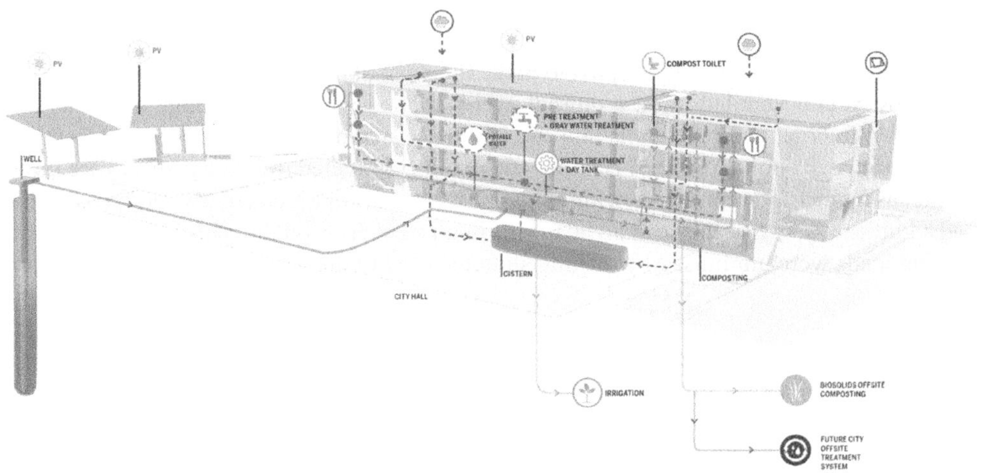

drains by gravity to the street sewer, along with miscellaneous sanitary wastewater. The leachate is handled offsite at the Hyperion Wastewater Treatment Plant, where it will be treated through anaerobic and aerobic biodigestion to create a sustainable source of heating and power for the wastewater treatment plant.

Many of the ideas the project team developed raised issues of code conformance, legality and land use regulations. As the ILFI asserts, 'Reaching the ideal for water use means challenging outdated attitudes and technology with decentralised site or district-level solutions that are appropriately scaled, elegant, and efficient' (Creighton, 2020).

The City Hall East design team ultimately worked to overcome significant obstacles posed by building, health and water codes in order to ensure the project met both the spirit and the Water Petal. The design team developed creative solutions to meet schedule and budget requirements, while limiting risk associated with non-traditional code compliance processes. Ultimately, the city's leaders persevered and secured a permit that authorises California's largest installation of composting toilets and the first building in California to treat rainwater for potable use on site, including drinking water.

In 2018, California Governor Jerry Brown signed a bill – 'Jerry Brown sets permanent water conservation rules for Californians' (Rogers, 2018) – to develop statewide regulations for on-site water reuse; a step that enables local governments like Santa Monica to take the lead in increasing water reuse for a more sustainable future.

'The Santa Monica City Services Building is an inspiring model of green design aimed at efficient operations, a healthy and productive workplace, and positive contributions to one of America's most successful municipalities' (Living Future, n.d.b).

Focus area (learning points)

VISION AND TECHNOLOGY

The Living Building Challenge is an attempt 'to raise the bar for building standards from doing less harm to contributing positively to the environment.' It demands that a building actually makes positive contributions to its environment as opposed to merely lessening its negative impact.

The Santa Monica City Hall East is a three-storey civic office building designed to meet the standards of the Living Building Challenge in the pursuit of full Living Building Challenge certification.

This case study presents an office building designed with the most rigorous performance standards in the built environment, surpassing the highest LEED certification requirements. The innovative design measures in energy and water were described in detail, offering valuable insights into how net-zero water can be achieved within a dense urban context.

The Santa Monica City Hall East has been celebrated as one of America's greenest city hall buildings (Stamp, 2020).

7.4.2 Empire State Building decarbonisation, New York

The Empire State Realty Trust (ESRT) is a real estate investment trust that owns and manages a diverse portfolio of office, retail and multi-family assets across Manhattan and the greater New York metropolitan area. The Trust owns the Empire State Building (ESB) (Figure 7.27) – the 'World's Most Famous Building', the Empire State Building Observatory (Empire State, 2024). It is rated as Tripadvisor's top attraction in the US. The ESB was the world's first skyscraper to stand at more than 100 stories with 1454 ft (443 m) from base to antenna.

Figure 7.27 A closer view of the ESB facade (photo credit to budgetplaces.com is licensed under CC BY-SA 2.0)

In response to New York City's Local Law 97, a progressive climate legislation that includes fines for missing emissions targets for buildings, ESRT developed a bold programme that would help define replicable and scalable solutions to address its buildings' energy demands and supply, and running cost savings. The Trust decided to prioritise on decarbonisation of the ESB as the first priority.

ESRT in partnership with the governmental body, New York State Energy Research and Development Authority (NYSERDA) commissioned a study to develop pathways to achieving net zero operational carbon emissions by 2030 for the ESB and 2035 for the rest of the commercial portfolio (*The Washington Post*, 2021).

The pathways were derived from a series of assessments of a range of interventions applied to their operational carbon reductions across ESRT's large portfolio of buildings, evaluated against the client's target date, and rated by cost and carbon. Alongside this, a series of financially robust plans were established to facilitate immediate action and define incentives to support energy solutions across ESRT's portfolio.

The work started with an evaluation of the existing asset conditions, which informed the business case for strategic decarbonisation. This requires thinking beyond a traditional energy audit approach or simple payback analysis. The process adopted techno-economic assessment including business-as-usual costs and risks against the costs and added value of phased decarbonisation investments in the long term (Retrofit Playbook for Large Buildings, 2022). The decarbonisation costs were estimated to be US $40.7 million for a '$CO_2$ mild' scenario. When accounting for the risks and running cost and incentives, the analysis yields a Net Present Value of US $4.3 million as net difference between the present value of cash inflows and outflows over a period of time.

Starting with the iconic Empire State Building, the team produced the *Empire Building Playbook: An Owner's Guide to Low Carbon Retrofits*, which was published by New York State Energy Research and Development Authority (NYSERDA) (NYSERDA, n.d.; Retrofit Playbook for Large Buildings, 2022). Developed with a focus on high-rise commercial buildings, the Playbook has had a far-reaching effect on the rest of building assets in New York and beyond, as the ESB shares many characteristics with other Manhattan skyscrapers from the 1930s. The findings and experience are equally applicable or easily adaptable to other building typologies. The *Empire Building Playbook* was unveiled at an event at the ESB in April 2022, attended by former President Bill Clinton, NYS Governor Kathy Hochul, NYC Mayor Eric Adams, NYSERDA President Doreen Harris, and a range of state and local officials (Casimira, 2022).

ESRT completed a detailed decarbonisation plan for ESB, with support funding from NYSERDA. The ESB set the bar for successful deep energy retrofits beginning in 2009, and has achieved 54% carbon emissions reduction to date through a range of interventions. Further improvements will be required to achieve net zero emissions.

Deep retrofit measures

Building envelope

- Improvements to the insulation of the walls and the frames and sashes of the windows.
- Upgrades to its 6514 windows.

Fossil fuel free

- Electrification of heating systems, adopting heat pump solution.
- Reducing existing electricity consumption.

Energy efficiency and management

- Energy recovery, to recover waste energy from the building's cooling loop to generate hot water for heating and to recover energy from exhaust airstreams to temper incoming outdoor air.
- 68 elevators were upgraded, with regenerative lift generated electricity for re-use.
- Control optimisation for existing systems.
- Integrating existing systems onto the Building Management System and implementing high performance sequences.

One of the key innovations is how the facade systems were upgraded in the building. Ninety-six per cent of original windows and frames have been taken out and modified with insulation and films (Figure 7.28), carried out at the window factory workshop that was created on the fifth floor of the building, without a need to be transported away. This significantly reduced waste and transport emissions.

The result is that 54% carbon emissions reduction have been achieved to date through this range of interventions. After the deep retrofit of the ESB building, ESRT is implementing the developed

Figure 7.28 Window assembly for upgrading of existing windows (recreated from information in Kennett (2010))

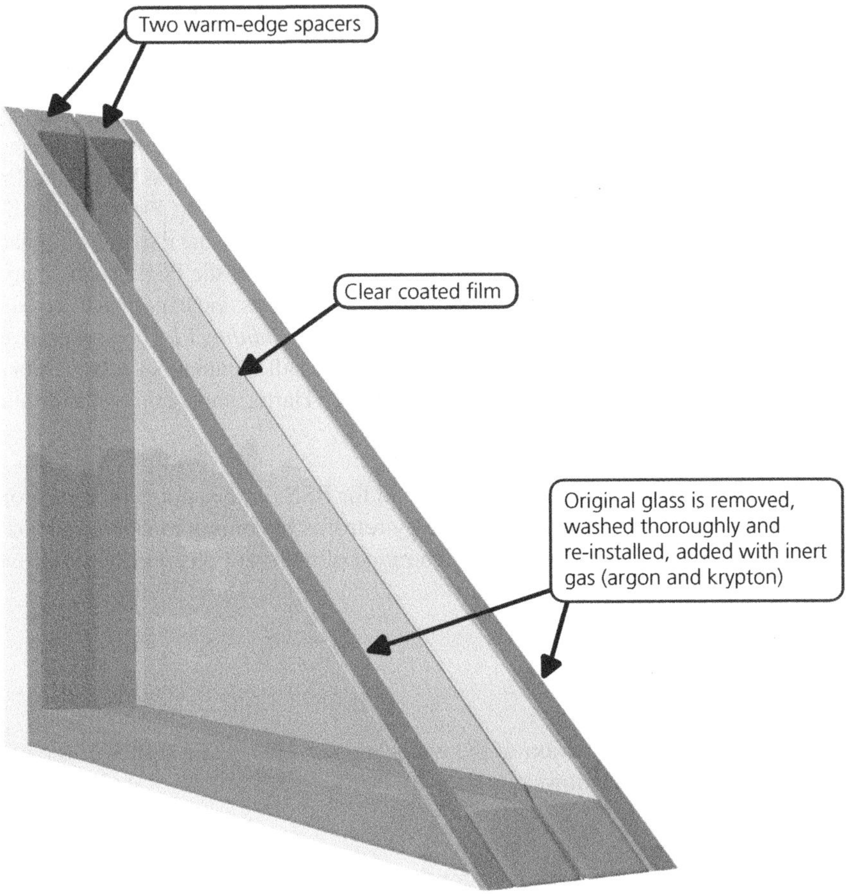

strategies across their portfolio, aiming to meet zero operational carbon emissions by 2030 for the ESB and 2035 for the rest of the commercial portfolio (Buro Happold, n.d.b). The specific onsite targets are as follows.

- Energy Use Intensity: 2019 baseline at 84 kBtu/ft^2 (265 kWh/m^2), will be reduced to 50 kBtu/ft^2 (157.8 kWh/m^2) by 2035, representing a 40% reduction.
- Carbon emissions: 2019 baseline at 15 640 tCO_2e/yr, to be reduced to 3986 tCO_2e/yr by 2035, at a 75% reduction.

> 'The Empire State Building is as innovative today as it was the day it was built and serves as the international beacon of the possibilities within the built environment to prove the business case to reduce carbon emissions. This important partnership between New York State and commercial real estates leaders creates local jobs, drives technological innovation, improves our communities, and stands as an example for climate-friendly retrofits'. (Tony Malkin, Chairman, President, and CEO, Empire State Realty Trust; NYSERDA, n.d.)

Focus area (learning points) **ACTION**

Empire State Building (ESB) is one of the 'World's Most Famous Buildings' and was the world's first skyscraper to stand at more than 100 stories with 1454 ft (443 m) from base to antenna. In response to New York City's Local Law 97, Empire State Realty Trust (ESRT) developed a bold renewal programme to decarbonise and retrofit their existing building stocks with an initial priority on the ESB.

This case study showcases a suite of innovative technologies and sustainable retrofit processes suitable for many existing high-rise commercial buildings. It demonstrates that by developing pathways to achieving net zero operational carbon emissions by 2030 for the ESB and 2035 for the rest of the commercial portfolio, it has helped ESRT to tackle the decarbonisation challenges and streamline the renewal programme. The pathways were derived from a series of techno-economic assessment of a range of interventions applied to their operational carbon reductions across ESRT's large portfolio of buildings, evaluated against the client's target date, and rated by cost and carbon.

This project was showcased at the September 2024 London International conference Council on Tall Buildings and Urban Habitats – 'New or Renew: Addressing the Density Dilemma' (CTBUH, 2024).

7.4.3 BIG U, New York

The vision for the BIG U emerged in 2024 from the aftermath of Hurricane Sandy 2012 and began with the Rebuild by Design Hurricane Sandy Competition launched by the Hurricane Sandy Rebuilding Task Force and the US Department of Housing and Urban Development (HUD). As the winning competition proposal, located in Lower East Side of New York, the BIG U was developed to protect Lower Manhattan from floodwater, storms and other impacts of a changing climate around the low-lying topography of Manhattan, spanning from West 57th Street, to The Battery, and East 42nd Street, home to approximately 220 000 residents and the core of a US $500 billion business sector that influences the world's economy (Rebuild by Design, n.d.).

Hurricane Sandy devastated not only the financial district, but also impacted 95 000 low-income, elderly and disabled city residents. Infrastructure within the 10-mile perimeter was damaged or destroyed, transportation and communication were cut off, and thousands were without power or running water. Backed by New York City, in collaboration with New York State Battery Park City Authority, the funded plans encompass 10 continuous miles of protection tailored to respond to individual neighbourhood typology as well as community-desired amenities. The project has gone through various phases of planning, design and implementation across multiple community boards.

The BIG U concept (Figure 7.29) was designed and led by Bjarke Ingels Group (BIG) with One Architecture, Starr Whitehouse, James Lima Planning + Development, Green Shield Ecology, AEA Consulting, Level Agency for Infrastructure, ARCADIS and Buro Happold. The BIG U is a prime example of 'hedonistic sustainability' – a term created by BIG's founder Bjarke Ingels to demonstrate how the sustainable development and the pursuit of pleasure can and, indeed, should co-exist – 'Living in a sustainable city meant increasing people's enjoyment, not sacrificing it', claimed by BIG.

BIG's hedonistic sustainability concept could be seen in their Copenhagen Harbour Bath project where Copenhagen residents could jump into clean water in the heart of the city by creating a clean port, instead of driving a long way to get to the beaches (Wright, 2025). Instead of

Figure 7.29 Overview of the BIG U plans (image credit to BIG-Bjarke Ingels Group)

thinking of sustainable living as an idea of compromise, hedonistic sustainability perfectly unites sustainability with human pleasure and well-being.

For the BIG U, the project's public realm component has been critical to overall delivery, not only delivering high quality style but also building consensus around taking major climate adaptive actions, while at the same time helping to deliver other social and environmental benefit. BIG's hedonistic sustainability approach helps to speed up the processes and advance the crucial resiliency projects by addressing local concerns and delivering benefits for multiple stakeholders.

BIG U has attracted further investment with the overall budget of US $2.7 billion. The initial federal government's allocation of US $335 million was to fund the first phase of the BIG U's construction for the delivery of access improvements, community-orientated programming and ecological enhancement within the local area as part of the East Side Coastal Resiliency Project. Since then, further investment was successfully leveraged to deliver wide benefits (Rebuild by Design, n.d.).

The proposal breaks the area into sections named as East Side Coastal Resiliency, Brooklyn Bridge Montgomery Coastal Resilience, Lower Manhattan Coastal Resilience and Battery Park City Resilience projects.

Figure 7.30 BIG U – designed to protect New York against extreme storms (photo credit to BIG-Bjarke Ingels Group)

Each section comprises a physically separate flood-protection zone, isolated from flooding in the adjacent zones, yet each also serves as a platform for integrated social and community planning. The phased projects have compartments that act in concert to protect and enhance the city, while each compartment's proposal is designed to function effectively on its own (Rebuild by Design, n.d.).

- The East Side Coastal Resiliency project consists of a number of phases (NYC,2025). A proposed bridging berm protects the area from storm surges and rising sea levels, and offers waterfront access for relaxation, socialising and enjoying river vistas by providing pleasant, accessible routes over the highway into the park (Figure 7.31). Additionally, salt-tolerant trees and plants provide a resilient urban habitat. East Side Coastal Resiliency uses a series of berms, flood walls, flood gates (Figure 7.30) and raised parklands to create a continuous 2.25 mile barrier to protect 110 000 residents of the Lower East Side from future coastal flooding (Michaels, 2022).

In August 2022, this project was selected to receive an 'Envision Gold Award' for sustainability from the Institute for Sustainable Infrastructure (ISI) (NYC, 2022) The Envision Sustainable Infrastructure Framework assesses project sustainability across five categories: quality of life, leadership, resource allocation, natural world, and climate and resilience. East Side Coastal Resiliency was cited by ISI for creating a cohesive and comprehensive response to climate resiliency goals, protecting the community, while enhancing public space and amenities, and developing a robust community engagement programme that considers the diversity of stakeholders. The East River Park amenities opened in May 2025.

Figure 7.31 BIG U – waterfront design against extreme storms (photo credit to BIG-Bjarke Ingels Group)

- The Brooklyn Bridge Montgomery Coastal Resilience project is a combination of flood walls and deployable flip-up barriers along the waterfront from the Brooklyn Bridge to Montgomery Street, was designed to protect the neighbourhood from projected 100-year storm surges in the 2050s, while maintaining access and visibility to the waterfront. Deployable walls attached to the underside of an elevated highway can flip down to mitigate flooding. Decorated by neighbourhood artists, the panels would create an inviting ceiling when not in use, while integrated lighting would transform a currently menacing area into a safe community destination. The project began construction in 2022.
- The Lower Manhattan Coastal Resilience project, building on the 2019 Lower Manhattan Climate Resilience Study, this programme comprises four projects: Brooklyn Bridge Montgomery Coastal Resilience, The Battery Coastal Resilience, Battery Park City Resilience, and the Financial District and Seaport Climate Resilience Master Plan. The design weaves an elevated path with a series of upland hills to form unique landscapes. The plan envisioned transforming the existing coast guard building into a new maritime museum or environmental education facility. The Battery Coastal project rebuilds and elevates the wharf promenade in The Battery by approximately 6 ft to protect it from projected sea level rise into the year 2100.

Wagner Park (Figure 7.32) was opened to public in July 2025. After an 18-month closure with a significant climate proof upgrade as part of South Battery Park City Resiliency Project, the park demonstrates that resilience and beauty can coexist—offering strengthened protection, along with a front-row seat to the harbour breeze (Ratliff, 2025).

Figure 7.32 Wagner Park re-opened in July 2025: a reimagined resilient waterfront destination (photo credit to Battery Park City Authority)

Focus area (learning points)

ACTION

The BIG U is one of the most ambitious storm resiliency projects in the world, yet this continuous U-shaped network of elevated parks, public lidos, covered walkways, cycle paths and sports facilities effectively creates the newly formed waterfront community spaces, public amenity for leisure and enjoyment, built upon flood plains populated with salt-tolerant flora and canopies covering the walkways which are designed to flip down to create instant storm barriers when required.

This case study demonstrates that by marrying architectural style, engineering resilience infrastructure with system thinking it is possible to create community spaces and bring joy to the city. Sustainable living should not be seen as a compromise, rather it can bring social and economic benefits. The project is setting a new standard for the concept of hedonistic sustainability (as was called by BIG) for futureproofing the built environment.

The project won multiple awards and 'Waterfront Edge Design Guidelines' (WEDG) verification from the Waterfront Alliance, recognised for its ability to reduce risk from coastal hazards, emergency preparedness plan, community engagement, quality of public access areas, waterfront greenway connectivity and inventive design (WEDG, n.d.).

7.4.4 Toronto Thermal Comfort Guidelines

The City of Toronto's City Planning Division commissioned an urban thermal comfort study aiming to inform their climate action. The city aims to encourage people to spend more time outdoors and be more physically active, while also addressing the needs of people experiencing homelessness and other disadvantaged populations who are living outdoors. This commission forms a part of the Heat Relief Strategy, serving as one of the foundational climate action resources.

Ensuring the quality and comfort of the city's public spaces has gained heightened significance due to their role in serving and providing enjoyment to all city dwellers. As climate change intensifies extreme heat events, which are occurring more frequently, it is crucial to integrate thermal comfort considerations into the planning of the public realm. This approach is essential for building resilience in the city and adapting to changing climatic conditions.

Toronto's City Planning Division contracted Buro Happold and Dialog (Ontario Association of Architects, 2023) to lead the study based on the former's extensive experience with similar plans for large metro areas such as New York, Los Angeles and Minneapolis. The scope of the study includes a review of the city's sun and shadow policies to ensure they achieve adequate opportunities for access to sunlight in the public realm. This study also combines the wind, sunlight and other parameters together to assess the quality of the public realm and shared outdoor amenity spaces (Figure 7.33). Thermal comfort recommendations have been developed through the study and provide design direction for thermal comfort in the public realm to be implemented through new and updated guidelines, standards and policies. This study also provides guidance on evaluating sunlight conditions in the public realm (Thermal Comfort Study, 2024).

The City of Toronto's City Planning Division established a core client team and relevant stakeholders to steer the direction of project, including City Planning (Urban Design, Community Planning and Strategic Initiatives and Policy Analysis), Environment and Energy, Parks, Forestry and Recreation, Toronto Public Health and Transportation Services.

The study examined how planned design elements, solar access, shade, urban greenspace, wind, water and cool materials could be utilised to ensure the quality and comfort of Toronto's public spaces. The work provided valuable lessons to urban planners, developers and other stakeholders, as the country's largest metropolitan area prepares a strategy to effectively mitigate the growing number of extreme heat days experienced each year. It highlighted the need for cities to integrate thermal comfort considerations into the planning of the public realm. The City Planning Division formally adopted the Thermal Comfort Guidelines at the Planning and Housing Committee on January 23, 2025 and the study materials were released in February 2025 (City of Toronto, 2025).

The Thermal Comfort Study focuses on several factors, including solar access, shade, the urban forest, wind, water and cool materials. The study identified the seasons and specific times when people feel comfortable or uncomfortable, as well as their preferred outdoor spaces for leisure activities. The survey was initiated in July 2022 and received responses from 556 individuals over the course of 35 weeks.

Key findings include the need to address the requirements for maintaining a healthy tree canopy in urban areas. Specific conditions, such as sunlight, moisture and healthy soil, are necessary for urban trees to thrive. Thermal comfort recommendations developed through the study and will

Figure 7.33 Toronto's skyline (photo credit to M Kuhn is licensed under CC BY-NC-SA 2.0)

inform public-realm design and be incorporated into future guidelines, standards and policies. The study's unique approach assesses exterior thermal comfort and cold weather comfort, offering a blueprint from global urban climate hazard mitigation. One direct aim for the guidelines is to increase the use of public outdoor spaces, reduce the number of heat-related illnesses that occur each year, and improve the greater mental health and social wellbeing of previously afflicted communities year-round (Niland, 2022).

The study significantly informs Toronto's response to heat and climate change, leading eventually to the creation and implementation of new comfort guidelines included as part of a broader Heat Relief Strategy. The study focuses on creating official guidelines, impacting city planning policies, and potentially informing new legislation. The initiative responds to the rising frequency and intensity of extreme heat days in Toronto. It aims to redefine the city's 'comfort guidelines', influencing property development and emphasising public health amidst rapid climate change. This rigorous, analytical approach far exceeds the basic guidelines produced by many other cities, introducing a data-driven approach for evaluating city conditions and the effectiveness of interventions (Construction Canada, 2024).

Focus area (learning points) **GOVERNANCE**

As climate change intensifies extreme heat events, the Toronto City Government wishes to develop a heat relief strategy serving as one of the foundational climate action plans. Central to this plan, they commissioned an urban thermal comfort study aiming to integrate thermal comfort considerations into the planning of the public realm.

This case study exemplifies government-led approaches to building resilience in the city and adapting to changing climate, especially for public spaces. The urban Thermal Comfort Study provides a foundation for new design guidelines, standards and policies that will shape how the city addresses extreme heat in the public realm.

The Thermal Comfort Guidelines (City of Toronto, 2025) was formally adopted in 2025. The Toronto Government is planning for training sessions for staff in the City Planning, Parks, and Forestry and Recreation divisions and testing the guidelines through two active studies before incorporating them into the City of Toronto Development Guide.

7.5. Australia

Buildings are typically designed with specific energy performance targets in mind. However, these targets are often not attainable and inaccurate in reality, leading to 'performance gap' due to the discrepancy between the intended and actual energy consumption. Australia devised a scheme, National Australian Built Environment Rating System (NABERS), to provide a simple, reliable and comparable sustainability measurement for the built environment. This scheme covers across building sectors including offices, hotels, shopping centres, apartments and data centres.

Since NABERS was introduced in 1999, there has been significant uptake of the NABERS scheme in Australia, supported by the Australian Government policies for making building energy efficiency visible. The NABERS annual report stated that 5000 ratings have been certified in 2024 (NABERS, 2024). The value of adopting NABERS is proven by 'what gets measured gets managed', with building owners having demonstrated energy use reduction by an average of 30–40% over a 10-year period (Smith, 2022). A NABERS rating helps building owners and operators to accurately measure, understand and communicate the environmental performance of their buildings, while identifying areas for cost savings and future improvements.

NABERS rating system has shown to have driven financial growth and solid returns in portfolios across building sectors. It has directed corporate procurement by clearly demonstrating their operational performance and sustainability achievements (Martin and LGIM Real Assets, 2021). One Australian study showed that businesses that strive for a 4.0–6.0 star rating achieve longer leases, lower vacancies and a higher valuation, compared with those with a 0–3.5 star rating (see Figure 7.34; Smith, 2022).

The UK Government introduced a UK version of the NABERS scheme in 2020, with the intention of increasing transparency regarding the building operational energy performance. This scheme is designed to play major part in tackling performance gap and driving better buildings and financial growth in real estate (Smith, 2022).

The selected NABERS case studies from Australia show the impact of using NABERS to drive for sustainability performance and transformative growth in real estate.

Figure 7.34 NABERS rating system drives financial growth in real estate (Martin and LGIM Real Assets, 2021; Smith, 2022) (reproduced with kind permission of RICS)

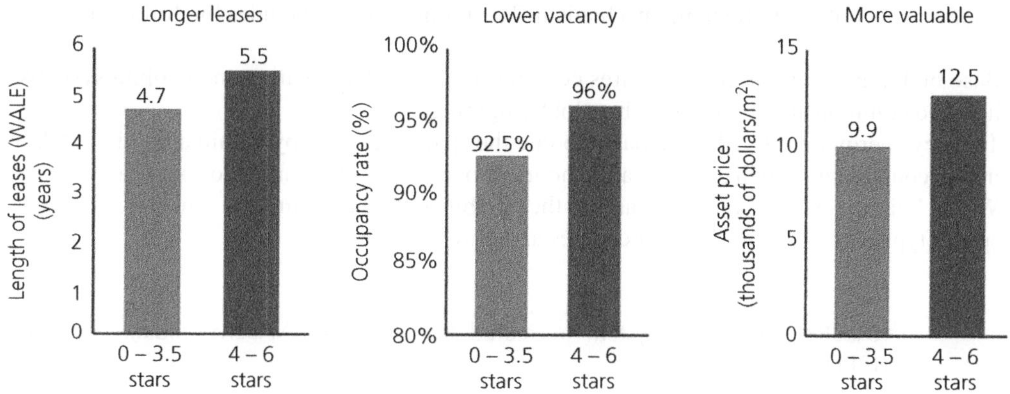

*Source: Real Investment Analytics, Australian Green Office Property indicators 2020 Q2. Figure represent the 5-year average.

7.5.1 NABERS

A NABERS rating (NABERS, n.d.a) helps building owners to accurately measure, track and communicate the environmental performance and progress of buildings. It uses star scoring system to benchmark the performance (Figure 7.35). It also identifies areas for savings and improvements. The scheme includes NABERS Energy, NABERS Water, NABERS Indoor Environment and NABERS Waste. NABERS ratings are valid for 12 months. This annual model helps ensure that the rating represents a building or workplace's current operational performance. Recently, NABERS has partnered with Climate Active to provide a Carbon Neutral certification.

Benefits of a NABERS rating can be summarised in four areas.

- Compare: NABERS ratings form the basis of a robust target and strategy. They provide a fair benchmark for a building's energy and water efficiency, enabling comparison across the industry.
- Validate and communicate: NABERS provides cost-effective, transparent and independent validation of sustainability data, ensuring confidence in reporting it publicly and communicating it simply.
- Track and improve: NABERS star ratings are an essential framework for annual reporting and sustainability strategies. With annual ratings, performance improvements and reduced energy and emissions can be demonstrated over time.
- Competitive edge: NABERS enables building owners to communicate simply and with confidence that their building has a lower environmental impact and running costs to prospective buyers, renters, procurement teams or hotel guests. NABERS ratings can be used to underpin sustainable finance, providing lenders with robust data and verified environmental outcomes. Almost 80% of Australia's commercial office buildings have achieved NABERS ratings, highlighting the system's extensive uptake and its strong influence across the sectors.

Comparing with other energy rating methods like Energy Performance Certificate (EPC) or BREEAM or LEED Energy credits, NABERS provides greater accountability and transparency in the rating process – for example, by breaking down to landlord and tenancy responsibility, and by incorporating a continuing tracking mechanism. The rating process includes the following.

- Base building energy rating: Measures central services such as heating and cooling systems, lifts and common area lighting such as lobby lighting.
- Tenancy energy rating: Allows tenants to rate the space they occupy within a building, taking into account factors such as office area, hours of occupancy and number of occupants.
- Whole building energy rating: Measures the combined base building and tenant occupied space (typically when one tenant occupies an entire building).

Figure 7.35 NABERS star scoring system (recreated from NABERS (NABERS, n.d.a). Copyright NABERS – nabers.gov.au)

Case study 1: Knox civic centre in Victoria (NABERS, n.d.b.)

One of the best ways to reduce the operational cost of an office building is to improve its energy efficiency. If the office building is more than ten years old, it is highly likely that making some modest updates and changes will make a significant difference to its energy efficiency, often generating cost savings that allow these measures to pay back rapidly. This case study shows a clear example of how targeted energy efficiency upgrades can enhance a building's NABERS rating while reducing overall energy cost.

The two-storey Knox civic centre in Victoria was built in 1998 and is typical of 'large low-rise' office buildings. It has a capacity of 255 people and a floor area of 6174 m^2. Simulations conducted by the University of New South Wales (UNSW) found that energy savings of 63% would be achievable if a package of cumulative energy efficiency upgrades are implemented. The case study is based on simulations of an office building's performance following implementation of a suite of energy efficiency measures and the findings indicate that energy savings of up to 63% are possible, which would improve the NABERS rating from 2.0 stars to 3.5 stars.

The key upgrades are shown in Figure 7.36.

Figure 7.36 Key recommendations from NABERS on Knox civic centre in Victoria (NABERS, n.d.b) (Copyright NABERS – nabers.gov.au)

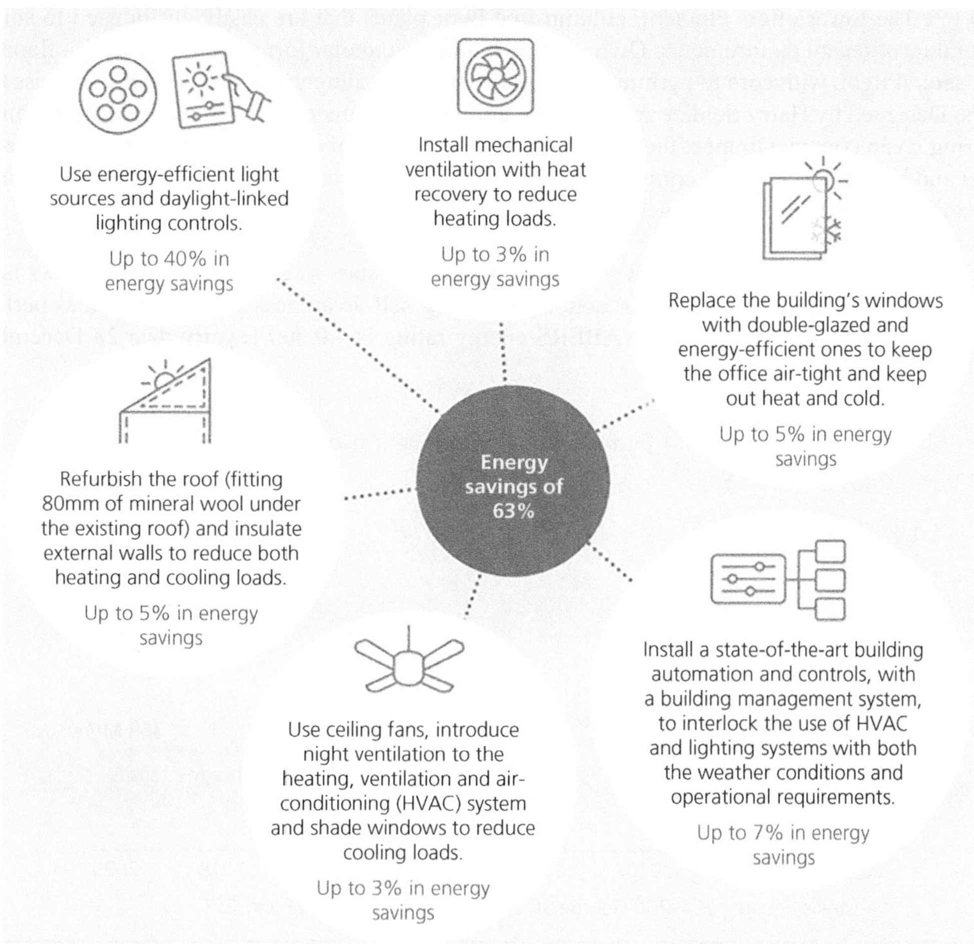

This study also provides recommendations relating to two factors that will have a significant impact on the building's performance.

(*a*) The operational parameters including hours of operation, set temperatures for heating and cooling, natural ventilation patterns and control of artificial lighting.
(*b*) The building's microclimate – that is, the extent to which it is shaded by buildings or natural features, the thermal storage of surrounding areas and how much heat is reflected or absorbed by the building's exterior.

Case study 2: Australia Square Tower, 264–278 George Street, Sydney NSW 2000 (NABERS, n.d.c)

Australia Square Tower is an A Grade building, comprising 48 levels with typical floorplates of 1100 m^2. The tower offers efficient, column-free floor plates that are easily customised to suit a wide range of tenant requirements. Owing to the buildings circular form, each floor is also flooded with natural light, with core to perimeter depths of 10 m providing up to 70% sightlines across the floors. Designed by Harry Seidler architect, the building was conceived with adaptability in mind, ensuring it can continue to meet the needs of future generation of occupants on a corporate, operational and human level. On its completion, Australia Square became Sydney's first skyscraper and the world's tallest lightweight concrete building.

Australia Square achieved a NABERS Energy rating of 5.0 stars in 2020. Figure 7.37 shows how the NABERS rating helped drive for action by raising self-awareness of environmental performance (Flores, 2022). The current NABERS energy rating is 4.0 star (expiry date 28 December 2025).

Figure 7.37 NABERS journey for performance improvement (Flores, 2022) (Copyright NABERS – nabers.gov.au)

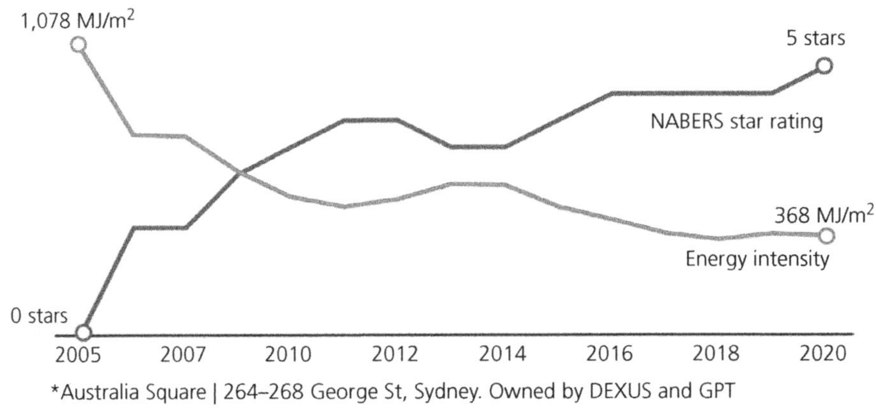

Australia Square has been rated under Energy, Waste, Water and Indoor Environment (NABERS, n.d.b). These NABERS ratings are based on scores and weighted according to the results of a building occupant satisfaction survey.

NABERS Energy rating 4.0 star: energy intensity 442 MJ/m^2. Total GHG emissions with renewable electricity per m^2 is 6.35 kgCO_2e/m^2 p.a. Total renewable electricity contribution 78% of total site energy.

NABERS Waste 3.5 star: recycling rate 39.00%, waste intensity 9.55 kg/m^2 p.a.

NABERS Water 4.0 star: total water use 18430.99 kL p.a., water intensity 0.59 kL/m^2 p.a.

NABERS Indoor Environment 5.0 star: thermal comfort 88.80%, indoor air quality 87.30%, acoustics 67.5%.

Focus area (learning points) **ACTION**

Australia has been a global leader in addressing the buildings' 'performance gap' through the development of the NABERS scheme to help building owners and operators to accurately measure, understand and communicate the environmental performance of their buildings while identifying areas for cost savings and future improvements. Since its launch in 1999, NABERS has successfully helped building owners to reduce energy use by an average of 30–40% over a ten-year period (Smith, 2022). It has been shown to stimulate financial growth and deliver solid returns in portfolios across building sectors.

This case study provides an overview of the NABERS scheme with its processes and associated benefits. Two NABERS project examples are included in this case study covering a civic centre refurbishment and a large modern commercial office, to illustrate a range of intervention measures and tangible benefits. All information presented in this case study has been sourced from NABERS (n.d.b, n.d.c).

As NABERS rating is valid for 12 months, this annual model helps ensure that the rating represents a building or workplace's current operational performance. NABERS ratings, including NABERS Energy, NABERS Waste, NABERS Water and NABERS Indoor Environment, help stimulate meaningful action by increasing organisations' self-awareness of their environmental performance.

7.6. Asia

Asia covers a large zone of diverse geographic and economic development areas. Asia Pacific is a centre of growth and urbanisation. As Munish Datta, Chair of Cambridge Institute for Sustainability Leadership (CISL)'s Business and Sustainability Programme (Datta and Cambridge Institute for Sustainability Leadership, 2024) put it, Asia is becoming a driving global force in terms of opportunity, innovation and resources. Asian countries are making substantial investments in renewable energy infrastructure, realising its significant potential for solar, wind and hydro to achieve dual benefits of securing energy supplies and reducing GHG emissions.

At the UN General Assembly in September 2020, China president Xi Jinping announced the country's carbon dioxide emissions 'Dual Carbon' goal to peak before 2030 and carbon neutrality before 2060. China's State Council released a document outlining its plan to achieve peak carbon dioxide emissions before 2030, aiming for carbon neutrality by 2060, with key targets including a 65% reduction in carbon dioxide emissions per unit of GDP by 2030 compared with 2005 levels (NDRC, 2021). Since then, China has developed a series of policy incentives for built environment, construction and manufacturing industries to take actions towards the carbon reduction goals. Renewably powered growth in China has accelerated in recent years, demonstrating how the Asian countries are meeting rising energy demand, driven by rapid urbanisation and industrialisation, through clean and renewable energy sources. China, in particular, is approaching its ambitious 2030 targets for additional wind and solar capacity as early as 2025 (Datta and Cambridge Institute for Sustainability Leadership, 2024).

One case study is chosen to illustrate the technology advancement led building decarbonisation in China.

Japan's sustainability leadership is reflected in its substantial number of RE100 commitments and Science-Based Targets net zero pledges, ranking second globally (RE100, n.d.). The RE100 Leadership Awards recognise the climate leadership and their work in accelerating the global transition to 100% renewable electricity. Out of 500 corporate respondents from Japan, 83% of Japanese business leaders identified sustainability as a top priority, according to Schneider Electric's 2023 Sustainability Survey (Schneider Electric, 2024). Japan has embraced international green building certifications and developed its own standards to encourage sustainable practices, named the Comprehensive Assessment System for Built Environment Efficiency (CASBEE). Given Japan's susceptibility to natural disasters, there is a growing emphasis on disaster-resilient green architecture. The Great East Japan Earthquake Memorial Museum in Fukushima, designed by architect Kengo Kuma, exemplifies this approach. The museum incorporates earthquake-resistant features and utilises local materials, embodying a commitment to both sustainability and resilience in the face of adversity (Yllah, 2024). The meticulous balance between tradition and modernity in Japanese green architecture offers valuable lessons for creating environmentally conscious buildings without compromising on functionality or beauty. The lessons learned from Japan's architectural endeavours serve as a testament to the transformative power of embracing sustainability as a guiding principle for shaping the future of built environment.

The Jikka House has been chosen as a case study to showcase Japan's innovative design solutions inspired by nature.

The Future Of Green Buildings plan in Singapore demonstrates Singapore Government's commitment and has contributed to the city state's remarkable sustainability progress in recent years. To improve the nation's commitments under the UN's 2030 Agenda for Sustainable Development and the Paris Agreement, the plan set out ambitious and specific targets for the next ten years, positioning the island state to meet its long-term net zero emissions goals. The Building and Construction Authority (BCA) introduced the Singapore Green Building Masterplan (SGBMP) (Building and Construction Authority, 2021) as a component of the Singapore Green Plan 2030 (Singapore Green Building Council, 2020). The SGBMP set out three primary goals, known as '80-80-80 in 2030'. By 2030, they aim to have 80% of new developments (by gross floor area) as Super Low Energy (SLE) buildings; it will accelerate efforts to green 80% of buildings by that date; and by 2030, will achieve an 80% increase in energy efficiency for the best-in-class green buildings.

Singapore is already demonstrating strong leadership in its efforts to create a greener urban environment in the years following COP26. This not only provides Singapore with a distinct and visually pleasant image, but the greener urban environment has also demonstrated its ability to alleviate urban heat issues, supporting sustainable water management and enhancing biodiversity in the city. Numerous projects have been developed to address these critical challenges and to provide sustainable design solutions. The Oasia Hotel Downtown (Goyal, 2023) is presented as a case study that exemplifies this vision of embracing nature, harnessing technology, and fostering a symbiotic relationship between humans and the environment.

According to the World Bank, future sea levels are projected to increase up to 100 cm by the year 2100. For centuries, bamboo has been widely used for traditional floating house structure by Indonesian people and Manobo houses in Philippines. Bamboo is a fast-growing vegetation and it can be found easily across the south-east Asia region – for example, rumah lanting are a type of floating settlement commonly found in most river streams in Kalimantan. These floating houses are designed as adaptive structures that mitigate the impacts of the environmental disasters typically experienced in traditional West Kalimantan area (Putro and Zain, 2021). In the race to create climate resilient cities, there have been a number of ambitious and innovative floating city endeavours and developments around the world.

The Netherlands have long been home to floating parks, a floating dairy farm (Floating Farm, 2025), and the floating district of Waterbuurt in Amsterdam that consists of nearly 100 individual floating homes moored on Lake Eimer (Spooky, 2020). The Floating Office Rotterdam building, headquarters of the Global Centre on Adaptation (Musmanni, 2021), is a carbon-neutral, climate-resilient building designed by Powerhouse Company with a three-storey structure that floats on the Rijnhaven port in Rotterdam, designed to rise with water levels.

Oceanix Busan, a project endorsed by the United Nations, was designed by architectural firms BIG and Samoo, and the technology company Oceanix (Rubio, 2023). The project has already been in development for more than a decade and is scheduled to be built between 2025 and 2028. Its aim is to build a prototype of a sustainable floating city near the Busan Metropolitan City of the Republic of Korea (UN-Habitat, 2022).

Among such developments, Maldives Floating City is the biggest floating city design using advanced and pioneering floating technology from the Netherlands. The Maldives Floating City has been chosen as a case study to illustrate how the modular approach to floating city development could potentially set a new template for future climate resilient cities.

Dubai's sustainability plan (Government of Dubai Environmental Authorities, n.d.), encompassing the Dubai 2040 Urban Master Plan (Government of Dubai, n.d.) and the Dubai Clean Energy Strategy (UAE, 2024) aims to transform Dubai into a global hub for clean energy and a green economy, with targets to produce 75% of its energy from clean sources by 2050 and increase green spaces across the city (HiDubai Newswire, 2023). Expo City Dubai, built on the site of Expo 2020, aims to be a net zero carbon city by 2050, integrating sustainability into its operations and events, with a focus on clean energy, water conservation and biodiversity protection. Expo City Dubai, serving as a blueprint for sustainable urban living, is committed to making this a reality (Expo City Dubai, 2020). Museum of Future and Terra are among the inspirational buildings showcased at Expo City Dubai. Terra is selected as a case study as it uniquely integrates architecture and climate responsive design into the public spaces and museum interior.

In summary, the five selected case studies are

- Swire Taikoo Li Sanlitun Beijing, China
- the Jikka House in Shizuoka
- Oasia Hotel Downtown, Singapore
- Maldives Floating City
- Terra Dubai Expo, Dubai.

7.6.1 Swire Taikoo Li Sanlitun Beijing, China

Established in Hong Kong in 1972, Swire Properties has become one of the leading real estate operators and property developers in Asia. Swire has joined the Hong Kong Green Building Council Advancing to net zero initiative and committed to Science-Based Targets (Swire Properties, 2022).

Swire has developed their sustainability strategy 2030, an extract of which is shown below.

- Align with their 1.5°C Science-Based Target.
- Operational Carbon Scope 1 and 2 – absolute GHG emissions to be reduced by 25% by 2025 and 46% by 2030 from a 2019 baseline.
- Tenant Carbon Scope 3 downstream leased assets, GHG emission intensity to be reduced by 28% per square meter by 2030 from a 2018 baseline.

- Upfront embodied carbon for new development, Scope 3 capital goods GHG emission intensity to be reduced by 25% per square metre by 2030 from a 2016–2018 baseline.
- 2050 net zero carbon vision.

Priority actions on energy aspects are as follows.

- Energy demand reduction – define, measure and progressively reduce asset level energy usage intensity (EUI).
- Advancing net zero – using low carbon innovative technologies.
- Generation – adopting renewable technology to generate renewable electricity and thermal energy.

Swire Properties recognised that renovation projects would increasingly become a strategic focus in Hong Kong and mainland China, particularly in the first and second tier cities. They recognised that maximising the effective use of the existing building and enhancing the energy-saving performance would become an increasingly critical area of focus. They also realised that this shift would introduce new opportunities and challenges for the majority of real estate developers in the industry.

Swire Taikoo Li Sanlitun is chosen as a case study to illustrate Swire's commitment to piloting innovative technologies and sustainable renovation practices.

After taking over the Taikoo Li Sanlitun project in Beijing, Swire Properties decided to gradually transform the existing system in the northen zones into an energy-efficient system over a five-year period from 2018 to 2023. Taikoo Li Sanlitun (Figure 7.38) has since been transformed into not only a vibrant urban commercial core area and destination for local and international shopping experiences, but also a benchmark for sustainability and innovation in China. The intelligent upgrading and transformation of this project has enabled Swire Properties to achieve cost reduction and energy efficiency improvements, while also allowing the public to directly experience the benefits of low carbon, smart building technologies, helping to promote a broader understanding of green, low-carbon, healthy and environmentally friendly development.

Building photovoltaic energy storage and direct current flexible technology (PEDF system)

One of the innovative systems installed in Taikoo Li Sanlitun, is the use of the PEDF system pilot located in the Taikoo Li Sanlitun North shopping mall. 'Photovoltaic storage, direct and flexible' technology refers to the application of photovoltaic (PV) power generation, energy storage, direct current (DC) power distribution and flexible energy consumption in the built environment. The pilot project covers two independent buildings, Building No. 15 and the Red Pavilion, and reserves space for the expansion of the DC microgrid in the future. The PEDF system includes PV panels, energy storage equipment, DC fan panel, DC motor for chillers at the end of the supply line, DC sockets, lamps and other future ready equipment. This integrated configuration enables the system to play a leading role in energy conservation and emission reduction while meeting the building's electricity demand alongside the main grid.

In 2020, Swire Properties launched this PEDF microgrid technology project in partnership with Tsinghua University alongside the supply chain Beijing Zhijian to improve the flexible use of electricity in buildings. The aim of the project is to achieve flexible dispatching of power grids and efficient utilisation of resources, and achieve dual control of energy conservation and carbon

Figure 7.38 Swire Taikoo Li Sanlitun Beijing (photo credit to Mei Ren)

reduction using innovative green technology. Seven hundred thin-film solar PV panels have been installed in Taikoo Li Sanlitun North, generating 20 000 kWh of clean energy per year.

The Taikoo Li PEDF system has reduced energy consumption, improved energy utilisation efficiency and reduced carbon emissions through flexible dispatching and optimised regulation of building energy loads and equipment; specifically with the following details.

(*a*) Through the visible, manageable and controllable PV storage direct flexible system master control platform, the real time energy consumption and other information of the system is centrally displayed, providing big data strategy to support the flexible regulation of building energy consumption and grid power load.
(*b*) Building managers can view the real time system with comprehensive data (power data, energy consumption, electricity bill, revenue savings etc.), enabling them to monitor and understand the overall operation of the system in real time.
(*c*) Create a cloud-edge-end coordination mechanism based on Artificial Intelligence of Things (AIoT) technology to achieve online interaction with the power grid and users.
(*d*) The system coordinates and controls each device through its intelligent flexible control algorithm, enabling flexible regulation and achieving seamless, intelligent integration of energy 'source, grid, load and storage'.

The development of PEDF buildings is an important technical advancement to achieve the transformation and upgrading of building energy systems towards the goal of carbon neutrality. It not only effectively enhances the building's own energy conservation and carbon reduction, improves the building's electricity efficiency, enables self-regulating and autonomous optimisation of building electricity, but also effectively helps alleviate the balancing pressure between electricity supply and demand.

The Beijing Sanlitun Taikoo Li project is the first of its kind, creating the benchmark demonstration of 'PV Energy storage and Direct and Flexible' commercial buildings, and supporting Swire Properties' broader objectives for energy conservation, low carbon and green transformation. Following the success of this project, the deployment of this technology has expanded, with larger installations now being implemented across other developments. This technology is promoted by the Chinese Government, institutions and manufacturers with the aim of advancing the national dual-carbon policy exploring scalable models for low carbon building development. An illustration of the PEDF system is shown in Figure 7.39 ((Liu *et al.*, 2023). With the rapid growth of renewable energy generation and increasing uptake of electric vehicles, the PEDF system represents a promising approach for enabling large scale integration of renewable energy, supporting the transition toward a carbon-neutral future.

Figure 7.39 A PEDF system integrating distributed photovoltaics, energy storages and a direct current distribution system (Liu *et al.*, 2023)

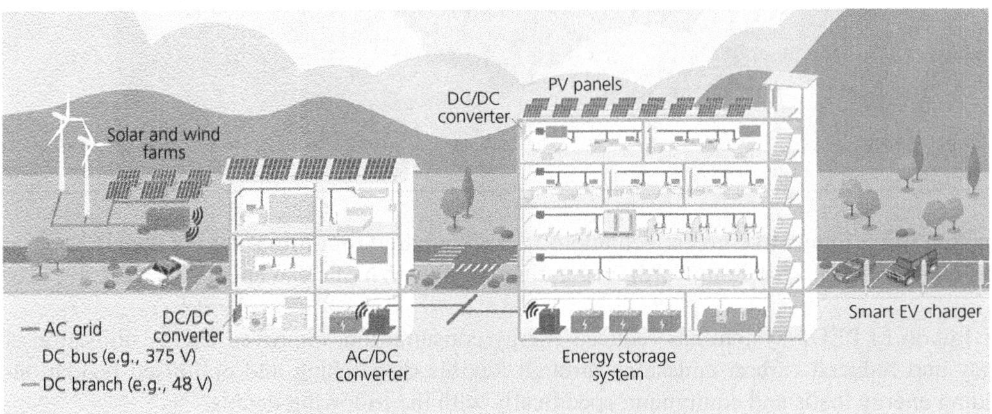

Summary

Apart from the PEDF installation, other sustainable renovation features for Beijing Sanlitun Taikoo Li (Swire Properties and AECOM, 2023) include the following.

- Chiller upgrade the original water ring heat pump system had an energy efficiency ratio (COP) value of less than 2 for more than 10 years. It has now been replaced with high-efficiency chiller using magnetic levitation chiller technology. The cooling system is transformed from a heat pump to a centralised maglev inverter chiller, high-efficiency cooling tower, full inverter water pump, a four-pipe water system, and simultaneous cooling and heating. The cost is approximately at RMB 150 million Yuan.
- Replace refrigerant from CFC with a lower environmental impact HFC-134a, which is harmless to the ozone layer and has reduced impact on the greenhouse effect.
- Use CFD technology to optimise cooling tower layout during its renovation, improving energy efficiency and reducing the heat island effect for nearby residential areas. The overall water-saving as well as acoustics performance of the cooling tower have been significantly improved.
- Smart meters – tenants billing system to drive accurate energy billing and encourage behaviour change.

- Indoor air quality – this is an essential part of improving customer experience. Taikoo Li Sanlitun has set up wall sensors in the public areas and elevator halls to measure and publicise the outdoor and indoor air quality parameters including temperature and humidity, PM2.5, PM10, carbon dioxide concentration, ozone, TVOC and formaldehyde. With reference to the corresponding requirements of the WELL standard, if the above data exceeds the standard threshold, the system will automatically turn on the fresh air ventilation equipment to purify the indoor air quality to the extent that it meets the requirements.
- The South and North districts first achieved LEED O+M Gold certification in December 2017, and in June 2022 achieved LEED O+M v4.1 Platinum certification with an overall score of 83 and 81, respectively, the highest level of certification available.

Focus area (learning points) **VISION AND TECHNOLOGY**

China has developed a series of policy incentives for built environment industry to take actions towards the carbon reduction goals including a 65% reduction in carbon dioxide emissions per unit of GDP by 2030 compared with 2005 levels (NDRC, 2021). Renewable energy generation has powered the growth in China in recent years.

Swire Taikoo Li Sanlitun in Beijing is chosen as a case study to illustrate the technology advancement-led building decarbonisation in China. Swire Properties is one of the leading real estate operators and property developers in Asia, having developed their Sustainability Strategy 2030 which is aligned with science-based targets. They identified that maximising the existing building and enhancing the energy-saving performance would become an increasingly critical area of focus. They also recognised that this evolving focus would introduce both new opportunities and significant challenges for real estate developers in the industry.

This case study described Swire's focus on pilot use of innovative technologies and sustainable renovation practice including Building PEDF system. Swire Properties launched this PEDF microgrid technology project in partnership with Tsinghua University with the supply chain Beijing Zhijian to improve the flexible regulation of electricity in buildings. This highly innovative PEDF system refers to the application of PV power generation, energy storage, DC power distribution and flexible energy asset. Its overarching objective is to achieve flexible dispatching of power grids and efficient resource utilisation in order to succeed in energy conservation and carbon reduction within built environment.

7.6.2 The Jikka House in Shizuoka, Japan

Japan is at the forefront of green architecture and sustainable construction practices, with Japanese architects increasingly embracing traditional and sustainable materials, harmonising heritage design principles with contemporary technologies. The Taizhou Contemporary Art Museum in Kyoto, designed by architect Kengo Kuma, exemplifies this approach. The museum features a facade constructed from bamboo, a locally abundant and renewable resource. This not only reduces the carbon footprint but also pays tribute to Japan's rich cultural heritage (Yllah, 2024).

The Jikka House in Shizuoka (Figure 7.40) is a fusion of the modern architecture with traditional Japanese homes. The house, designed by architect Issei Suma, maximises natural ventilation and integrates a turf-covered roof for insulation and visual harmony with the surrounding landscape.

Figure 7.40 The Jikka House (image by way of Issei Suma, photography by Takumi Ota)

While the Japanese word 'Jikka' literally means 'the house I was born in', it is more broadly understood as 'where my heart belongs'. Set on a mountain ridge in the rural district of Shizuoka in Japan, the Jikka House is a teepee-shaped private home for two friends in their sixties. The 100 m^2 cluster of tent-like structures accommodates a home and workplace. The space serves as a food delivery service for the elderly and a nursing care facility for people with disabilities.

The Jikka House is designed in a series of five huts, each with square-shaped plans of varying proportions to accommodate the individual programmes. The pointed huts are wrapped in light, slender wooden panels with large, arched windows that curve from the ground halfway up each structure until the roof starts to slope. These glass apertures provide ample natural light into each hut alongside small, circular openings at the top of each point. The interior is clad in smooth concrete and white-coated timber beams (Franklin, 2017) but sit lightly within the green landscape.

The living quarters include a shared bedroom with a bathroom in a compact hut featuring low ceilings. At the centre of the cluster sits the dining room, distinguished by its striking ceiling of eight metres, giving the space a more communal and elevated character. This central room serves as a gathering place where the house owner hosts guests from the local community and welcomes old friends. Within each hut, the building provides ample space and ease of circulation for the home owner and their visitors. The interior is not only visually striking but are also fully, wheelchair accessible for disabled guests.

The project was a finalist in the 2017 A+Awards in the Residential-Private House category, for its striking mixed-use dwelling.

Five years since its opening, International Longevity Centre Global Alliance (ILC)-Japan (ILC, 2022) interviewed the client – two older Japanese women who had moved from Tokyo to Izu Highland, a popular retreat for city dwellers, to build their private house and restaurant. Jikka has since evolved into a community hub, fostering social connection, and strengthening cooperative relationships among local residents. Nobuko and Sachiko hope that Jikka will eventually operate as an informal day care centre, embodying the principles of productive aging. The warm, genuine smiles of the local residents who were interviewed by ILC-Japan were particularly noteworthy, underscoring the positive social impact of the project. (ILC, 2022).

Focus area (learning points) **ACTION**

Japan is at the forefront of green architecture and sustainable construction practices using traditional and sustainable materials for creating environmentally conscious buildings without compromising on functionality or beauty.

The Jikka House in Shizuoka is a fusion of modern architecture with traditional Japanese homes. This case study showcases Japan's innovative design solutions inspired by nature. The house maximises natural ventilation and integrates a turf-covered roof for insulation and visual harmony with the surrounding landscape.

The project was a finalist in the 2017 A+Awards in the Residential-Private House category, for its striking mixed-use dwelling. This case study also provides an example of achieving the social benefit of the project's initial aims. Jikka has evolved into a community hub, fostering social connection and strengthening cooperative relationships among local residents.

7.6.3 Oasia Hotel Downtown, Singapore

Singapore's Green Mark scheme is a green building rating system operated by the BCA. The Green Mark scheme promotes the adoption of green building designs and technologies that improve energy efficiency and reduce buildings' impact on the environment. Compared with other countries' sustainability benchmark assessment method, Singapore's Green Mark has put more emphasis on greening of buildings and urban design.

Singapore's Urban Redevelopment Authority (URA) introduced the Landscaping for Urban Spaces and High-Rises (LUSH) programme in 2009 to capitalise on development as a means of introducing more greenery into the city. In its latest version, it introduces a Green Plot Ratio (GPR) standard as a 'measure of the density of greenery within a site' (URA, 2017).

Singapore has reinvented itself from early aspiration of becoming a 'garden city' to a vision of a modern 'city in a garden' (Treepedia, 2018), a shift highlighted by the collaboration between the Massachusetts Institute of Technology (MIT) Senseable City Lab and World Economic Forum's Global Future Council on Cities and Urbanisation; Singapore has the highest density of greenery based on a list of 17 cities (Treepedia, 2018) including Sydney, Miami, London and Paris.

The Oasia Hotel Downtown (Goyal, 2023) in Singapore, designed by WOHA, is a prominent example of biophilic design and a green building, recognised for its extensive use of greenery and its 'living facade'. The hotel is certified under Singapore's Green Mark scheme and has received numerous awards for its sustainable design and architectural excellence.

Figure 7.41 A view of Oasia Hotel Downtown (photo credit to Mei Ren)

The Oasia Hotel Downtown, shown in Figure 7.41, comprises abundantly landscaped sky terraces, interleaved in naturally ventilated breezeway atria between room blocks. This case study presents an example of land use intensification in a tropical climate (Wong *et al.*, 2018). The high-rise typology in Singapore has been evolved from the Western model to better suit the local culture and climate.

Cities often feel harsh and isolating when densely packed development leaves little spaces for greenery or shared outdoor spaces. City dwellers are confined living in the towers leading increasingly insular lives with minimal contact with nature. Oasia Hotel Downtown re-imagined the modern tower – from a vertical city to a garden city, as building blocks for living, with greenery and amenities that support and contribute to the overall urban environment. By incorporating greenery beyond the ground floor, buildings can function as biophilic environments that visually and emotionally engage both their inhabitants and the surrounding public.

Located in the heart of Singapore's Central Business District (CBD) and overlooking the historic Tanjong Pagar district, Oasia Hotel Downtown stands as a striking example of architecture design and engineering excellence. The tower is 191 m high and rises up from the tree lined streets. Constrained within a tight site of 47 m by 47 m, the building comprises offices, a hotel and club. The building is designed with L-shaped blocks, with each block served by vertical circulation at the corners. This opens up the centre of each block as a breezeway atrium. Sky terraces are interleaved between each stack, offering generous landscaped amenity decks. By dividing the tower into vertical segments, the sky terraces provide not only multiple elevated grounds for greenery and amenity, but also create human scale environments in the sky. Figure 7.42 shows a view from the sky terrace located on the 12th floor, and Figure 7.43 for the green wall.

For the L-shaped blocks, the sky terraces serve as large overhangs, directly shading the terrace, open-sided, the atria are cross ventilated from all directions, aided by a natural funnelling effect. Each atrium achieves an approximate 1:1 height to depth ratio, affording a bright and airy environment with daylight, cross ventilation, coupled with evaporative cooling from water and shading from the greenery. The atria are comfortable, sheltered open spaces with natural light and fresh air, without being air-conditioned.

Unlike other high-rise towers, the Oasia Hotel Downtown does not have a solid mass or a full-height curtain wall. The stacked blocks and terraces are enveloped by 25 000 m^2 of expanded aluminium mesh screens. About 1800 prefabricated fiberglass planters are located on every level, abutting the screen so that creeping vines are only required to climb 3–5 m before overlapping with the next tier of planters. Landscaping is used as the architectural environment filter and material palette, not merely an applied aesthetic layer. As an envelope, the greenery and the screen function as filters that provide shade, reduce heat, dampen noise, cut out glare and dust, and improve

Figure 7.42 A view of Oasia Hotel Downtown from the 12th floor sky terrace (photo credit to Mei Ren)

Figure 7.43 A view of Oasia Hotel Downtown's green wall mesh (photo credit to Mei Ren)

air quality. The screen is a composition of five colours to mimic the natural colour variation in plants, and would eventually recede behind greenery and appear as a background of varying colour accents like flowers within the vertical foliage.

Twenty-one species of creepers were selected and distributed across the facades based on their sunlight requirements, rate of growth, density of coverage, texture and colour. Some species produce coloured flowers that attract birds and insects at different times of the year. Together with 33 different species of trees and shrubs planted on the sky terraces, the living tower accommodates a total of 54 species of plants. To ensure the practicality and ease of maintaining the greenery, cat ladders and catwalks are incorporated alongside the planters so that maintenance personnel have safe and direct access without relying on ropes. All planters are watered by an automatic irrigation system to avoid wastage (Wong *et al.* 2018).

Within the building, people are surrounded by nature. Externally, the vertical green screens and landscaped sky terraces provide a welcome counterbalance to the surrounding urban fabric, creating moments of respite within the dense city environment. This building is a prototype for liveability and sustainability.

Focus area (learning points) **ACTION**

Singapore Green Building Masterplan (SGBMP) has three primary goals known as '80-80-80 in 2030'. By 2030, they aim to have 80% of new developments (by gross floor area) to be Super Low Energy (SLE) buildings. It will accelerate efforts to green 80% of buildings by that date. By 2030, achieving an 80% increase in energy efficiency for the best-in-class green buildings. Comparing with other countries' sustainability benchmark assessment method, Singapore's Green Mark has put greater emphasis on greening of buildings and urban design.

Singapore has reinvented itself from early aspirations of becoming a 'garden city' to a vision of a modern 'city in a garden'. This vision affords Singapore not only with a distinct and visually pleasant image, but also a greener urban environment that has demonstrated its ability to alleviate urban heat issues, increase biodiversity in the city and support sustainable water management.

Oasia Hotel Downtown is selected as a case study to represent this vision of embracing nature, harnessing technology and fostering a relationship between humans and the environment. This case study showcases a series of innovative technologies and practices for achieving the sustainability goals.

The building with its vertical green 'living wall' screens and landscaped sky terraces provides a prototype for liveable and sustainable city living.

7.6.4 Maldives Floating City

The Maldives consists of approximately 1200 coral sand islands that form the world's lowest-lying nation, with 80% of the country less than 1 m above sea level. Around 200 of the islands are inhabited, while 100 more are used as hotel islands. The capital, Malé, is one of the world's most densely populated cities. As sea levels continue to rise, the risk of the Maldives becoming engulfed by water increases. The Maldives Floating City is a project designed to address this risk. The innovative solution will serve as an integrated tourism model, including houses, hotels, shops, restaurants and a marina aiming to house around 20 000 people in a series of floating housing units linked by walkways. The project is a private-public partnership between Dutch Docklands and Maldives Government. Dutch Docklands was founded by architect Koen Olthuis at Dutch architectural firm Waterstudio and building contractor Paul van de Camp (Benkö, 2022). The Maldives Floating City will have several thousand low rise buildings. Located in a lagoon 15-min boat ride from Malé and a 10-min boat ride from the international airport, it will float in a 200-hectare lagoon in the Indian Ocean, where it will be able to adapt to the changing sea level (Ancheri, 2022).

The Maldives Floating City (Figure 7.44) is an innovative housing project being developed in a lagoon using the latest floating technology from the Netherlands. The floating city will comprise 5000 housing units, tethered to the lagoon floor and linked together to create a safe and comfortable environment (Government of Maldives and Dutch Docklands Maldives, 2021). The floating city represents a major commitment that embraces sustainable and modular urbanisation. The 'island city' consists of individual units rather than a giant mega-structure which can be cost prohibitive.

The new city is envisioned as being 'close to nature', with 'roads' and canals. Drawing inspiration from nature, the structures give it the appearance of a real brain coral and reflect a similarly 'efficient organisational' logic. Dutch Docklands also feels that the shape of the Maldives Floating City will raise the international profile of the Maldives as a centre for coral conservation.

Figure 7.44 An image showing the completed city (courtsey to Waterstudio)

The project emphasises sustainable living and aims to be a car-free zone, with navigation by way of canals such as boat taxi, natural white-sand roads, and electric buggies and scooters (Figure 7.45). Everything is designed from an environmental perspective. The city will be equipped with a smart energy grid powering the city solely with renewable energy.

The city consists of thousands of units using a modular system (Figure 7.46). The modular system ensures quality, standardisation, short construction time, cost control and efficient maintenance. The new development is conceived as a flexible city with an intelligent power grid that can respond to changing demands and weather conditions. Power comes mainly from solar energy generated on the island city itself. The aim is to use as little energy as possible, with any surplus energy generated being fed into the grid. Cold deep-sea water will be pumped into the lagoon as an alternative to air conditioning (Benkö, 2022).

Apart from the sustainable, innovative technologies, ecological best practice is used to replenish and protect the marine ecosystem. The unique city creates 'blue habitats' both above and below the surface of the water, which in turn encourage coral growth. This is because artificial coral reefs made from glass foam are fitted underneath the structures. The surrounding coral reefs serve as natural breakwaters for the city, stabilising it and providing added security for its inhabitants.

The supply and distribution of goods, storage and controlled waste treatment and disposal are designed to be as resource and environmentally friendly as possible. In addition, the shadow impact of the hexagonal floating structures ensures that enough sunlight reaches the seabed. The anchoring systems must also be rigid yet flexible enough to allow the city to withstand extreme outside forces. Up to 20 000 people could live in the Maldives Floating City in the future.

The island state's government gave the project its approval in 2022 (Maldives Floating City, 2022). The marine engineering, technical details and specifications for the Maldives Floating City

Figure 7.45 An image showing the urban street (courtsey to Dutch Docklands Maldives / Waterstudio)

Figure 7.46 Brain coral shaped city with flexible and modular urbanization (courtesy to Waterstudio)

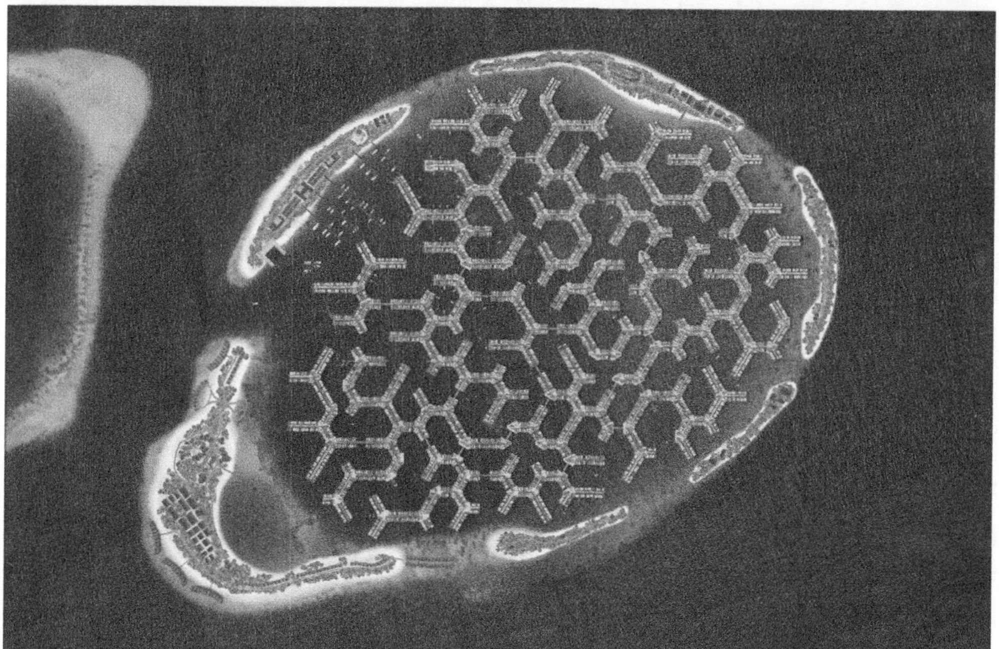

development have been completed, and construction began after remaining government approvals were granted. The modular city construction started in January 2023 and will take 4–5 years to complete with the anticipated completion in 2027. The first residential units are already on view and ready in 2024 (Figure 7.47).

Maldives Floating City aims to be the 'backbone' of future floating cities. The developers at Dutch Docklands and the architects at Waterstudio hope this will allow the Maldivians to present themselves not as climate refugees but as climate innovators (Benkö, 2022). With its colourful low-rise buildings, the Maldives Floating City presents a viable approach to living on the water with resilient, environmentally friendly floating structures. This approach is becoming increasingly pertinent as global sea level is rising by around 3.1 mm a year. Mohamed Nasheed, President of the Maldives from 2008–2012, Speaker of Parliament and CVF Ambassador of Ambition, explains:

> 'This Maldives Floating City does not require any land reclamation, therefore has a minimal impact on the coral reefs. Our adaption to climate but work with it, as the Maldives Floating City proposes. In the Maldives we cannot stop the waves, but we can rise with them: change mustn't destroy nature'. (Government of Maldives and Dutch Docklands Maldives, 2021)

Figure 7.47 Maldives Floating City, photographed from space in 2024 (courtsey to Maxar Technologies – (Versano, 2024))

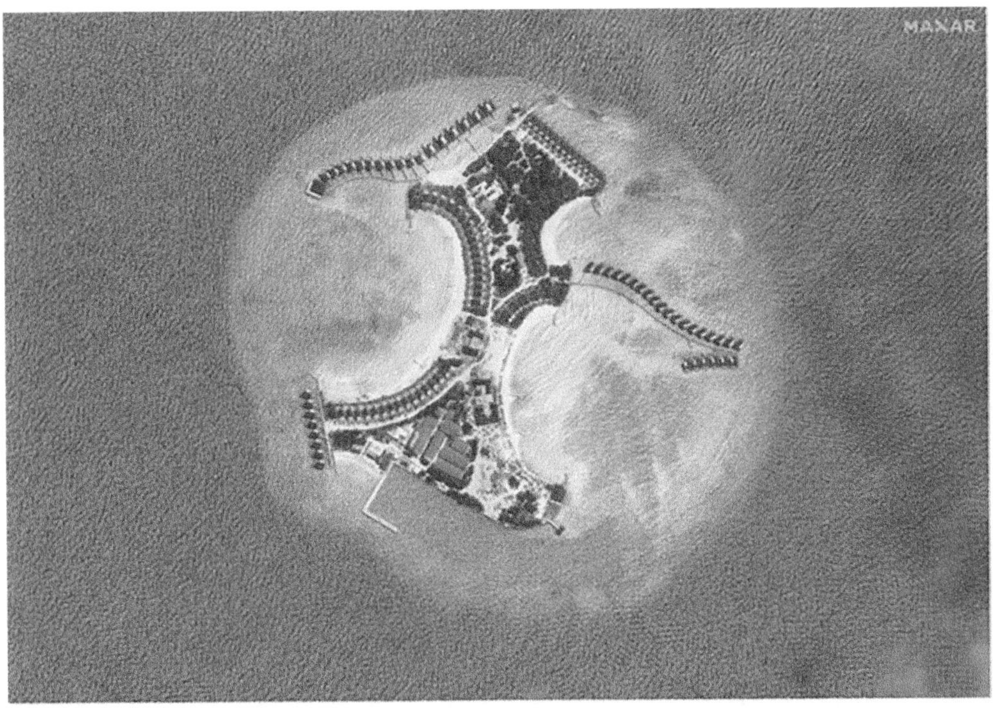

Focus area (learning points) **ACTION**

It has been predicted by the World Bank that future sea levels will be increased by up to 100 cm by the year 2100. The Maldives island is the world's lowest-lying nation with 80% of the country less than a metre above sea level. In order to future proof Maldives's living, in partnership with Dutch Docklands, the Maldives Government initiated Maldives Floating City project using private-public partnership designed to address this climate risk. The innovative solution will serve as an integrated tourism model, including houses, hotels, shops, restaurants and a marina aiming to house around 20 000 people in a series of floating housing units linked by walkways.

This case study provides an insight to innovative floating housing development in a lagoon using the latest floating technology from the Netherlands. The 'island city' consists of individual units using a modular system rather than a giant mega-structure, which can be cost prohibitive. The modular system ensures quality, standardisation, short construction time, cost control and efficient maintenance. The new development is conceived as a flexible city with an intelligent power grid and resource efficient networks that can respond to changing demands and weather conditions.

The floating city represents a major commitment that embraces sustainable and modular urbanisation and sensitive to surrounding pristine habitats. The modular city construction started in January 2023 and will take 4–5 years to complete with the anticipated completion in 2027. When it is complete, Maldives Floating City has the potential to serve as a template for future climate resilient floating cities.

7.6.5 Terra Dubai Expo 2020

The World Expo showcases innovations in culture, technology, healthcare and learning. Extending from the 1083-acre site's central plaza, three unique districts were dedicated to the Expo's core themes: sustainability, mobility and opportunity. Each of these districts features an inspirational pavilion that not only houses a range of exhibits, but also embodies the pertinent theme. The Sustainability Pavilion, Terra (Figure 7.48), gave an opportunity for the UAE to showcase innovations in energy generation and water management for the region and convey an inspiring message about the natural world and technology application while providing an immersive educational experience. Built for client Emaar Properties, this Pavilion is a permanent structure that would become a sustainability museum after the Expo.

The pavilion's core building features an approximately 8000 m^2 exhibition space that includes exhibition galleries, an auditorium, a courtyard and a reservoir. The structure consists of a podium with various levels. The pavilion sits partially below ground, meaning it is generally cooler than the ambient temperature. Energy is saved by burying much of the occupied spaces and providing thick, insulated walls with minimal glazing. The excavated waste was used in the landscaping around the building.

Sustainability stood front and centre in the design brief and all design thinking for the pavilion. The building itself has an optimised building form, high envelope performance and orientation to provide shading and reduce the cooling demand. Branching out from a central support zone, a 130 m-wide elliptical cantilevered canopy overarches the pavilion. This provides shade over the atrium and exhibition spaces to reduce solar gain, while enabling passive cooling to the building and reducing energy demand. Secondary steel catwalks are positioned above the primary steel structure of the canopy and simultaneously function as rainwater collectors, which is then treated and used in the building exhibitions. Surplus rainwater from the canopy is distributed throughout the landscape to recharge the ground water table by way of percolation tanks. Between these catwalks sit the PV panels to generate power (Grimshaw, n.d.).

To realise the complex design of the canopy, the team applied advanced computational methods for coordinating and optimising the architecture and the structural solution, which led to a significant reduction in materials use. In alignment with the project's sustainability requirements, the structural steel incorporates 40% recycled content.

Energy Trees

The most identifiable aspect of the pavilion is the distinctively shaped steel canopy, which features more than 1055 solar panels (8000 m^2), to maximise the energy generated on site by renewable sources. These series of bespoke PV 'Energy Trees' not only provide shade, but also track the sun's azimuth to provide additional electricity. Grimshaw architect and Buro Happold structure engineering design team developed these unique structures, which will contribute to the pavilion's ability to produce an expected 4.3 GWh of electricity a year through solar panels with an aspiration to achieve net zero energy during the legacy phase of the building's life.

Water Trees

This project employs water harvesting, water recycling and innovative materials in a context that is unique to the region. The Water Tree enables a passive method of dew harvesting, generating 30 litres a day on average, taking advantage of the rapid change in temperature overnight where the inner

Figure 7.48 Photos showing Terra (reproduced with kind permission of Buro Happold (Buro Happold, n.d.c), photo credit to Craig Palmer, copyright Buro Happold)

cone surface becomes cooler to generate droplets of water that gather and run down to a collection point, irrigating the landscape below. The pavilion aims to produce up to 21 000 litres of water a day, extracted from atmospheric humidity and recycling grey and black water. Design calculations showed a water demand reduction of 92% against a baseline building including landscape. It treats 100% of its wastewater, captures and infiltrates 100% of storm water on site, and employs a district-scale strategy that recovers wastewater from neighbours for treatment and reuse within the project (Grimshaw, n.d.).

Summary

- The Sustainability Pavilion achieved LEED Platinum, LEED Zero – Energy and Water certification.
- Energy Use Intensity: 226 kWh/m^2/year (regulated loads), a 42% reduction from an equivalent new building.
- Total site PV energy generation is 4 GWh, 100% renewable in Legacy mode.
- Aspirational net zero water in Expo and Legacy modes.
- The Pavilion is a small-scale water treatment plant. It supplies 100% of its water use from on-site sources, by treating 100% of its wastewater, capturing 100% of storm water, using a district-scale water strategy as well as harnessing water from the air (Water Trees).

Focus area (learning points) — **VISION AND TECHNOLOGY**

Dubai has developed various sustainability policies and incentives aiming to transform Dubai into a global hub for clean energy and a green economy. Expo City Dubai, built on the site of Expo 2020, aims to be a net zero carbon city by 2050, integrating sustainability into its operations and events, with a focus on clean energy, water conservation and biodiversity enhancement. It stands as a compelling exemplar of sustainable urban living for Dubai and beyond (Expo City Dubai, 2020).

Terra is one of the landmark inspirational buildings showcased at Expo City Dubai. Sustainability stood front and centre in the design brief and overall design process. The Terra case study showcases a net zero carbon development with high quality public spaces and museum interior by harmonising architecture with climate responsive design.

Terra case study provides a series of innovative technologies and practices for achieving LEED Platinum, LEED Zero – Energy and Water certification, aiming to achieve net zero energy and water at legacy operation mode.

7.7. Africa

Africa faces a complex web of sustainability and economic challenges and opportunities, requiring multiple approaches to tackle climate change impacts, involving governments, businesses and communities (Lui, 2024). Integrating ESG principles into financial markets is crucial for promoting sustainable investment and development in Africa.

Key challenges include the following.

- Climate change – African countries are disproportionately affected by climate change, facing issues such as droughts, floods and rising sea levels. African countries' abundant natural resources, including minerals and land, present both opportunities and challenges for sustainable development, requiring responsible management.
- Social and economic development – Africa's economic growth needs to address social inequality, poverty and access to resources and opportunities. The impact of conflicts and diseases also hinder sustainability efforts. African countries have limited access to green finance for green projects. Their lack of necessary infrastructure and regulatory hurdles often complicate efforts.

Africa accounts for approximately 6% of global energy demand, with more than half of this coming from its buildings. Given that Africa's population is expected to reach 2.4 billion people by 2050, with 80% of this growth occurring in cities, sustainability needs to be a core principle of all future buildings. These solutions are highlighted in the UN Environment Programme (UNEP)'s *2022 Global Status Report for Buildings and Construction* (UNEP, 2022), launched at the UN Climate Conference (COP27) in Egypt (United Nations, 2022). The report focused on how Africa can manage this urban growth and increase the resilience of its housing stock while avoiding an increase in GHG emissions. Africa has significant potential for green business development, including sustainable agriculture, renewable energy and circular economy initiatives. Developing sustainable urban areas through green building practices, efficient transportation and access to clean water and sanitation is vital.

As part of UN Environment Programme News and Stories (UNEP, 2024), African countries are leaning to traditional building practices to tackle some of the challenges they face.

Locally adapted sustainable design, construction practices and materials coupled with rich in renewable energy sources, such as solar and wind, with nearly half of the planet's total renewable energy potential, and innovation represent great opportunities for both mitigation and resilience in Africa's rapidly growing building stock. Inspiration for climate-resilient building can be found throughout African history, from Eswatini's beehive huts to the Drogon cliff villages of Mali to the mud-brick mosques of West Africa (UNEP, 2024). Africa is rich in natural, sustainable materials such as adobe, laterite, termite mound soil, timber, stone, bamboo, sand and dry vegetation. Traditional construction techniques include rammed earth, sun-dried bricks, compressed earth blocks, wattle and daub, cob, timber-framed construction, sandbag construction and thatched roofs. Ensuring the use of sustainable materials is particularly critical in a context where a significant proportion of population lives in overcrowded informal settlements (UN-Habitat, 2020) that are especially vulnerable to the impacts of climate change.

The UK's Prince William founded the Earthshot Prize (The Earthshot Prize, n.d.a) to champion the game-changers, the makers, the creatives and the leaders in sustainability. The Earthshot Prize is a global environmental challenge, recognising innovative solutions that address five key areas, one of which is 'Clean Our Air', which aims to achieve clean, healthy air for everyone by 2030. Mukuru Clean Stoves (Mukuru Clean Stoves, n.d.) was the Earthshot Prize winner 2022 (World Economic Forum, 2022) for developing a stove that burns processed biomass, reducing pollution by 70% compared with traditional cookstoves. Mukuru Clean Stoves have demonstrated a method whereby ingenious people can create sustainable businesses by innovation to 'Eradicate Household Air Pollution in Africa!' (Mukuru Clean Stoves, n.d.).

Two case studies are selected to illustrate how African countries and people took on the sustainable development opportunities to tackle the challenges they are facing.

- Lycée Schorge Secondary School.
- Mukuru Clean Stoves.

7.7.1 Lycée Schorge Secondary School, Burkina Faso

Located on the outskirts of the Burkina Faso town of Koudougou, the Lycée Schorge Secondary School showcases an exemplar case study of creating a sustainable building by mixing traditional techniques with new materials.

Completed in 2016, Lycée Schorge (Kéré, n.d.a) has a built area of 1660 m^2 designed by Berlin-based Burkinabè-founded architecture firm Kéré Architecture, and commissioned by Stern Stewart Institute and Friends.

The school consists of nine modules arranged around a central courtyard, protecting the central space from wind and dust. Each module is built out of locally sourced laterite, which is cut into bricks and left in the sun to harden.

The walls of each module are built out of locally sourced laterite stone, which lends them their striking deep red colour. When first extracted from the earth, laterite can be easily cut and shaped into bricks, which are then left in the sun to harden. The material provides an excellent source of thermal mass, absorbing the daytime heat and radiating it at night.

A secondary facade, made of locally sourced eucalyptus wood, wraps around the classrooms like a transparent fabric and creates a variety of shaded intermediary spaces between itself and the classrooms to protect students from daytime high temperatures (Figure 7.49). The classrooms' ceilings, made of perforated plaster vaults, diffuse indirect sunlight in order to improve the light quality while avoiding the heat otherwise caused by direct radiation. Wind towers located at the back of each classroom allow hot air to escape, thus helping to further lower the internal temperature. The sculptural forms of these towers stand out above the main body of the building, creating a landmark in its surroundings (Figure 7.50).

In order to minimise costs and reduce material waste, the school's furniture is made from local hardwoods and steel offcuts from the roof construction.

The building forms part of a series of school and community projects by Kéré Architecture following the studio's acclaimed success in designing Gando Primary School where creative use of abundantly available clay enabled the production of structurally robust bricks. Gando Primary School (Kéré, n.d.b) was completed in 2001, awarded with Aga Khan Award for Architecture 2004 and Global Award for Sustainable Architecture 2009.

These examples demonstrate that such techniques can help prevent the need for air conditioning, lessen the carbon footprint of their buildings, and draw on traditional construction techniques that minimise the need for the long-range transport of building materials and concrete production.

Figure 7.49 Walls at Lycee Schorge Secondary School, Burkina Faso (photo credit to Sierd Hoekstra licensed under CC BY-SA 4.0)

Figure 7.50 The Schorge Secondary School, Burkina Faso (photo credit to Sierd Hoekstra licensed under CC BY-SA 4.0)

Quality, sustainable housing is essential for strengthening the resilience of vulnerable populations to climate crisis. Rising temperatures and expanding access to electricity are expected to drive a sharp increase in air-conditioning use, making cooling a major source of future residential energy demand across Africa. Designing new homes to maximise natural and passive cooling is therefore crucial for reducing energy use and ensuring safe, comfortable living conditions in a warming climate.

Focus area (learning points) **VISION AND TECHNOLOGY**

African countries face complex sustainability and economic challenges. However, there are untapped resources in the continent presenting strong opportunities for sustainable development. Leading practitioners in Africa are adopting traditional building practices to tackle some of the challenges, using locally adapted sustainable design, construction practices and materials coupled with rich in renewable energy sources.

This case study of Lycée Schorge Secondary School shows an exemplar approach of a sustainable building by mixing traditional techniques with new materials. It showcases a set of innovative yet locally available materials and processes for creating climate resilient community facilities.

The case study demonstrates that the strong vision and ingenuity of adapting locally available resources can deliver quality and sustainable buildings that help vulnerable populations to build greater resilience to the impacts of the climate crisis.

7.7.2 Mukuru Clean Stoves, Kenya

In Africa, 700 million people use cookstoves that burn firewood or charcoal, which are harmful both to humans and the environment. Open fires and traditional stoves expose families to dangerous air pollution often exceeding World Health Organization (WHO) clean air guidelines by up to 50 times. Household air pollution is linked to over 1 million deaths annually, with women and children disproportionately affected due to prolonged exposure while cooking with inefficient stoves.

Rather than burning solid fuels, Charlot Magayi created Mukuru Clean Stoves in 2017, which use processed biomass made from charcoal, wood and sugarcane, reducing pollution by 70% compared with a traditional cookstove. They also cost less to buy and to run than traditional cookstoves.

Mukuru Clean Stoves (Mukuru Clean Stoves, n.d.) has the vision – 'To Eradicate Household Air Pollution in Africa!' – and their mission is to provide the most reliable cooking asset to help people to fight against household air pollution and malaria. Mukuru Clean Stoves uses recycled metal to manufacture improved cookstoves and agricultural waste to manufacture mosquito repellent fuel.

Charlot Magayi's first job was selling charcoal within the slum community. Her young family suffered from respiratory tract infections and her daughter suffered a severe burn injury from a traditional stove when she was two years old. After Charlot learnt about the health hazards caused by charcoal and other harmful solid fuels, as an enthusiast of science and social studies, she invented Mukuru Clean Stoves and wants to inspire fellow women to lead the fight against household air pollution in Africa. Mukuru Clean Stoves, a social enterprise that produces clean, affordable and reliable cook stoves targeting under-served markets to help mothers keep their children safe, save on fuel consumption and reduce household air pollution. A female-founded business with mostly female staff and distribution agents, Mukuru is empowering women to make a living by making a difference.

Mukuru Clean Stoves also put affordability and accessibility at the heart of their business as this remains significant barriers, preventing millions of households from making the switch to cleaner, more sustainable cooking solutions.

Two years on since Mukuru Clean Stoves won the Earthshot 2022 prize, Mukuru's progress has been incredible, they have reached more than 600 000 households, employing more women, expanding into new countries. Mukuru Clean Stoves (Figure 7.51) are now sold and distributed across many African countries including an expansion into Ghana and Nigeria. Ikea Social Entrepreneurship Partnership (Ikea Social Entrepreneurship, n.d.) distributes Mukuru to Kenya, Uganda, Ethiopia and Rwanda.

Charlot Magayi aims to develop a cleaner stove that burns ethanol to save around US $10 million in fuel costs while reducing the health risks linked to traditional cooking fuels. She plans to reach one million customers within three years and expand to ten million people across Africa within the next decade (The Earthshot Prize , n.d.b).

Figure 7.51 Mukuru Clean Stoves (Mukuru Clean Stoves, n.d.)

Focus area (learning points) **ACTION**

Africa's complex sustainability and economic challenges requires a new approach to tackle climate change impacts, involving governments, businesses and communities.

Mukuru Clean Stoves was Earthshot Prize winner 2022 for the 'Clean our Air' category in recognising their effort and ingenuity to achieve clean, healthy air for everyone by 2030. Founded by the UK's Prince William, the Earthshot Prize aims to champion the game-changers, the makers, the creatives and the leaders. Mukuru Clean Stoves has developed a stove that burns processed biomass, reducing pollution by 70% compared with traditional cookstoves.

This case study of Mukuru Clean Stoves demonstrates that by empowering local businesses and communities, Mukuru Clean Stoves has not only created a cleaner and more sustainable cooking solution that improves household air quality and health, but also keep affordability and accessibility at the heart of their business.

REFERENCES

Adelfio M, Hamiduddin I and Miedema E (2020) London's King's Cross redevelopment: a compact, resource efficient and 'liveable' global city model for an era of climate emergency? *Urban Research & Practice* **14(2)**: 180–200, 10.1080/17535069.2019.1710860.

Ancheri S (2022) Maldives is building the world's first floating city. *Conde Nast Traveller*, 5 July. https://www.cntraveller.com/article/maldives-floating-city#:~:text=Everything%20you%20need%20to%20know%20about%20Maldives%20Floating%20City&text=Located%20a%2015%2Dminute%20boat,designing%20architecture%20to%20withstand%20floods (accessed 24/01/2026).

Andrés-Doménech I *et al.* (2021) Sustainable urban drainage systems in Spain: a diagnosis. *Sustainability* **13(5)**: 2791, 10.3390/su13052791.

Battersea Power Station (2024a) Battersea Power Station awarded gold winner for the Heritage category at 2024 FIABCI World Prix. *Battersea Power Station*, 31 May. https://batterseapowerstation.co.uk/news/battersea-power-station-fiabci-prix-dexcellence-awards-2024/#:~:text=Battersea%20Power%20Station%20Awarded%20Gold%20Winner%20For,At%202024%20FIABCI%20World%20Prix%20d'Excellence%20Awards.&text=Battersea%20Power%20Station%20was%20also%20awarded%20World,completed%20phase%20Circus%20West%20Village-%2C%20in%202019 (accessed 22/06/2025).

Battersea Power Station (2024b) *Masterplan Exhibition for Phase 3C of the Battersea Power Station Masterplan and the Completion of the Pedestrianised High Street, Electric Boulevard.* https://batterseapowerstation.co.uk/content/uploads/2024/04/BPSDC_P3C_-exhibition-boards-web-version.pdf (accessed 23/01/2026).

Battersea Power Station (2024c) *The Masterplan.* https://batterseapowerstation.co.uk/about/building-battersea-the-masterplan/ (accessed 23/01/2026).

Benkö (2022) *The world's first floating city. ubm magazine.* https://www.ubm-development.com/magazin/en/maldives-floating-city-climate-innovators-at-work/#:~:text=%23city%20planning,The%20world's%20first%20floating%20city,are%20used%20as%20hotel%20islands (accessed 24/01/2026).

BLT Design Awards (2023) *EDGE Suedkreuz Berlin – Winner - BLT Built Design Awards.* https://bltawards.com/winner/blt/2023/21329/0/#:~:text=Winner%20%2D%20EDGE%20Suedkreuz%20Berlin%20%2D%20single%20winner (accessed 08/07/2025).

BNIM (n.d.) *Case Study No. 1: Santa Monica City Hall East.* https://issuu.com/bnim/docs/designing_20for_20zero_20carbon/s/15514702 (accessed 08/07/2025).

Building and Construction Authority (Singapore) (2021) *Green Building Masterplans. Updated July 2022.* https://www1.bca.gov.sg/buildsg/sustainability/green-building-masterplans (accessed 23/01/2026).

BRE (n.d.) *London Borough of Camden gets BREEAM's highest bespoke building rating.* https://bregroup.com/case-studies/london-borough-camden-breeams-highest-building-rating (accessed 22/01/2026).

Buro Happold (2022) *Chobham Manor Phase 1 Post Occupancy Evaluation study.* https://live-qeop.pantheonsite.io/sites/default/files/attachments/Chobham%20Manor%20Phase%201%20POE.pdf (accessed 23/01/2026).

Buro Happold (n.d.a) *EDGE Suedkreuz – Buro Happold.* https://www.burohappold.com/projects/edge-suedkreuz/ (accessed 08/07/2025).

Buro Happold (n.d.b) *Empire State Realty Trust Deep Energy Retrofits – Buro Happold.* https://www.burohappold.com/projects/empire-state-realty-trust-deep-energy-retrofits/ (accessed 08/07/2025).

Buro Happold (n.d.c) *Terra – the Sustainability Pavilion – Buro Happold.* https://www.burohappold.com/projects/terra-the-sustainability-pavilion/ (accessed 24/01/2026).

Camden Council (2006) *Section 106 Agreement for Kings Cross Central.* https://kxdf.wordpress.com/wp-content/uploads/2014/01/the-section-106-agreement-for-kings-cross-central.pdf#:~:text=(M)%20The%20Council%20considers%20it%20expedient%20in,of%20this%20Agreement%20are%20of%20public%20benefit (accessed 22/01/2026).

Casimira C (2022) Retrofitting for a Greener Future: The Empire Building Decarbonisation Playbook. *Reos Partners*, October. https://reospartners.com/blog/empire-building-playbook-decarbonization (accessed 08/07/2025).

City of Toronto (2025) *Thermal Comfort guidelines: for large area studies, public realm capital projects, and large site developments – final report. Thermal Comfort Guidelines, report*, pp. 1–13. https://www.toronto.ca/legdocs/mmis/2025/ph/bgrd/backgroundfile-252013.pdf. https://www.toronto.ca/city-government/planning-development/planning-studies-initiatives/thermal-comfort-study/ (accessed 23/01/2026)

Clark WC and Harley AG (2020) Sustainability science: toward a synthesis. *Annual Review of Environment and Resources* **45(1)**: 331–386, 10.1146/annurev-environ-012420-043621.

Construction Canada (2024) *Study Aims to Impact Toronto's Thermal Comfort for Smarter Urbanism - Construction Canada.* https://www.constructioncanada.net/study-aims-to-impact-torontos-thermal-comfort-for-smarter-urbanism/ (accessed 07/06/2025).

Creighton H (2020) *Sustaining a City, Sustaining Hope: Water and the Santa Monica City Hall East.* https://trimtab.living-future.org/living-building-challenge/sustaining-a-city-sustaining-hope-water-and-the-santa-monica-city-hall-east/ (accessed 23/01/2026).

CTBUH (2024) Council on Tall Buildings and Urban Habitat International Conference. https://2024.ctbuhconference.com/ (accessed 23/01/2026).

Datta M and University of Cambridge Institute for Sustainability Leadership (2024) *Asia's Crucial Role as a Leader in Global Sustainable Business.* https://www.cisl.cam.ac.uk/news/blog/asias-crucial-role-leader-global-sustainable-business (accessed 07/06/2025).

Dean E (2022) *Designing for Zero Carbon Volume 1.* Southern California Edison. https://calbem.ibpsa.us/wp-content/uploads/2022/03/Designing-for-Zero-Carbon-Volume-1.pdf (accessed 23/01/2026).

Directorate-General for Research and Innovation (2020) *100 Climate-Neutral Cities by 2030 – by and for the Citizens: Report of the Mission Board for Climate-neutral and Smart Cities.* https://research-and-innovation.ec.europa.eu/knowledge-publications-tools-and-data/

publications/all-publications/100-climate-neutral-cities-2030-and-citizens_en#details (accessed 24/01/2026).

DOGA (2021) *Kristian Augusts gate 13.* https://doga.no/aktiviteter/dogas-priser/doga-merket-design-arkitektur/vinnere-av-doga-merket/kristian-augusts-gate-13/ (accessed 08/07/2025).

Dowson M (2023) *Placemaking and Performance: Chobham Manor Housing POE.* https://www.cibsejournal.com/case-studies/placemaking-and-performance-chobham-manor-housing-poe/ (accessed 28/09/2025).

Empire State (2024) *News – The-Empire-State-Building-Observatory-1-Attraction-in-the-World-in-Tripadvisor-2024-Travelers-Choice-Awards.* https://investors.esrtreit.com/news/Press-Releases/news-details/2024/The-Empire-State-Building-Observatory-1-Attraction-in-the-World-in-Tripadvisor-2024-Travelers-Choice-Awards/default.aspx#:~:text=The%20Empire%20State%20Building%20Observatory%20is%20located,which%20is%20the%20only%20360%2Ddegree%2C%20open%2Dair%20observatory (accessed 23/01/2026).

EPA (US Environmental Protection Agency) (2024) *EPA Sustainability Progress.* US EPA. https://www.epa.gov/greeningepa/epa-sustainability-progress#:~:text=EPA%20works%20to%20demonstrate%20leadership,to%20fund%20facility%20efficiency%20investments (accessed 24/01/2026).

EUmies (n.d.) *EUmies Awards – Kristian August Street 13.* https://eumiesawards.com/heritageobject/kristian-august-street-13/#work-details (accessed 23/01/2026)

Expo City Dubai (2020) *Sustainability & Green Initiatives.* https://www.expocitydubai.com/en/sustainability/ (accessed 24/06/2025).

Floating Farm (2025) *Home – Floating Farm.* https://floatingfarm.nl/ (accessed 27/08/2025).

Flores C (2022) *Driving sector-wide energy performance improvements in buildings. NABERS UK Energy for Offices.* https://www.betterbuildingspartnership.co.uk/sites/default/files/media/attachment/NABERS%20Energy%20for%20Offices%20Final%20Slide%20Deck.pdf (accessed 23/01/2026).

Franklin S (2017) *Woodland Retreat: JIKKA Might Be Japan's Most Beautiful New Home.* https://architizer.com/blog/inspiration/stories/issei-suma-jikka/ (accessed 23/01/2026).

Garcia R (2022) *A Climate-Neutral Valencia – Eurocities.* https://eurocities.eu/stories/a-climate-neutral-valencia/ (accessed 23/01/2026).

Good Design Awards (2023) *Green Good Design Sustainability Awards 2023: EDGE Suedkreuz Berlin.* https://www.good-designawards.com/green-good-design-award-details-page.html?award=51128 (accessed 23/01/2026).

Government of Dubai (n.d.) *Dubai 2040 Urban Master Plan.* http://dubai2040.ae/en/dubai-2040-urban-master-plan/ (accessed 24/01/2026).

Government of Dubai Environmental Authorities (n.d.) *Environment & Sustainability.* https://www.dubai.ae/environment-sustainability#:~:text=Commitment%20to%20Sustainability,from%20clean%20sources%20by%202050 (accessed 08/07/2025).

Government of Maldives and Dutch Docklands Maldives (2021) *Maldives Floating City: The World's First Floating Island City.* https://maldivesfloatingcity.com/wp-content/uploads/2021/03/20210314-Press-release-MFC-ENGLISH-N.pdf (accessed 23/01/2026).

Goyal J (2023) Top 10 Leading Green Buildings in Singapore for 2025. *Novatr*, 12 December. https://www.novatr.com/blog/green-architecture-buildings-singapore (accessed 23/01/2026).

Grimshaw (n.d.) *Terra — The Sustainability Pavilion Expo 2020 Dubai.* https://grimshaw.global/projects/culture-and-exhibition/dubai-expo-2020-sustainability-pavilion/ (accessed 25/08/2025).

HiDubai Newswire (2023) *Dubai's Sustainable Initiatives.* https://focus.hidubai.com/dubais-sustainable-initiatives-pioneering-environmental-efforts-in-the-middle-east/#:~:text=

Dubai's%20Clean%20Energy%20Strategy%20sets%20a%20clear,for%20clean%20energy%20and%20the%20green%20economy (accessed 25/08/2025).

IKEA Social Entrepreneurship (n.d.) *Mukuru Clean Stoves*. https://www.ikeasocialentrepreneurship.org/en/social-enterprises/mukuru-clean-stoves (accessed 24/01/2026).

ILC (International Longevity Centre Global Alliance) (2022) *Jikka House: An Innovative Senior Living in Japan*. https://www.ilc-alliance.org/care/jikka-house-an-innovative-senior-living-in-japan/ (accessed 25/06/2025).

Johnson T (2024) Valencia flooding. Engineering solutions could have saved lives. *New Civil Engineer*. https://www.newcivilengineer.com/latest/valencia-flooding-engineering-solutions-could-have-saved-lives-04-11-2024/ (accessed 23/01/2026).

Kéré (n.d.a) *Lycee Schorge Secondary School*. https://www.kerearchitecture.com/work/building/lycee-schorge (accessed 23/01/2026).

Kéré (n.d.b) *Gando Primary School*. https://www.kerearchitecture.com/work/building/gando-primary-school-3 (accessed 23/01/2026).

Kennett S (2010) *Empire State Building Window Refurb: The Empire Strikes Back*. https://www.building.co.uk/news/empire-state-building-window-refurb-the-empire-strikes-back/5005104.article (accessed 23/01/2026).

Kings Cross (n.d.) *Sustainable Buildings*. https://www.kingscross.co.uk/sustainable-buildings (accessed 22/01/2026).

KoozArch (2022) *Kristian August gate 13. Reimagining the Built Environment*. https://www.koozarch.com/interviews/kristian-august-gate-13-reimagining-the-built-environment (accessed 12/07/2025).

Kristian Augusts gate 13 – Mad arkitekter (n.d.) https://mad.no/prosjekter/kristian-august-gate-13 (accessed 12/07/2025).

Lafortune G *et al.* (2024) *European Elections, Europe's Future and the SDGs: Europe Sustainable Development Report 2023/24*. https://s3.amazonaws.com/sustainabledevelopment.report/2024/europe-sustainable-development-report-2023-24.pdf (accessed 24/01/2026).

Liu X *et al.* (2023) Photovoltaics and energy storage integrated flexible direct current distribution systems of buildings: definition, technology review, and application. *CSEE Journal of Power and Energy Systems* [preprint], 10.17775/cseejpes.2022.04850.

Lui A (2024) The role of companies in shaping Africa's sustainability and climate resilience. *Clyde & Co*, 4 October. https://www.clydeco.com/en/insights/2024/10/shaping-africa-sustainability-climate-resilience#:~:text=Africa's%20sustainability%20and%20climate%20resilience%20challenges,practices%20to%20safeguard%20the%20environment (accessed 24/01/2026).

Living Future (n.d.a) *Living Building Challenge Basics – Living Future*. https://living-future.org/lbc/basics/ (accessed 05/05/2025).

Living Future (n.d.b) *Certified Case Studies Living Building Challenge*. https://living-future.org/case-studies/program/lbc/#:~:text=The%20Santa%20Monica%20City%20Services,and%20productive%20workplace%2C%20and%20positive (accessed 23/01/2026)

Maldives Floating City (2022) *Ministry of Tourism Permits Dutch Docklands Maldives to Run Floating City Development Based on the Integrated Tourism Model*. https://maldivesfloatingcity.com/wp-content/uploads/2022/06/PRESS-RELEASE-Ministry-of-Tourism-permits-Dutch-Docklands-Maldives-to-run-Floating-City-Development-based-on-the-Integrated-Tourism-Model.pdf (accessed 23/01/2026).

Martin R and LGIM Real Assets (2021) *The Investment Case for Net Zero Buildings*. https://prod-epi.lgim.com/landg-assets/lgim/real-assets/_files/the-investment-case-for-net-zero.pdf (accessed 24/01/2026).

Michaels I (New York City) (2022) *First 16-Ton Movable Flood Gate Installed at East Side Coastal Resiliency*. https://www.nyc.gov/site/ddc/about/press-releases/2022/pr-022422-ESCR-Gate.page (accessed 28/09/2025).

Mukuru Clean Stoves (n.d.) https://mukurustoves.org/ (accessed 24/01/2026).

Musmanni GD (2021) *GCA Moves to World's Largest Floating Office, a Model of Self-Sufficient Climate-Resilient Design*. https://gca.org/gca-moves-to-worlds-largest-floating-office-a-model-of-self-sufficient-climate-resilient-design/ (accessed 23/01/2026).

NABERS (2024) *NABERS Annual Report 2023–2024 major Achievements*. https://nabers.info/annual-report/2023-2024/major-achievements (accessed 23/01/2026).

NABERS (n.d.a) NABERS homepage. https://www.nabers.gov.au/ (accessed 23/01/2026).

NABERS (n.d.b) *Improving energy efficiency and reducing costs for your office building - example from Victoria: Knox Civic Centre, Victoria*. https://www.nabers.gov.au/sites/default/files/2023-02/The%20easy%20NABERS%20guide%20for%20Office%20-%20Victoria%20-%20case%20study.pdf (accessed 23/01/2026).

NABERS (n.d.c) *NABERS: Find a current rating*. https://www.nabers.gov.au/ratings/find-a-current-rating?premise=10&search=Australia±Square±Complex%2C±Sydn%2C±264-278±George±Street%2C±SYDNEY±NSW±2000&details=1 (accessed 11/07/2025).

NDRC People's Republic of China (Department of Resource Conservation and Environmental Protection) (2021) *Action Plan for Carbon Dioxide Peaking before 2030*. https://en.ndrc.gov.cn/policies/202110/t20211027_1301020.html (accessed 23/01/2026).

Niland J (2022) *Buro Happold and Dialog Unveil New Urban Thermal Comfort Study for Toronto*. https://archinect.com/news/article/150413745/buro-happold-and-dialog-unveil-new-urban-thermal-comfort-study-for-toronto (accessed 28/09/2025).

NYC (New York City) (2022) *East Side Coastal Resiliency Project Receives 'Envision Gold Award' from the Institute For Sustainable Infrastructure*. https://www.nyc.gov/site/ddc/about/press-releases/2022/pr-080122-ESCR-ISI.page (accessed 24/01/2026).

NYC (New York City) (2025) *East Side Coastal Resiliency 2025 Phasing*. https://www.nyc.gov/assets/escr/downloads/pdf/20250516_ESCR_2025-Phasing.pdf (accessed 07/06/2025).

NYSERA (New York State Energy Research and Development Authority) (n.d.) *Planning for Resource Efficient Decarbonization. A Phased Approach to Eliminating Greenhouse Gas Emissions from Tall Buildings in Cold Climates: A Holistic Approach, Combined with a Realistic Phasing Plan, Can Make Decarbonization Technically and Economically Feasible*. https://www.nyserda.ny.gov/All-Programs/Empire-Building-Challenge/Empire-Building-Challenge-Projects. https://www.nyserda.ny.gov/About/Publications/Featured-Case-Studies/Empire-State-Realty-Trust (accessed 23/01/2026).

Omnis (n.d.) *Kristian Augusts Gate 13 receives Norwegian State Award for Building Quality*. https://omnisusa.com/projects/kristian-augusts-gate-13/#:~:text=Kristian%20Augusts%20Gate%2013%20Receives,good%20and%20innovative%20building%20processes (accessed 08/07/2025).

Ontario Association of Architects (2023) *News & Insights – DIALOG and Buro Happold Release New Study as Component of Toronto's Climate Action Efforts*. https://oaa.on.ca/whats-on/news-and-insights/news-and-insights-detail/DIALOG-and-Buro-Happold-release-new-study-as-component-of-Torontos-climate-action-efforts (accessed23/01/2026).

OPDC (Old Oak and Park Royal Development Corporation) (2022) *Passive Energy Performance, Daylight and Overheating in High-Density Development, Supplementary Planning Document*. Old Oak and Park Royal Development Corporation. https://www.london.gov.uk/sites/default/files/passive_energy_performance_daylight_and_overheating_in_high_density_development_spd_0.pdf (accessed 02/07/2025).

Policy Circle (2024) Spain floods: Climate change preparedness is a moral imperative. *Policy Circle*, 3 November. https://www.policycircle.org/environment/spain-floods-climate-change/ (accessed 20/08/2025).

PPIAF (Public-Private Infrastructure Advisory Facility) (2015) Railway Reform: Toolkit for Improving Rail Sector Performance Case Study: London King's Cross. The World Bank, pp. 453–456. https://www.ppiaf.org/sites/ppiaf.org/files/documents/toolkits/railways_toolkit/PDFs/RR%20Toolkit%20EN%20New%202017%2012%2027%20CASE8%20LONDON.pdf (accessed 23/01/2026).

Putro J and Zain Z (2021) Active and passive adaptation of floating houses (Rumah Lanting) to the tides of the Melawi river in West Kalimantan, Indonesia. *Geographica Pannonica* **25(2)**: 72–84, 10.5937/gp25-30422.

Ratliff L (2025) Wagner Park reopens with $1.7 billion Climate-Ready upgrade. *Time Out New York*, 1 August. https://www.timeout.com/newyork/news/this-beloved-nyc-waterfront-park-just-got-a-1-7-billion-makeover-073125 (accessed 28/09/2025).

RE100 (n.d.) Climate Group RE100. https://www.there100.org/re100-members (accessed 24/01/2026).

Rebuild by Design (n.d.) *Project Pages: The Big U*. https://rebuildbydesign.org/work/funded-projects/the-big-u/ (accessed 07/06/ 2025).

Retrofit Playbook for Large Buildings (2022) *Empire State Building*. https://retrofitplaybook.org/resource/empire-state-building/ (accessed 15/04/2025).

Rogers P (2018) Drought or no drought: Jerry Brown sets permanent water conservation rules for Californians. *The Mercury News*. https://www.mercurynews.com/2018/05/31/california-drought-jerry-brown-sets-permanent-water-conservation-rules-with-new-laws/ (accessed 23/01/2026).

Royal Academy of Engineering (2023) *2023 Major Project Award for Sustainability*. https://raeng.org.uk/news/major-project-award-for-sustainability-2023#:~:text=The%20award%20is%20made%20to,and%20business%20quarter%20for%20London%2C (accessed 22/01/2026).

Rubio I (2023) Oceanix Busan: The dream of building a floating city to tackle climate change. *EL PAÍS English*, 17 June. https://english.elpais.com/science-tech/2023-06-17/oceanix-busan-the-dream-of-building-a-floating-city-to-tackle-climate-change.html (accessed 23/01/2026).

Salat S and Ollivier G (2017) *Transforming the Urban Space through Transit-Oriented Development: The 3V Approach*. World Bank. https://hdl.handle.net/10986/26405 (accessed 24/01/2026).

Schneider Electric (2024) *Best Practices to Drive Successful Sustainability Initiatives in APAC*. https://perspectives.se.com/blog-stream/advancing-sustainability-best-practices-to-drive-successful-sustainability-initiatives-in-the-multicultural-asia-pacific-and-japan-region (accessed 24/01/2026).

Schröder C (2022) *Truly Outstanding: The Edge Suedkreuz Office Complex in Berlin – DGNB Blog English*. https://blog.dgnb.de/en/edge-suedkreuz-berlin/ (accessed 23/01/2026).

Silva JMC *et al.* (2024) The sustainability of development pathways and climate change vulnerability in the Americas. *Ecological Economics* **220**: 108164, 10.1016/j.ecolecon.2024.108164.

Singapore Green Building Council (2020) *Singapore Green Building Masterplan – Singapore Green Building Council*. https://www.sgbc.sg/sgbmp/ (accessed 08/07/2025).

Smith R (2022) Creating a clearer picture of office energy efficiency. *RICS*, 17 January. https://ww3.rics.org/uk/en/journals/property-journal/creating-a-clearer-picture-of-office-energy-efficiency.html#:~:text=The%20new%20Energy%20Efficiency%20for%20Offices%20scheme%20from%20NABERS&text=Real%20Investment%20Analytics%2C%20Australian%20Green%20Office%20Property%20indicators%202020%20Q2%20 (accessed 24/01/2026).

Spooky (2020) *Waterbuurt – Amsterdam's Floating Neighborhood.* https://www.odditycentral.com/architecture/waterbuurt-amsterdams-floating-neighborhood.html (accessed 28/09/2025).

Stamp E (2020) This is one of America's greenest city hall buildings. *Architectural Digest*, 27 August. https://www.architecturaldigest.com/story/one-americas-greenest-city-hall-buildings (accessed 23/01/2026).

Suzuki H *et al.* (2015) *Financing Transit-Oriented Development with Land Values: Adapting Land Value Capture in Developing Countries.* World Bank Group, Washington DC. http://documents.worldbank.org/curated/en/468551468165570019 (accessed 24/01/2026).

Swire Properties (2022) *Net-Zero Roadmap.* https://sd.swireproperties.com/2022/en/performance-environment/climate-change/net-zero-roadmap (accessed 24/01/2026).

Swire Properties and AECOM (2023) *Interpretation of Commercial Energy-Saving Renovation Technology of Taikoo Li Sanlitun, Beijing.* http://m.mep-link.com.cn/h-nd-342.html (accessed 24/01/2026).

The Earthshot Prize (n.d.a) *The Earthshot Prize – Urgent Optimism to Repair our Planet.* https://earthshotprize.org/ (accessed 24/01/2026).

The Earthshot Prize (n.d.b) *Mukuru Clean Stoves – Earthshot Prize Winner 2022.* https://earthshotprize.org/winners-finalists/mukuru-clean-stoves/ (accessed 24/01/2026).

The London Plan (2021) *The London Plan: The Spatial Development Strategy for Greater London.* Greater London Authority. https://www.london.gov.uk/sites/default/files/the_london_plan_2021.pdf (accessed 07/07/2025).

Treepedia (2018) *Treepedia: Exploring the Green Canopy in Cities Around the World.* http://senseable.mit.edu/treepedia (accessed 24/01/2026).

UAE (2024) *Dubai Clean Energy Strategy.* https://u.ae/en/about-the-uae/strategies-initiatives-and-awards/strategies-plans-and-visions/environment-and-energy/dubai-clean-energy-strategy (accessed 24/01/2026).

UNEP (UN Environment Programme) (2022) *2022 Global Status Report for Buildings and Construction: Towards a Zero-Emissions, Efficient and Resilient Buildings and Construction Sector.* DTI/2482/PA. https://www.unep.org/resources/publication/2022-global-status-report-buildings-and-construction (accessed 24/01/2026).

UNEP (2024) *Traditional Building Practices Offer Sustainable Solutions as African Cities Grow.* https://www.unep.org/news-and-stories/story/traditional-building-practices-offer-sustainable-solutions-african-cities (accessed 24/01/2026).

UN-Habitat (2022) *UN-Habitat and Partners Unveil OCEANIX Busan, the World's First Prototype Floating City | UN-Habitat.* https://unhabitat.org/news/27-apr-2022/un-habitat-and-partners-unveil-oceanix-busan-the-worlds-first-prototype-floating (accessed 23/01/2026).

UN-Habitat (2020) *Fighting COVID-19 in Africa's Informal Settlements.* https://unhabitat.org/news/08-jun-2020/fighting-covid-19-in-africas-informal-settlements (accessed 24/01/2026).

United Nations (2022) *COP27: Delivering for People and the Planet | United Nations.* https://www.un.org/en/climatechange/cop27 (accessed 24/01/2026).

URA (Urban Redevelopment Authority) (Singapore) (2017) *Updates to the Landscaping for Urban Spaces and High-Rises (LUSH) Programme: 3.0.* https://www.ura.gov.sg/Corporate/Guidelines/Circulars/dc17-06 (accessed 23/01/2026).

Versano C (2024) Satellite image shows construction of world's first floating city, *Newsweek*, 14 June. https://www.newsweek.com/maldives-floating-city-satellite-photos-construction-1897098 (accessed 24/01/2026).

Washington Post (2021) Climate solutions. https://www.washingtonpost.com/graphics/2020/climate-solutions/empire-state-building-emissions/ (accessed 23/01/2026).

WEDG (n.d.) Verified Projects: East Side Coastal Resiliency Project. https://wedg.waterfrontalliance.org/verified-projects/ (accessed 23/01/2026).

Wong MS *et al.* (2018) *Oasia Hotel Downtown, Singapore: a tall prototype for the Tropics, Council on Tall Buildings and Urban Habitat (CTBUH).* https://global.ctbuh.org/resources/papers/download/3790-oasia-hotel-downtown-singapore-a-tall-prototype-for-the-tropics.pdf (accessed 28/09/2025).

World Economic Forum (2022) *How the Winners of the Latest Earthshot Prize Will Scale Environmental Solutions.* https://www.weforum.org/stories/2022/12/earthshot-prize-winners-climate-change/ (accessed 24/01/2026).

Wright E (2025) *What Is Hedonistic Sustainability? BIG's Take on Fun-Injected Sustainable Architecture Arrives in New York.* https://www.wallpaper.com/architecture/what-is-hedonistic-sustainability (accessed 24/01/2026).

Yllah @Freewill.Inc (2024) Japan's role in the Rise of Green Architecture and Sustainable Construction Practices. *Medium*, 29 February. https://medium.com/sustainable-eco-society/japans-role-in-the-rise-of-green-architecture-and-sustainable-construction-practices-f9f2d58b081f#:~:text=The%20Taizhou%20Contemporary%20Art%20Museum,harmony%20with%20the%20surrounding%20landscape (accessed 24/01/2026).

Navigating an Evolving Sustainable Built Environment: Delivering Net Zero Carbon and Climate Resilience

Mei Ren and Rachel Waggett
ISBN 978-1-83662-853-8
https://doi.org/10.1108/978-1-83662-850-720261012

Chapter 8
What will the future hold for the built environment?

8.1. Two major driving forces

Earlier chapters have discussed some of the available design tools and frameworks (Chapter 5) and near-term opportunities for sustainability integration into the built environment (Chapter 6), many of which are already being trialled or included in leading projects. Chapter 7 provides some examples of places where innovative measures have been made reality, showcasing the possibilities and demonstrating their practicality and impact on society and wellbeing.

Rapid advances in materials engineering are already being demonstrated and will gather pace in the short term. These advances will provide new, lower-impact materials that have reduced environmental impact, or are stronger, lighter, easier to manufacture and reuse. Similarly, decarbonisation (particularly of exceptionally difficult areas such as steel and concrete manufacture), clean water and food production will change the way urban environments are planned and managed. Energy generation is undergoing a seismic change, with the shift away from fossil fuel to renewable energy already underway and accelerating. Changes to transportation systems, especially in cities and urbanised areas, are already starting to revolutionise the way people move around cities and urban places. This is helped to a great extent by mobile and satellite-enabled technology applications that enable on-demand personal transportation as well as access to streamlined incentives for public transportation. There is no doubt this trend will accelerate as connectivity systems become even more widespread and affordable.

But these are all relatively near opportunities – what of the further future? In designing and constructing the built environment, a legacy is being created that will likely be in existence for decades, so how can this be planned for, when future needs are uncertain? Nobody has a crystal ball to see the future clearly, as the BBC's Tomorrow's World in 1995 when predicting 30 years ahead, failed to predict the wide spread of big tech and social media (Fraser, 2025). On reflection, Monty Don, who was one of the presenters on that Tomorrow's World programme 30 years ago, commented: 'what we have really learnt since then is that mankind has a habit of making things worse, particularly environmentally, and we have to work with nature rather than try modify and control it' (Fraser, 2025).

Looking towards the future 30 years, it is clear that there are major forces underway that will change and disrupt the way people currently live, work, learn and enjoy daily life. The built environment is undergoing major shifts with climate change and artificial intelligence (AI) as two of the major forces. By exploring some of the trends, this chapter discusses a glimpse into possible futures. For certain, those working in the built environment must also change to adapt to the climate change and AI advancements that will surely happen.

Climate change

Earlier chapters have described the realisation that the global climate is changing as a result of human activities, primarily the rapid release of carbon dioxide (and other greenhouse gases (GHGs), such as methane) primarily due to burning of fossil carbon as fuel and deforestation. The realisation of the need to act and the setting of global targets to try and arrest and reverse the impacts have only been determined within the last decade, showing just how long it has taken to reach a consensus on the existence of the problem, let alone the solutions to resolve it.

The Paris Agreement (United Nations, 2015) discussed in Chapters 1 and 3, is a commitment that sought to agree a goal to limit the increased global temperature and avoid dangerous impacts for global stability. The agreement stated that it would hold the increase in global average temperature to 'well below 2°C above pre-industrial levels' and pursue efforts 'to limit the temperature increase to 1.5°C above pre-industrial levels'.

Since the Paris Agreement, observed global average temperatures have continued to rise, and at the time that the latest Indicators of Global Climate Change report was published, are 1.24°C higher than pre-industrial levels (Forster *et al.*, 2025). Over land areas, where people live, the temperature increase is even greater, at 1.79°C. 2024 was the first calendar year with a global mean temperature of more than 1.5°C above the pre-industrial level (World Meteorological Organization, 2025). While this was only a single year (and therefore does not, in itself, breach the Paris Agreement target) it comes on top of a decade of record-breaking temperatures both on land and in the oceans. Surface temperature is only a small part of the story, with much of the accumulated heat being stored in the oceans, leading to marine heatwaves that damage undersea ecosystems and exacerbate sea level rise. The Indicators of Global Climate Change report (Forster *et al.*, 2025) also found that sea level has increased by around 26 mm over the 6 years to 2024, more than double the long-term rate.

Sea level rise has numerous impacts; quite apart from the obvious implications for coastal and island communities, it drives more damaging coastal erosion and storm surges. Sea level rise is not globally uniform and varies regionally. The low-lying coastal zone is home to around 680 million people, projected to reach more than one billion by 2050. Extreme sea level events that are historically rare (likely to happen only once in 100 years) will occur at least once per year at many locations by 2050. Although well designed coastal protection is very effective in reducing expected damages and cost efficient for urban and densely populated regions, it is generally unaffordable for rural and less wealthy areas (IPCC, 2022).

The changing climate frames some of the biggest challenges humanity faces: decarbonisation, clean water, food. Rapidly evolving and increasingly powerful modelling systems are now allowing better replication of earth systems, meaning that the likely impact and ramifications for a warming planet can be better estimated. It is also becoming possible to estimate more localised effects, rather than global averages – for example, NASA using their satellite data to model future wave action and sea level rise impacts utilising highly accurate topography. This is important, because the spatial aspects of how to cope with a different climate in the future will be vital to targeting and building in resilience to the built environment.

AI

We are living in an era of greater technological change than humanity has ever witnessed. The pace of change itself is accelerating (Kurzweil, 2024), as each advancement in technology makes it easier to design the next stage of their own evolution, as being seen by those generative AI systems that can write code, systems that develop systems. The underpinning technologies are advancing at an explosive exponential rate. McKinsey found in 2023 that generative AI alone stands to add up to £4.4 trillion to the global economy annually (McKinsey and Company, 2023).

This will have a profound significance for the future technology and on humanity. In the years ahead, human lives will be transformed by technologies that have not yet been created or even imagined at this point of time. This exponential growth has no finishing line, we anticipate new invented systems, not expected by most computer scientists, to come to market all the time.

Professor Richard Susskind (Susskind, 2025) articulated the following five hypotheses offering diverse perspectives on its potential trajectory regarding AI's future development:

- Hype hypothesis – AI has inherent limitations and will not have substantial improvements to our current AI.
- GenAI + hypothesis – predicts that current generative AI systems will be significantly improved and become widely integrated into everyday life. AI will be fully integrated into the operations of most organisations securing notable efficiency and productivity gains.
- AGI hypothesis – Artificial General intelligence (AGI) would possess intelligence and self-awareness equal to those of humans. AGI could achieve near-human-level intelligence by 2035.
- Superintelligent hypothesis – AI surpasses human intellect in all domains.
- Singularity hypothesis – envisions a future where humans merge with AI, immensely augmenting both intelligence and consciousness, as predicted by Ray Kurzweil (Kurzweil, 2024).

While technological advancement will continue, AI evolution in the future will depend on social, regulatory and ethical considerations, leaving AI's future uncertain, especially for a 'superintelligent' and 'singularity' future.

With the exponential technology advancement with Large Language Models (LLM) on which programs such as Gemini or ChatGPT are based, these applications are now already being used widely for summarising, drafting and comparing documents, but also for visualising, analysing documents to producing arguments, conducting research and answering questions. With reasonable confidence, a move towards GenAI + by 2030 can be predicted.

Susskind (2025) believes that AI impact on our daily life has been overstated in the short term; however, the long-term impact on humanity is hugely understated. This belief was evidently based on the scale of investment, market demand, significant amount of interest and energy from start-up community. Susskind (2025) considers that AGI systems are technically feasible and we must not merely dwell on the existing known generative AI technology. The impact of those currently inconceivable yet invented technologies would have far greater impact on us in the 2030s. How do we live with AI as double-edged sword – containing AI and protecting humanity from AI? Each of us, both personally and professionally, needs to think systematically about the risks looking beyond the current available technologies and develop most effective ways of imposing our own preferences. Susskind strongly recommends all government bodies and businesses to develop a strategy for an AGI future, to avoid future threats of uncertainty in AI development. It is likely that over time, all organisations will reach similar conclusions, therefore early consideration will provide greater safeguards as well as greater opportunities to reap any benefits.

8.2. Future living

Burca's *Construction Voices* article (De Burca, 2024), paints a picture of future living as being vertical, smart and with new urban mobility. Predictions for future trends include increased focus on sustainability, adaptable living spaces and the integration of urban agriculture in high-rise structures.

Urbanisation is transforming cities worldwide as populations grow and land becomes scarce. It is predicted that more than four out of six people will be living in urban areas by 2050, in sharp contrast to one in six in 1900 (Miles, 2023). As vertical cities become more feasible, they offer potential solutions to overpopulation and environmental challenges by providing high-density urban housing development, prioritising efficient land use while addressing sustainability, affordability and community engagement challenges.

High-density living refers to housing developments that maximise the number of units within a given area, often featuring tall buildings with shared amenities and integrated public spaces. Key characteristics include efficient layouts, creative use of vertical space, and a focus on fostering social connections. In an era of diminished land availability and scarcity of land as a resource, increased urbanisation and vertical growth of communities will be likely. More developments similar to Singapore's Oasis type of vertical development are likely (see Chapter 7, Section 7.6 for Asia–based case studies).

How else to alleviate and tackle the challenges in future high density living? Future living will be underpinned by technology that should embrace sustainability. Embedding and harnessing the use of key technologies would shape urban design in smart cities to include Internet of Things (IoT) devices, AI-driven systems and data analytics platforms. These technologies enable features such as automated lighting, climate control, shared electric vehicle charging stations, real-time energy monitoring and community Wi-Fi networks to be available to all.

There is also the opportunity to look below the land surface, exploiting the cooler and resilient spaces underground, that are better protected from extreme weather. Subsurface opportunities could also include better, more efficient storage of resources such as energy and water, hydrogen and brine using geothermal technologies.

8.2.1 Future cities

Sir James Bevan (Bevan, 2023) outlined an optimistic vision for the city of the future, at his keynote speech at Imperial College, to be a clean, green, climate-positive dream.

> Our Future City would not just stop being part of the climate problem. It would also be a major part of the solution. Its green areas – parks, woodland, grasslands, flowerbeds, football pitches – would all be acting as carbon sinks, taking damaging carbon out of the atmosphere and so reducing the extent of climate change. In its design and its infrastructure our city would be perfectly adapted to living safely and well in a climate-changed world. It would have flood defences that protected people from the worst that the violent weather caused by a changing climate could fling at it. It would have power and transport systems designed to cope just as well with periods of high temperature and drought as with record-breaking rainfall. Better still, its trees and plants would not just take carbon out of the atmosphere but cool the air and provide habitats for wildlife. Our city would not just be liveable: it would be beautiful.

There are already projects emerging that provide a glimpse of how this vision can become a reality, for example, BIG U – Hedonistic Sustainability (see Chapter 7, Section 7.4 for US-based case studies). More of this type of future cities practices will emerge that offer multiple benefits to people, with sustainability and climate conscious solutions at their core.

In the rapidly changing climates, sea levels will be approximately 1 m higher in 2100 (in the UK). Coastlines will be subject to higher tidal ranges and stronger wave action. Water scarce environments will be commonplace, storage of water and filtration of waste water for crop irrigation and non-potable water use will be necessary. Storage of surface water in heavy deluges will be required to be built into all urban areas, to retain water for use in times of drought and to prevent flooding.

Futuristic living predicted by Porcelanosa (2023) features ten trends, not surprisingly including sustainable features and smart homes using the connected IoT. They predict that car-free streets with sky bridges will be future norm for transport in future cities. Subterranean shuttles will also revolutionise product delivery (more will follow British tech scale-up Magway, drawing investors like grocery delivery organisation Ocado). There will be more mixed-use neighbourhoods with mini-centres a short walk from residential buildings. Buildings will provide multi-purpose spaces, with inter-generational living and adaptable spaces in which people to work, learn, relax and connect.

Responding to the need for ecological networks and green space, nature-connected forest-cities will be developed, with green walls, rooftop allotments, pocket parks and tree lined streets. Big Data and the IoT will make cities more accessible, inclusive, and safer for all (Miles, 2023). Every individual house and building will become a power plant with renewable energy sources connected, creating an energy revolution similar to the communication network revolution. On landscapes and gardens of tomorrow, both farming and gardens will change radically with greater appreciation and understanding of the interconnected ecosystems and the importance of insects and indigenous species of plants and animals, adopting a strategy to work with nature rather than try to modify and control it (Fraser, 2025).

As part of the Future Cities project led by the Institution of Civil Engineers (ICE, n.d.), funded by the research councils, Professor Chris Rogers and Dr. Ellie Cosgrave discussed the future of cities and how to tackle some of the world's biggest engineering challenges. They predict that while the twentieth century has seen engineers get to grips with the technical needs of the environment, the challenge of the twenty-first century will be on the more complex areas of social justice and inclusion.

It was found that Songdo in South Korea, generally credited as the world's first smart city where buildings have automatic climate control and automated access, where water, waste and electricity systems track and respond to the movement of residents, provided a solution to endeavour to meet the needs of growing urban populations. However, a 'computerised' city such as this has attracted criticism as it is perceived to make a city where everyone is under surveillance.

The challenge for built environment professionals will be to remember that cities are much more than nuts, bolts and high-tech innovation. Urban infrastructure should serve the needs of citizens and communities – helping them to prosper. For an existing city, a future alternative may be using new technologies to integrate with existing systems, as well as putting the needs of citizens first when upgrading their environment.

Examples of this approach are already being embedded in existing cities, including using smart sensors to measure and find water leaks in real time, control traffic flow (smart motorways) or dim street lights when no-one is around. There will be many more opportunities to use such smart technologies to operate the city more efficiently and to the benefit of citizens.

Retrofitting of properties to meet new requirements will be particularly important. Materials to help retrofit easily – that is, technologies that can easily be incorporated into older properties without excessive cost and effort will be of vital importance in existing urban areas, helping citizens to cope with future climates by including cooling technologies, ventilation with heat recovery or climate resilient materials, for example.

Nature-based solutions will be prioritised to benefit humans, such as minimising urban heat island and provide shading, but at the same time to preserve the ability of surrounding natural biodiversity to thrive while continuing to provide food and materials needed by society.

8.2.2 Future home

FindMyHome (2024), presented four influencing factors in future home living. This includes sustainability and environmental awareness, the influence of technological developments (smart homes), flexible forms of living (combination of living and working) and demographic changes (ageing population and changing family structures).

Sustainability and green living will continue to play a big part in future homes, with more ecologically conscious features in interior design, such as more recycled materials as décor, resource and energy efficiency, the greening of roofs and façades in urban areas. Living spaces should be increasingly shifted towards sustainability and wellbeing. AI is increasingly moving into the home and its influence will continue to grow, with an increasing role for AI in personalised living and a significant role in home security. Technology will become seamlessly integrated with the surfaces that define our living spaces. It is predicted that there will be a fully automated interior design in 50 years, with individual pieces of furniture equipped with sensors or actuators that enable them to adapt to the resident's needs. An increase in wearable augmented reality (AR) is likely, such as cameras and sensors, working with the physical environment; AR providing hands-free access to real-time information, instructions and interactive experiences.

AI will certainly also play a major role in the healthcare sector in the future. Built environment designers will need to truly consider design for robotic access, which will become as mainstream as current designs for ease of human access, to help those with disabilities and protecting workforce welfare. How to approach designing for robotic access, such as for robotic care assistants and workers, is a question that will need to be resolved, to understand what requirements and needs they will need to operate successfully.

8.3. Future energy

Electricity demand is expected to rise enormously, driven by factors such as a growing use of AI technologies and electrification of buildings, transportation and industrial processes. Data centres and the rise of energy demand need to have power to meet the increased loads.

8.3.1 How could we do more with less in the future?

Self sufficiency of buildings and urban cities means generating power on site and storage of power using various new technologies. Energy demand management and control systems balance power across cities and urban areas by optimising storage. There will be many more distributed data centres and heat harvesting and storage.

Waste heat is everywhere. Those industries such as cement and steel utilise significant amounts of the energy generated globally, 60% of which is simply lost as waste heat. How could this vast amount of industrial waste heat be captured and reused to reduce emissions from these industries? Advanced thermovoltaic systems (ATS), an Earthshot 2024 award winning technology, offers a simple, safe and scalable solution to transform industrial waste heat into electricity for use in hard to abate industries, with the potential to save gigatonnes of carbon dioxide (The Earthshot Prize, 2025). ATS convert waste heat into electricity by passing it over small plates, manufactured using advanced materials and a smartphone-style assembly process. It can also harvest lower temperature levels of heat and generate electricity from any heat source, including geothermal energy. Their patented solution is also 98% circular, with components that can be reused, and cartridges that can be replaced as needed. The technology was titled as 'thermovoltaic', to imitate the familiar 'photovoltaic' when they discovered that their panels can capture ambient heat to generate power, without sunlight. The market potential for ATS technology is enormous, with the ability to transform energy use in heavy industries worldwide.

8.3.2 Learning from nature – does nature provide the best solutions for a resource-efficient future?

An exponential growth in renewable energy sources, particularly solar and wind, is already underway globally. This enormous energy transition, alongside advancements in energy storage and grid management technologies will happen in the future. Improvements in efficiency and technology are being made constantly and, increasingly, researchers are looking to nature for inspiration. In a revolutionary breakthrough for renewable energy, the world's first photovoltaic leaf (PV-leaf) has been developed by researchers at Imperial College London (Ravate, 2023). This leaf-inspired technology mimics natural leaf transpiration to maximise the efficiency of solar panels whose performance can be impacted by overheating. The new design produces three types of output: electricity, freshwater and heat. The PV-leaf uses a biomimetic transpiration layer that mimics natural leaf structures and combines vascular fibre bundles and sponge-like cells, made from inexpensive materials like bamboo fibres as well as super-absorbent polymer hydrogels. This PV-leaf can also function by using seawater rather than freshwater, which increases its usage in coastal and arid areas. The technology, when fully developed, has the potential to be a low-cost and easily accessible solution for widespread implementation (Sanusha, 2025).

With the high demand for renewable energy, innovative solutions such as this can help tackle the challenges in providing for increased energy use, while making a rapid shift away from fossil fuel consumption. The ability of this technology to simultaneously address water sustainability supports the need to deliver 'more with less' by lessening our dependence on fossil fuels while also tackling the global scarcity of freshwater.

Other experimental techniques for energy generation offer promising future opportunities. There have been various trials using algae to produce energy thanks to algae's ability to convert sunlight into stored energy and their efficient carbon dioxide consumption makes them a promising renewable energy source. Although algae have the potential to become a significant contributor to the renewable energy mix, the technology is still at experimental stage. Further research, development and cost reductions are needed to overcome existing challenges, such as improving the efficiency of electron harvesting and scaling up the technology for practical applications to realise their full potential (Lejtenyi, 2024).

Nuclear energy using fusion technology could have a very significant impact on future energy generation. While nuclear fission involves splitting a heavy nucleus into smaller ones and is a relatively established technology with huge challenges, primarily centred around safety concerns, radioactive waste disposal and high initial costs, nuclear fusion is a promising energy source, producing significantly more energy with minimal radioactive waste. If successfully scaled up to commercial levels, it could produce endless amounts of clean energy without carbon emissions. The technology is currently still in the experimental stage as recreating the conditions of the sun on Earth is a monumental scientific and engineering challenge. It requires extreme temperatures and pressures to initiate and maintain for a sustained reaction (Chatzis and Barbarino, 2021). The UK government is now hoping to build the world's first fusion power plant in Nottinghamshire with operations beginning in the 2040s (BBC Science and Environment, 2024).

8.3.3 Will space play a part in this pursuit?

As noted by the National Space Society, space-based solar power has the potential to outperform all other energy sources combined. The process of space-based solar power technology includes using solar panels to capture solar energy in space, utilising reflectors or inflatable mirrors to focus solar radiation onto these panels and then beaming the energy to Earth through a microwave or laser. On Earth, this energy is received by a microwave antenna (a rectenna). Space-based solar power can deliver large amounts of energy with a minimal negative environmental impact. Although it is costly, it serves as an excellent source of clean energy with the potential to generate more power than the world currently consumes or is expected to need in the future. Various countries and organisations are actively researching and developing the technology with demonstration projects underway to test the feasibility of key components, such as wireless power transmission (development@everythingcreative.com, 2024).

Depending on how quickly global carbon emissions can be reduced, there are possible scenarios that could lead global temperatures to increase by several degrees by the end of the century (see Chapter 3 for more detail). There are uncertainties in how quickly, or by how much, temperatures could increase, depending on the sensitivity of the global climate systems that could cause climate 'tipping points' to be crossed over the next century. Effort in 'climate interventions' that could actively reduce temperatures globally or regionally over shorter timescales is gathering more momentum.

Space located solar radiation management or geoengineering, offers a potential method of climate modification mitigating global warming (David, 2023). It involves deploying sunshade devices in space to reflect a portion of the sun's incoming radiation back into space, thus reducing the amount of solar energy reaching Earth. Instead of directly addressing GHG emissions, space-based shading or other intervention technologies focus on reducing the amount of solar radiation reaching the planet's surface. This method offers global temperature control, fast implementation and a degree of reversibility. However, there are still huge technological hurdles and high costs involved to deploy a large-scale sunshade.

Several research groups from different countries involving collaboration between scientists, engineers and policymakers are working on developing prototypes, carrying out theoretical modelling and conducting experiments to test the feasibility of space-based shading technologies (ARIA, n.d.), as well as other geoengineering experiments, including into marine-cloud brightening (Gruenbaum, 2025). Researchers are clear that such technologies are not a replacement for carbon-reducing action but propose that such interventions might well be necessary in the future.

8.4. Food and materials

Global supply chains are becoming strained and disrupted, with increasing nationalisation of resources becoming popular (on-shoring) and emphasis on reduced reliance on other nations.

This implies more use of materials found on site or locally produced being valued. It also has land use implications, as there will need to be sufficient land area for materials to be produced (i.e. timber, clay, bamboo, hemp etc.) and food to be grown.

While food is not the main focus of this book, food supplies to sustain the global population is intertwined with the built environment and will greatly influence future sustainable designs. Feeding a growing global population requires a multi-faceted approach that includes increasing food production, reducing waste and shifting towards more sustainable agricultural practices. Specifically, the United Nations projects a 50% increase in food production will be needed by 2050, and some estimates suggest even a 70% increase (Ranganathan, 2018). This will necessitate addressing issues like land use, water conservation and climate change impacts.

Co-location of cheap, plentiful energy generation and industrial product manufacture such as lab-grown proteins, fermentation products as well as manufactured materials and goods will grow with more new industrial centres. Food growing factored into all spaces will provide local resilience and fresh produce in a globally restricted marketplace. Vertical farming, also known as controlled environment agriculture, is a promising approach to address future food security challenges by growing crops indoors, often in stacked layers, using innovative hydroponic farming technologies with artificial lighting and nutrient delivery systems. It offers potential benefits like increased production efficiency, localised food systems and reduced reliance on traditional farming methods and resources. The biggest indoor vertical farm in Dubai, can produce 2–3 tonnes daily of leafy vegetables, for example (Gavine, 2022).

Lab grown protein, while not yet a certainty, is attracting significant attention, with some projections suggesting it could be a more sustainable alternative to traditional meat production. Solutions to deal with challenges and uncertainties surrounding its widespread adoption, including cost and safety concerns, are yet to be fully explored.

8.5. Future humanity

On a global scale, the world seems to be constantly on the brink of catastrophe, lurching from one crisis to the next. Even putting aside human-made financial crises and war conflicts, would the frequency and damages of natural disasters happening across the world cause climate anxiety? Is it valid to worry about the future of humanity at an existential level in years to come? One third of Pakistan was left underwater by floods in 2022 (BBC News, 2022). Wildfires of unprecedented scale rampage across America (Earth.Org, 2025) and Australia. Years of glacier retreat in Kazakhstan at a rate of about 25 m per year, is resulting in significant risk of a water crisis for millions (Medeu *et al.*, 2022). Lake Chad, once one of Africa's largest lakes offering a lifeline to nearly 40 million people, has decreased by 90% over the past 60 years as a result of over exploitation and climate change (UNEP, 2018).

How can built environment professionals respond to these increasing threats? It is hard to escape the sense of looming apocalypse compounded by the climate crisis and the potential risks posed by AI. Simply avoiding consideration of them feels wrong. While the understandable fear must not be dismissed – nor should the damages inflicted on the planet be ignored – still, hope and optimism is a reasonable response, because the tools to change are within grasp. By paying attention, improving understanding and appreciating the need for change, the sector and all participants can be better prepared to respond and rebuild.

8.5.1 AI challenges

The leaders of McKinsey, Bain and BCG (Boston Consulting Group) talked about their common challenges (Vlamis *et al.*, 2025) being AI and climate change. There are huge business opportunities for them to help their clients grapple with significant amount of impending change, to seize the opportunity, fundamentally redefine operating models, value chains and decision making. However, the impact on professions is huge. According to Deloitte's 2025 survey (Deloitte Global, 2025), employees seek to use cutting edge technology, with 57% already using generative AI daily for data analysis, creative work and project management. Using modern digital tools is becoming essential to the majority of jobs. McKinsey has navigated the AI revolution and created new adventures to exploit the use of AI, but this evidently comes with positive and negative impacts. McKinsey's AI arm, QuantumBlack, has supported One Ocean Foundation in their research into sea protection projects such as reseeding of ocean-floor meadows, by applying analytics and AI to depict exactly how businesses are working to protect the seas (McKinsey & Company, 2024), with huge positive potential. In the meantime, however, McKinsey has reduced its staff by 10% in 2025, in part contributed by AI driven new levels of efficiency (Young, 2025).

These are already uncomfortable questions about AI-empowered people and businesses, and the ability of companies and communities to remain competitive by building the systems that replace old ways of working.

Susskind (2025) recommended that the short-term focus on AI is on automation, improving efficiency and productivity, whereas the longer-term trends are innovation and elimination, replacing old workflows with AI enabled technologies, which would in turn create new job opportunities. This needs organisations to make radical structural changes through AI, in whatever sectors whether in the public sector or in business. Some organisations are considering building their own AI systems that will replace their old ways of working. The transformation goes beyond merely improving efficiency, their disruptive nature to existing business models must be vision based to succeed in their long-term future.

The power of AI to help humanity cope with (and prepare for) global climate change could be a huge advantage. Climate systems are very complex and weather patterns are increasingly unpredictable. AI is already starting to be used to improve emergency planning and crisis responses to flooding and wildfires, for example. There has also been an increased focus on the use of AI combined with satellite weather forecasting and prediction to a smaller geographic 'cell' level to allow early warning systems to be precisely targeted to people's geographic location. Early warning systems of this type could in future be linked to 'smart' protective features on buildings such as protective shutters, flood gates and walls, opening of water storage diversion facilities, retractable covered walkways to protect pedestrians and many other opportunities.

For built environment professionals, what is the future for us? Should the question be *how in the future will we solve the problems facing the built environment and our society?*

8.6. The role could you play?

One thing is certain at this time of great change. We all have a part to play to tackle the twin challenges, be prepared for the acceleration of AI technology and the need to embrace and accelerate sustainability opportunities, as we cannot afford to stand still. Frankly it is about survival – facing these two major forces and adapting to the disruption may be the only way for us.

There is no doubt that the climate is changing. AI has moved to a place of no return in areas of automation. We need to better prepare for future generations who will inhabit a world that has unknown implications for humanity. We cannot know the future, and we also cannot always control how the world will be changed, even though we are getting better at predicting the future climate patterns. We do know – and the journey that we have explored in this book proves – that human ingenuity is incredibly resilient and adaptive to change.

After all, sustainability is about our moral obligation to account for those who come after us, to avoid bequeathing a more catastrophic world to our children and grandchildren. People working as built environment professionals are constantly working to create structures and systems that will last many generations into the future; creating a legacy is not something unfamiliar to us.

We are increasingly aware that humanity depends on the network of all the organisms on the planet operating together to create us a place where we can survive and thrive. To sustain our communities and provide ourselves with a future, we need to think beyond bricks and mortar, beyond our own shelter and the other things that the built environment delivers. As described by the Doughnut Economics framework (Fanning and Raworth, 2025; see Chapter 4 for more detail), this is about creating the 'safe space' in which humanity can operate between the ecological ceiling and the social foundation. It is vitally important that we think beyond the immediate needs of building occupiers, or even beyond the needs of human communities and start to think instead about how we can preserve life for all of the organisms on the planet, to maintain the complex, interconnected ecosystems on which we rely.

This chapter is a reflection on the future. We have seen how at the end of the last millennium, built environment professionals listened to the scientific community and perceived the need for sustainable development as an opportunity and an exciting challenge. Despite obstacles that will continue to be presented to us, as built environment professionals we have a reason and impetus to remain optimistic and ambitious. Not all the ventures and opportunities that we try will be successful, but as a sector we must remain undaunted and keep moving forward at pace, without striving for absolute perfection before we choose to act. The tools to deliver a sustainable built environment are now in our hands.

REFERENCES

ARIA (Advanced Research + Invention Agency) (n.d.) *Exploring Climate Cooling*. https://www.aria.org.uk/opportunity-spaces/future-proofing-our-climate-and-weather/exploring-climate-cooling (accessed 25/08/2025).

BBC News (2022) *Pakistan Floods: One Third of Country Is Under Water – Minister*. https://www.bbc.co.uk/news/world-europe-62712301 (accessed 25/08/2025).

BBC Science & Environment (2024) *Nuclear Fusion: New Record Brings Dream of Clean Energy Closer*. https://www.bbc.co.uk/news/science-environment-68233330#:~:text=Nuclear%20fusion%20has%20produced%20more,explained%2C%20this%20is%20not%20straightforward (accessed 25/08/2025).

Bevan J (2023) *The Future City: a clean, green, climate-positive dream*. GOV.UK, 24 January. https://www.gov.uk/government/speeches/the-future-city-a-clean-green-climate-positive-dream (accessed 30/08/2025).

Chatzis I and Barbarino M (2021) What is fusion and why is it so difficult to achieve? *IAEA Bulletin*: **62-2**. http://www.iaea.org/bulletin/what-is-fusion-and-why-is-it-so-difficult-to-achieve (accessed 25/08/2025).

David L (2023) These scientists want to put a massive 'sunshade' in orbit to help fight climate change. *Space*, 19 December. https://www.space.com/sunshade-earth-orbit-climate-change (accessed 04/09/2025).

De Burca J (2024) The future is vertical: urban design trends in high-density living. *Constructive Voices*, 19 November. https://constructive-voices.com/the-future-is-vertical-urban-design-trends-in-high-density-living/ (accessed 04/09/2025).

Deloitte Global (2025) *2025 Gen Z and Millennial Survey*. https://www.deloitte.com/content/dam/assets-shared/docs/campaigns/2025/2025-genz-millennial-survey.pdf (accessed 25/08/2025).

development@everythingcreative.com (2024) *Space-Based Solar Power: The Future Source of Energy?* Quantarverse. https://quantarverse.com/space-based-solar-power-the-future-source-of-energy/#:~:text=What%20is%20Space%2DBased%20Solar,terms%20of%20research%20and%20funding (accessed 25/08/2025).

Earth.Org (2025) *17 largest wildfires in US History*. https://earth.org/worst-wildfires-in-us-history/ (accessed 25/08/2025).

Fanning AL and Raworth K (2025) Doughnut of social and planetary boundaries monitors a world out of balance. *Nature* **646**: 47–56, 10.1038/s41586-025-09385-1.

FindMyHome (2024) Living in the future. myhome blog, 22 February. https://myhome.at/en/living-in-the-future/ (accessed 25/08/2025).

Forster PM *et al.* (2025) Indicators of Global Climate Change 2024: annual update of key indicators of the state of the climate system and human influence. *Earth System Science Data* **17(6)**: 2641–2680, 10.5194/essd-17-2641-2025.

Fraser G (2025) *30 Years Ago Tomorrow's World Predicted 2025 – How Did It Do?* https://www.bbc.co.uk/news/articles/cjdn10yk0k1o (accessed 10/08/2025).

Gavine A (2022) *Emirates Flight Catering Opens World's Largest Vertical Farm*. https://www.aircraftinteriorsinternational.com/news/catering-onboard/emirates-flight-catering-opens-worlds-largest-vertical-farm.html#:~:text=Emirates%20Flight%20Catering%20opens%20world's%20largest%20vertical,the%20greens%20for%20many%20Emirates%20inflight%20meals (accessed 25/08/2025).

Gruenbaum O (2025) Mirrors in space and underwater curtains: can technology buy us enough time to save the Arctic ice caps? *The Guardian*, 25 June. https://www.theguardian.com/environment/2025/jun/25/climate-geoengineering-arctic-ice-melting-mirrors-space-underwater-curtains-technology-solar (accessed 25/08/2025).

ICE (Institution of Civil Engineers) (n.d.) *Future Cities – Design and Build Cities That Really Meet the Needs of Growing Urban Populations*. https://www.ice.org.uk/what-is-civil-engineering/infrastructure-projects/future-cities (accessed25/08/2025).

IPCC (Intergovernmental Panel on Climate Change) (2022) *The Ocean and Cryosphere in a Changing Climate: Special Report of the Intergovernmental Panel on Climate Change*. Cambridge, Cambridge University Press, UK, 10.1017/9781009157964.

Kurzweil R (2024) *The Singularity Is Nearer: When We Merge with AI*. Bodley Head, New York, USA and London, UK.

Lejtenyi P (2024) Algae offer real potential as a renewable electricity source, new Concordia research shows. *News – Concordia University*. https://www.concordia.ca/news/stories/2024/06/11/algae-offer-real-potential-as-a-renewable-electricity-source-new-concordia-research-shows.html (accessed 25/08/2025).

McKinsey & Company (2023) *What's the Future of Generative AI? An Early View in 15 Charts*. https://www.mckinsey.com/featured-insights/mckinsey-explainers/whats-the-future-of-generative-ai-an-early-view-in-15-charts (accessed 25/08/2025).

McKinsey & Company (2024) *With Gen AI, Helping Green Businesses Strengthen Our 'Blue Economy'*. https://www.mckinsey.com/about-us/new-at-mckinsey-blog/with-gen-ai-we-are-helping-green-businesses-strengthen-our-blue-economy (accessed 25/08/2025).

Medeu AR *et al.* (2022) Moraine-dammed glacial lakes and threat of glacial debris flows in South-East Kazakhstan. *Earth-Science Reviews* **229**: 103999, 10.1016/j.earscirev.2022.103999.

Miles E (2023) *What Will Cities Look Like in 2050?* https://www.countryandtownhouse.com/culture/future-cities-2050/ (accessed 30/08/2025).

Porcelanosa (2023) The future of houses: 10 ways homes of tomorrow are taking shape today. *Porcelanosa TrendBook*. https://www.porcelanosa.com/trendbook/uk/future-of-houses (accessed 19/01/2026).

Ranganathan J *et al.* (2018) *How to Sustainably Feed 10 billion people by 2050, in 21 Charts*. World Resources Institute. https://www.wri.org/insights/how-sustainably-feed-10-billion-people-2050-21-charts#main-content (accessed 25/08/2025).

Ravate B (2023) A new bio-inspired solar leaf design with increased harvesting efficiency. *Imperial News*. https://www.imperial.ac.uk/news/246833/bio-inspired-solar-leaf-design-with-increased/ (accessed 30/08/2025).

Sanusha S (2025) 'First photovoltaic leaf in history created — Infinite energy with 'living chips'. *ECO News*, 14 February. https://www.ecoticias.com/en/first-photovoltaic-leaf-in-history/11323/# (accessed 30/08/2025).

Susskind R (2025) How to think about AI: Guide for the Perplexed. In Oxford University Press eBooks, pp. 23–38, 10.1093/oso/9780198941927.003.0003.

The Earthshot Prize (2025) *Advanced Thermovoltaic Systems – Earthshot Prize Winner 2024*. https://earthshotprize.org/winners-finalists/advanced-thermovoltaic-systems/ (accessed 25/08/2025).

UNEP (UN Environment Programme) (2018) *The Tale of a Disappearing Lake*. https://www.unep.org/news-and-stories/story/tale-disappearing-lake (accessed 25/08/2025).

United Nations (2015) *Paris Agreement*. https://unfccc.int/files/essential_background/convention/application/pdf/english_paris_agreement.pdf (accessed 30/08/2025).

Vlamis K, Paradis T and Varanasi L (2025) *Meet the Leaders of MBB, the Consulting Giants Advising the World's Most Powerful CEOs*. https://www.businessinsider.com/mbb-leaders-consulting-firms-advising-leaders-and-ceos-2025-7 (accessed 25/08/2025).

World Meteorological Organization (2025) *State of the Global Climate 2024 WMO-No. 1368*. https://library.wmo.int/idurl/4/69455 (accessed 25/08/2025).

Young J (2025) McKinsey cuts 10% of global workforce amid slowdown in consulting demand. *Business Matters*. https://bmmagazine.co.uk/news/mckinsey-cuts-10-of-global-workforce-amid-slowdown-in-consulting-demand/#:~:text=McKinsey%20&%20Company%20has%20cut%20more,it%20remains%20committed%20to%20growth (accessed 25/08/2025).

Navigating an Evolving Sustainable Built Environment: Delivering Net Zero Carbon and Climate Resilience

Mei Ren and Rachel Waggett
ISBN 978-1-83662-853-8
https://doi.org/10.1108/978-1-83662-850-720261013

Index

Printed and bound by CPI Group (UK) Ltd, Croydon, CR0 4YY

19/07/2026

14924083-0001